THE WISCONSINAN STAGE

The Geological Society of America, Inc.
Memoir 136

The Wisconsinan Stage

Edited by

ROBERT F. BLACK

University of Connecticut

Storrs, Connecticut

RICHARD P. GOLDTHWAIT

The Ohio State University

Columbus, Ohio

H. B. WILLMAN

Illinois State Geological Survey

Urbana, Illinois

1973

Published by
THE GEOLOGICAL SOCIETY OF AMERICA, INC.
3300 Penrose Place
Boulder, Colorado 80301

*The publication of this volume has been made possible
through the bequest of
Richard Alexander Fullerton Penrose, Jr.*

Foreword

To the late Lewis M. Cline goes the credit for the initial impetus for this symposium volume. As Technical Chairman of the 1970 Annual Meeting of The Geological Society of America, he conferred with Black on possible topics for symposia at the Milwaukee meeting, and specifically questioned whether something on the Pleistocene would be timely. This led to the sanction by the Geomorphology Group of a symposium on the Pleistocene and to the appointment of the undersigned as a committee to carry out the wishes of the group. The committee selected the Wisconsinan Stage as a representative and amenable topic.

Because of timing and the availability of contributors, it was not possible for the committee to obtain anywhere near the number of papers desired to provide representative coverage of the Wisconsinan Stage in the Midwest and other areas of the world. This volume includes all but one of the six papers actually presented at the symposium in Milwaukee. Some of the nine additional papers were presented in part at the regular sessions of the Annual Meeting, and others are more recent offerings.

The committee recognizes both regional and topical gaps in this coverage of the Wisconsinan Stage. However, between papers on historical perspective for the Midwest and both polar regions, its scope is considered to be sufficiently broad to provide the reader with a reasonable concept of the development of thought on the Wisconsinan Stage, its present status, and to indicate some trends for the future. We wish to thank the individual contributors for their cooperation in the demanding and difficult task of bringing this volume to completion.

Robert F. Black
Richard P. Goldthwait
H. B. Willman

Contents

Foreword   vii

HISTORICAL

History of investigation and classification of Wisconsinan drift in north-central
United States . . . .*George W. White*   3

REGIONAL

Late Wisconsin fluctuations of the Laurentide ice sheet in southern and eastern
New England . . . . *Harold W. Borns, Jr.*   37
Wisconsinan history of the Hudson-Champlain Lobe . . . . . *G. Gordon Connally
and Leslie A. Sirkin*   47
Wisconsin glaciation in the Huron, Erie, and Ontario Lobes . . . . *A. Dreimanis
and R. P. Goldthwait*   71
The Erie Interstade . . . .*Nils-Axel Mörner and A. Dreimanis*   107
Wisconsinan climatic history interpreted from Lake Michigan Lobe deposits and
soils . . . . *John C. Frye and H. B. Willman*   135
Superior and Des Moines Lobes . . . . *H. E. Wright, Jr.,
Charles L. Matsch, and Edward J. Cushing*   153

TOPICAL

Glacial dispersal of rocks, minerals, and trace elements in Wisconsinan till, south-
eastern Quebec, Canada . . . . *W. W. Shilts*   189
Differentiation of glacial tills in southern Ontario, Canada, based on their Cu,
Zn, Cr, and Ni geochemistry . . . . *R. W. May and A. Dreimanis*   221
DeKalb Mounds: A possible Pleistocene (Woodfordian) pingo field in north-central
Illinois . . . . *Ronald C. Flemal,
Kenneth C. Hinkley, and James L. Hesler*   229
Tunnel valleys, glacial surges, and subglacial hydrology of the Superior Lobe,
Minnesota . . . .*H. E. Wright, Jr.*   251
Pleistocene-Holocene boundary and Wisconsinan substages, Gulf of Mexico
. . . . *John H. Beard*   277
Climatic fluctuations during the late Pleistocene . . . . *C. C. Langway, Jr.,
W. Dansgaard, S. J. Johnsen, and H. Clausen*   317
Climatological implications of stable isotope variations in deep ice cores from
Byrd Station, Antarctica . . .*Anthony J. Gow, Samuel Epstein, and Robert P. Sharp*   323
Index   327

ix

HISTORICAL

GEOLOGICAL SOCIETY OF AMERICA
MEMOIR 136
© 1973

History of Investigation and Classification of Wisconsinan Drift in North-Central United States

GEORGE W. WHITE

Department of Geology, University of Illinois, Urbana, Illinois 61801

ABSTRACT

Outwash deposits in the Ohio Valley that were thought to be alluvium were illustrated by Volney in 1803. Drake (1815) proposed an iceberg origin for Wisconsinan erratics in Ohio. Hitchcock (1841a) in Massachusetts reviewed the glacial theory of Agassiz favorably, and in Ohio, St. John (1851) wrote in detail on the subject. Ice sheet origin of drift was widely and generally accepted by 1865.

Multiple glacial advances, many of which later turned out to be Wisconsinan, had been noted by Lyell (1849), Whittlesey (1866), Worthen (1868), Orton (1870), Winchell (1873), and Newberry (1874). Chamberlin (1878) gave the first detailed description and analysis of the different ages of surface drifts in the Kettle Moraine region. After Chamberlin introduced the term Wisconsin (1894, 1895), Leverett (1899) divided the Wisconsinan into early, middle, and late, and Leighton (1931) assigned the Iowan to the Wisconsin and gave the names Tazewell, Cary, and Mankato to the early, middle, and late substages of Leverett. Rock- and time-stratigraphic terminology of the Lake Michigan Lobe was defined by Frye and Willman (1960), and the terms Altonian, Farmdalian, Woodfordian, and Twocreekan were introduced. Other geologists have recently used this classification to differentiate the drift of other lobes.

INTRODUCTION

The classification of glacial deposits in the central part of North America has progressed from the recognition of surficial material as "diluvium" or "drift," to recognition that drift was of glacial origin, and finally to the realization that the drift was deposited by more than one episode of glaciation. There then followed more detailed and refined divisions of the Pleistocene deposits, particularly of the later, mainly Wisconsinan deposits, which have the widest surface outcrop, are best preserved, and can be dated in actual years by isotopic methods.

Evidence for multiple glacial advances in the Pleistocene and in the Wisconsinan in North America is best exhibited and most studied in the Midwest. The stratigraphy and terminology are developed mainly from this region, and it is now the scene of active studies. For lack of space the history of classification of deposits of the Ontario Basin in the United States and Canada and of the Canadian part of the St. Lawrence Basin is not included in this paper. The classification in southern Ontario has been summarized by Goldthwait and others (1965) and is discussed by Dreimanis and Goldthwait (this volume, p. 71-105).

Very Early Notices of Drift

The earliest diagrams of Quaternary deposits in the Middle West, and indeed in America, were published by Volney in France in 1803 and in America in 1804. He described the material in the Ohio River terraces at Louisville (1803, p. 62, Pl. 1; 1804, p. 48, Pl. 1) and the terraces and their material at Cincinnati (1803, p. 90, Pl. 1; 1804, p. 68-69, Pl. 1). He recognized that these materials (Fig. 1) had been carried down the Ohio Valley from the northeast, but he did not know that they were glacial outwash.

The first suggestion in America of the influence of ice on origin of drift, and especially of erratics, was by Drake (1815, p. 75). In 1817 he described erratics in greater detail as "foreign and adventitious debris . . . consisting of granite . . ." (Drake, 1825, p. 133-135) and gave the first description (in the United States) of till, which he called "geest" (p. 135-137). He explained the origin of the erratics and geest as being from "large fields of ice in a region far beyond the lakes and floated hither by the same inundations that brought down and spread over the surface of this country the *geest* in which they are imbedded" (1825, p. 137). His diagram (p. 138) shows the "alluvial formation" in the Ohio Valley and the geest on the uplands. However, Drake's ice origin of these materials was too radical for general acceptance at the time, even by such an astute observer as Thomas (1819, p. 248).

Maclure, in a brief but important paper on American geomorphology, noted the presence of erratics in the Central Interior. He suggested that those in Ohio had been transported by icebergs in a once far more extensive Lake Erie (Maclure, 1823, p. 102).

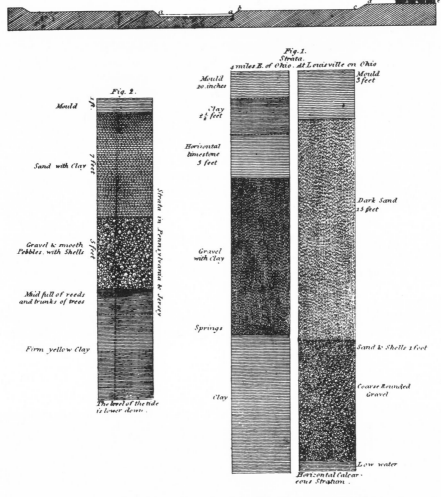

Figure 1. Volney's sections. Upper figure is of the Ohio River terraces which he observed at Cincinnati in August 1796, in "the history of which three distinct periods may be traced" (Volney, 1804, p. 69; Pl. 1). The two columnar sections at the right are of deposits in the Ohio River Valley and are thus the first sections of American Pleistocene deposits. The section on the left is from the Coastal Plain in the Philadelphia region.

Ice Sheet Origin of Drift

The ice sheet origin of drift in America was first seriously proposed and supported by close reasoning from field data by Hitchcock[1] (1841a, 1841b). Earlier (Hitchcock, 1832) he had extensively described drift; but upon reading Agassiz's *Etudes sur les Glaciers,* which had been published in October 1840, he immediately recognized that this theory explained

[1]Conrad (1839, p. 241-242) thought that "freezing of enormous lakes . . . converting them into immense glaciers" might explain erratics and striations in New York, but he did not go into detailed and extended explanations as did Hitchcock.

satisfactorily the features he had been studying in such detail (White, 1967, p. 212). The ice sheet concept was not enthusiastically accepted for several years, and Hitchcock himself soon retreated from the position of glacier transport he had taken and went part way back to iceberg origin of drift.

Many workers, such as Mather, Hitchcock, and Hall, in the newly founded geological surveys of Ohio, New York, and elsewhere, had in the late 1830s written extensively on drift and erratics, but suggested iceberg origin of the drift. These men, as well as others, as a result of Hitchcock's bringing Agassiz's theory to their attention so cogently, began to consider the glacial theory. However, for about 20 years after Agassiz and Buckland announced evidence of continental glaciation in Great Britain, glacial geology did not generate much interest in that country nor in America. Then, in Great Britain, the "years 1861 to 1865 saw a very remarkable development of interest in glacial geology" (Newbiggin and Flett, 1917, p. 157) in which James Geikie participated and soon became the leader in Britain and was widely recognized in the United States. Soon thereafter, as a proponent of multiple glaciation, he had a great effect on American workers. He and T. C. Chamberlin were in frequent communication, which culminated in Chamberlin's contribution to the classic third edition of the *Great Ice Age* (Geikie, 1894).

The first acceptance of ice sheet origin of drift by anyone of professional standing in the Midwest was in 1851, by Samuel St. John of Western Reserve College (White, 1967). This is not the place to trace in detail the acceptance of the glacial hypothesis in the central United States—it is sufficient to realize that by about 1870 the ice sheet origin of the drift was generally accepted, although the concomitant action of ice in bodies of water to explain what we now interpret as outwash deposits, made by water from melting ice, was more or less insisted upon by some until late in the nineteenth century.

In America, studies of the drift rapidly increased in the 1860s, as they did in Great Britain, and it was inevitable that, as field work became more detailed, the layers of the drift and the varying degrees of erosion and weathering on different surface drifts would be recognized.

PREVIOUS HISTORICAL STUDIES

One of the earliest reviews of the literature, with a comprehensive bibliography citing 500 earlier reports up to 1899, is that of Leverett (1901, p. 29-49). This important historical study deals mainly with the Erie Lobe and adjacent regions and includes some references to divisions of the post-Illinoian drift. He compiled another exhaustive bibliography (Leverett and Taylor, 1915, p. 33-54) which included scores of references on classification. Still later, Leverett (1932, p. 209) presented another, briefer history and bibliography.

In his review of the "Development of the glacial hypothesis in North

America,'' Merrill (1906, 1924) touched very briefly on the early recognition of multiple glaciation (his references were usually very sketchy), but had nothing to say about the Wisconsinan and its duration.

The paper by Thwaites (1927), ''The development of the theory of multiple glaciation in North America,'' traces year by year from 1873 to 1926 the appearance of reports on this subject. It emphasizes the ''Central District,'' but also includes historical summaries of the East, the western mountains, and Canada. It deals mainly with the whole Pleistocene classification, but does include some notices of reports on Wisconsinan divisions up to 1926. This fundamental study with its accurate references has been most useful in the preparation of this paper.[2]

Leighton (1964) summarized some of the history of discoveries of multiple drifts in midwestern North America and gave detailed consideration and references to his most recent classification of Wisconsinan deposits (p. 124-133).

The brief but excellent paper by Flint (1965) deals with the history of classification of the Pleistocene but not with the divisions of the Wisconsinan. His Table II is an accurate summary of the origin of stage names in the central United States.

Charts in recent reports by Frye and Willman (1963, Fig. 1) and by Frye and others (1968, p. E4) show the evolution of Wisconsinan classification. A report on a part of the Lake Michigan Lobe by Frye and his associates (1969, p. 2-5) discusses the history of classifying and naming of Pleistocene units (mainly late Pleistocene) in Illinois. The references and maps are valuable parts of this report. An even more extensive bibliography was included by Willman and Frye (1970) in their definitive bulletin on Pleistocene of Illinois.

The history of discovery and classification of Wisconsinan drifts in Wisconsin is included in a series of papers in elaborate guides prepared for the Pleistocene field trips before and after the 1970 Annual Meeting of The Geological Society of America in Milwaukee. The historical summaries are particularly detailed for southeastern Wisconsin by Laska (in Black and others, 1970, p. B3-B7, C3-C14, E3-E13) and by both Black and Bleuer for south-central Wisconsin (in Black and others, 1970, p. G1-G11, H5, J6-J8). The history of the Two Creeks buried forest is summarized by Black (1970, p. 21).

THE INQUA VOLUME—1965

The great Quaternary volume published in 1965 on the occasion of the Seventh Congress of the International Association for Quaternary

---

[2] The paper by Thwaites (1927) is not readily accessible and in our opinion should be reprinted to be available to interested researchers. It might be noted that Keyes (1928) made lengthy and strident objections to Thwaites's paper, mainly on the basis that multiple glaciations were doubtful and had not been proved, and that the merits of the proponents of a ''lone Great Ice Age'' had not been recognized.

Research (Wright and Frey, 1965) contains chapters descriptive of the
Pleistocene geology of various parts of North America. Most of them
emphasize Wisconsinan deposits; all the chapters have some historical
references, but some have many more than others.

In the northern Great Plains—Montana, North Dakota, and South Dakota—Lemke and others (1965, p. 20, 21-25) summarized the evidence
for "six distinct and separate glacial advances during Wisconsin time,"
related them to moraines, shown by a useful map, but did not name
the tills. Advance 1 is the Altonian; the others are Woodfordian. Wright
and Ruhe (1965) summarized the classification and extent of the complex
Wisconsinan drifts in Minnesota and Iowa from four major ice lobes
and included a useful series of maps. (In this volume, Wright and others
discuss the current classification.)

In Illinois and Wisconsin the Wisconsinan divisions were described
by Frye and others (1965) following the classification introduced earlier
by Frye and Willman (1960). Elaboration of this classification has been
recently presented by Frye and others (1968, 1969), and Willman and
Frye (1970). Wayne and Zumberge (1965) presented a table showing
the Wisconsinan divisions in Indiana after Wayne (1963) and the geologic-
climatic units after Gooding (1963).

The Wisconsinan deposits of the Huron, Erie, and Ontario Lobes,
long differentiated into substages called early, middle, and late Wisconsin, and later given rock-stratigraphic names by a number of workers,
are reviewed and summarized by Wayne and Zumberge (1965), Goldthwait
and others (1965), and Muller (1965). The recognition at Sidney, Ohio,
of a buried soil more than 22,000 years old below three tills and above
at least two earlier tills (Forsyth, 1965, Fig. 2) has provided the basis
for elaboration of Wisconsinan classification in the eastern part of the
Miami Lobe and hopefully for correlation with deposits of the Scioto
Lobe. The term Sidney (?) Interval has been used for the time represented
by the formation of the Sidney paleosol (Goldthwait and others, 1965,
p. 91).

The rock units of the Erie Lobe and their age assignments are dealt
with in detail in this volume by Dreimanis and Goldthwait.

It should be pointed out that ice sheets in much of the Central Lowlands
advanced into low regions and generally downslope. In contrast, in the
Allegheny Plateau in eastern Ohio, northwestern Pennsylvania, and New
York, the Wisconsinan ice sheets, as well as the earlier ones, advanced
upslope and, therefore, to not nearly as great a distance. Consequently,
in the plateau the various tills are exposed at the surface only in narrow
belts, some of which are less than 5 mi wide, causing severe mapping
problems. However, drifts in the subsurface can now be correlated with
surface drifts in these narrow belts so that classification is more certain
(White, 1969; White and others, 1969). Also in the plateau, organic deposits
that can be used for C-14 dating are much more scarce than in the
Central Lowlands. Fewer than six localities in the plateau, aside from

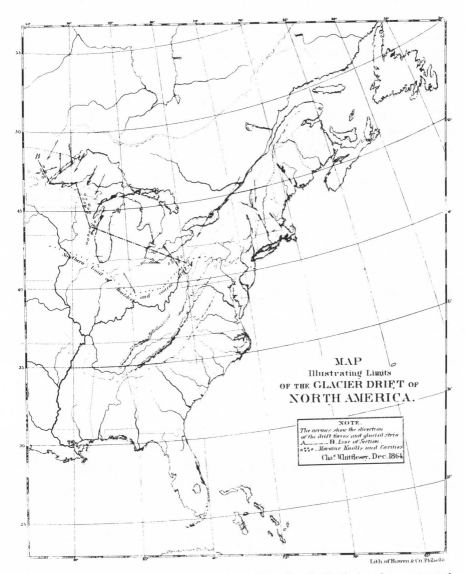

Figure 2. A very early (earliest?) map of glacial boundary in North America constructed by Charles Whittlesey in 1864. It shows "Southern limits of boulders and coarse drift" in a line extending from New Jersey to Iowa. Original map is 8 in. wide (Whittlesey, 1866).

flood plain and postglacial peat-bog materials, have so far provided organic material for dates.

MULTIPLE DRIFTS 1849–1878

Before 1868 there had been a few notices of differences in ages of the drift, but in the years 1868 to 1878 came the more general recognition of multiple drift deposits and the realization that they could only be explained by one or more glaciations that followed one or more interglacial times. Criteria other than buried organic layers came to be recognized by 1880.

Charles Lyell in his second visit to the United States in 1845-1846 had sensed that "the northern drift, however, is by no means all of the same age" (Lyell, 1849, vol. 2, p. 265). However, it was not until much later that buried organic layers were reported.

In one of the most important early papers, Whittlesey (1866) described at length the morphology of the drift and its different units. His catalog of the many sections that showed "buried timber" (p. 13-15) is the earliest precise description of interglacial (or interstadial) material. Whittlesey's paper also contains the earliest map (dated December 1864) to show the drift boundary in the United States (Fig. 2). The map also shows "striae," and "Moraine Knolls and Cavities."[3]

Another very early notice of buried organic deposits that separated drift sheets was by Worthen in Illinois (1868, p. 75-87). He described buried organic deposits below upper drift. Although realizing their importance, he did not speculate on their origin, saying that "wherever these beds are penetrated in sinking wells or are otherwise exposed, a careful examination should be made . . . for any organic remains . . . as these would no doubt throw some light on their true origin, and the conditions under which they have been deposited" (p. 87). Some of these organic deposits may have been Sangamonian in age, but all those checked have turned out to be Farmdalian, and thus separated two substages of Wisconsinan drift as now recognized (Willman, 1971, oral commun.).

Although the forest bed in southeastern Ohio was known before 1870 and was described in detail by Orton (1870), the explicit recognition in the Midwest of multiple glacial and interglacial deposits that required more than one episode of glaciation to explain their origin was in 1873. Orton (1873, p. 430) first used the term "the *interglacial stage*" in America in his description of the "forest bed" in southwestern Ohio and southeastern Indiana. At the location he studied, the forest bed lies between outwash deposits that he recognized as requiring two glacial stages (Fig. 3). This is the earliest illustration of a deposit specifically labelled "interglacial." It is quite likely an interstadial deposit of Wisconsinan-Farmdalian age and hence records the first division of Wisconsinan glacial materials.

Winchell (1873, p. 61) suspected that some thin weathered material in southeastern Minnesota was "the remains of a previous glacial sheet." He suggested "vegetable growth" had been removed "toward the north" by later ice advance, but that in the southeast at places it was buried "beneath the debris from the hillsides."[4]

Newberry (1874, p. 3) recognized a widespread forest bed in Ohio

---

[3] Stewart and MacClintock (1969, p. 55-56) have summarized early observations of evidence of multiple glaciations in Vermont.

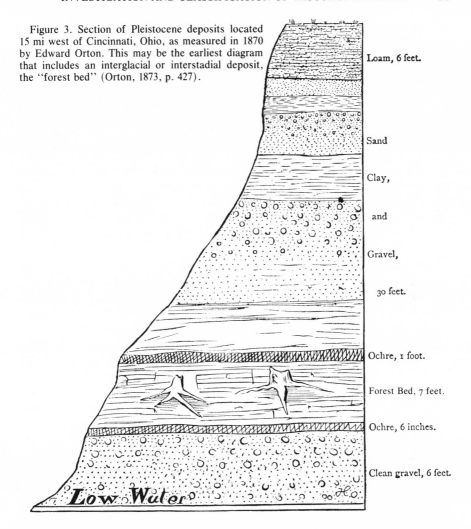

Figure 3. Section of Pleistocene deposits located 15 mi west of Cincinnati, Ohio, as measured in 1870 by Edward Orton. This may be the earliest diagram that includes an interglacial or interstadial deposit, the "forest bed" (Orton, 1873, p. 427).

Loam, 6 feet.

Sand

Clay,

and

Gravel,

30 feet.

Ochre, 1 foot.

Forest Bed, 7 feet.

Ochre, 6 inches.

Clean gravel, 6 feet.

Low Water

resting upon "boulder clay" and overlain by drift, mainly "water transported material" (p. 7). Although Whittlesey's map of 1864 is the first to show the drift border, Newberry's (Fig. 4) is the earliest map to show the drift border in a more detailed way. In his earlier papers (Newberry, 1870), there is reference to drift margins and to different kinds of drift.

McGee (1878) referred to Newberry's report on the forest bed and asserted that a similar bed is to be "observed at many points in Illinois, all through southwestern Indiana, and at many localities in Wisconsin, throughout northeastern Iowa, in Canada, and in many other places" (p. 339). He gave illustrative sections and stated that "the forest

---

[4]Winchell's small but significant five-page paper included distinctions between meltwater deposits in valleys sloping away from the ice and those sloping toward the ice, and also an analysis of superglacial deposits let down when the glacier melted.

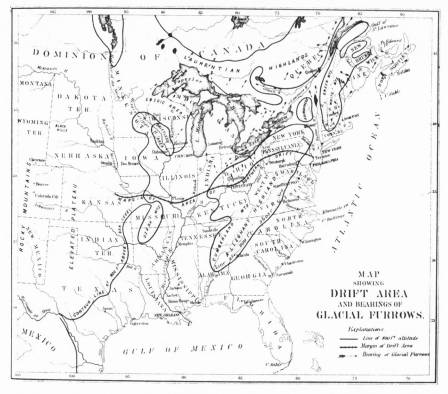

Figure 4. Newberry's map prepared about 1870 of "Drift area and bearings of glacial furrows." Note "Driftless Area." Original map is 9-1/4 in. wide (Newberry, 1874, opposite p. 76).

bed is overlaid by true glacial drift, and hence must be of interglacial age" (p. 341).

Winchell (1875, p. 185) thought that buried peat in Mower County, Minnesota, "seems to mark a period of interglacial conditions." He said that patches of much weathered drift in Fillmore County, on the margin of the Driftless Area, "are believed to belong to a glacial epoch that preceded the epoch that produced the great drift sheet of the Northwest," and an "interglacial epoch separated them." In 1878 he reported on wood deposits "between the two drift periods" in Goodhue County (Winchell, 1878, p. 44).

Chamberlin, who was beginning his work in Wisconsin about the same time, was familiar with the discoveries in Ohio and Minnesota. In his chapter on the Quaternary deposits in *The Geology of Eastern Wisconsin* (1877, p. 196–246), he ascribed the multiple layers of drift to "alternating advance and retreat of the ice-mass" (p. 214). Many of his diagrams illustrate multiple layers of drift, especially those in his Plate 8. The relations of ground-water resources to the drift layers were well illustrated in a series of diagrams. A separate paper that elaborated on these ideas, and which was Chamberlin's first paper entirely on glacial subjects, was his report on the Kettle Moraine (1878), a revision of a paper he had presented orally in 1875. In this he clearly distinguished between the drift in the Kettle Moraine and the drift outside on the basis of different development of drainage and of weathering rather than on stratigraphic grounds. However, the drift outside the Kettle Moraine also is largely Wisconsinan. After describing the Kettle Moraine and tracing it (Fig. 5), both far to the west and far to the east, he summarized his ideas about the different ages of material within and outside the moraine: "*outside* of the Altamont Moraine" is a "fairly established drainage system," in contrast to "the much less established drainage inside the moraine" (p. 16). He recognized moraines buried beneath later deposits (p. 17, 19, 29). He assumed "the moraine was, therefore, formed *after the retreat of the glacier had commenced and marks a certain stage of its subsequent history*" (p. 30). The moraine was ascribed not to a halt, but to an "advance of the ice-mass" (p. 31). It is "*a definite historical datum line . . .* separating the formations on either hand by a chronological barrier" (p. 33–34). Thus, almost 100 years ago, as a result of rigorous analysis of field data, the divisions within the last glacial stage were recognized, and a map was presented to show different ages of drift which are Wisconsinan in age.

It is significant that Newberry, Winchell, Orton, and Gilbert had all worked in Ohio in the first part of the 1870s and had published reports in the Ohio Survey volumes. They were familiar with each other's discoveries, and those of Winchell in Minnesota and McGee in Iowa strongly supported the Ohio findings. Chamberlin had seen the drift in Ohio and correlated the Ohio findings with his own studies in Wisconsin.

These men supposed that the drift above and below the "interglacial deposits" was derived from separate ice advances, but they did not distinguish between stages and substages. They were, however, the first to recognize multiple deposits in the late Pleistocene, that is, within what we now call Wisconsinan.

MULTIPLE DRIFTS AND THEIR CLASSIFICATION, 1878–1894

After the definitive paper on the Kettle Moraine by Chamberlin (1878), an increasing number of reports of interglacial deposits and of multiple tills appeared, as recorded by Thwaites (1927). Only a few can be selected for mention. Intertill deposits, some interglacial and others interstadial, were reported by McGee and Call (1882) in Iowa and by Upham (1883, 1884) in Minnesota. They recognized both stratigraphic evidence and degrees of weathering, and varying amounts of valley cutting where different drifts were exposed at the surface.

McGee and Call (1882) also recognized loess between the drift sheets and stated that this proved readvance of the ice sheet "not during an

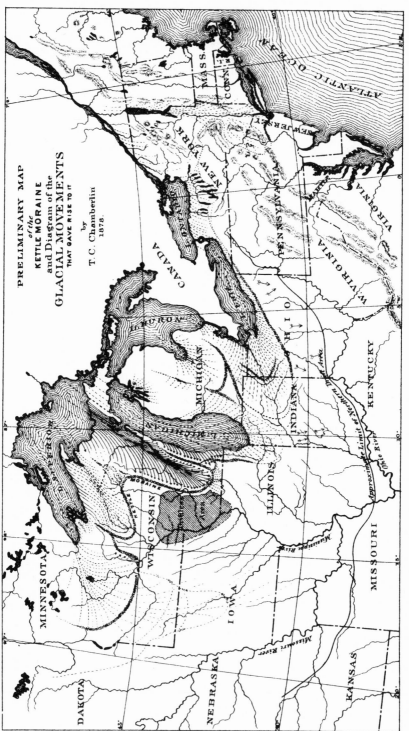

Figure 5. Chamberlin's map of 1878, showing "Approximate limit of northern drift area," the "Driftless Area," and "Kettle Moraine." This is the first map to show explicitly drift of more than one age, because the Kettle Moraine formed the margin of drift definitely younger than that between it (Kettle Moraine) and the drift limit. Original map is 7-1/2 in. wide (Chamberlin, 1878, p. 10).

independent ice-period, but during a temporary halt and slight readvance of the slowly retreating ice-sheet." Here was an early recognition of stratigraphic divisions of the drift of the last glacial stage.[5] In a classic report McGee (1891, p. 472-547) described "upper till" separated from "lower till" by a forest bed or other organic material, and included many diagrams and maps. However, by this time he had interpreted the forest bed as interglacial (p. 576-577).

Chamberlin was becoming more certain about two glacial advances in Wisconsin, both of which are now known to be of Wisconsinan age. He developed this concept in increasing detail and certainty in a series of papers. Chamberlin (1882, p. 93) recorded the "extensive range of terminal moraines" extending from Pennsylvania to Dakota, beyond which was earlier drift.[6] *"The moraine marks a second glacial advance separated from the former by a considerable interval of time"* (p. 94). This was an elaboration of the extramorainal drift he had first described in 1878.

The classic paper by Chamberlin on "The terminal moraine of the second glacial epoch" (1883) is one of the most important for the advancement of Pleistocene classification. It also contains a history of the study of glacial deposits in North America up to that time, and the references to notices of moraines and drift are very valuable. Chamberlin traced the great series of moraines from the Atlantic Ocean to Dakota (p. 310) and stated that they do not lie at the drift margin (p. 314). The extramorainic drift was recognized as older on the basis of more erosion and of more advanced development of the drainage, greater weathering, and thinner and less continuous drift (p. 347). He considered that in places an earlier moraine had been overridden by a later one (p. 347).

Another more elaborate classification of the Quaternary epochs was that of Chamberlin and Salisbury (1885, p. 212). In this classification "the Earlier glacial epoch" was divided into two sub-epochs separated by an "interglacial sub-epoch." The following "Later glacial epoch," which followed the "Chief interglacial epoch" [Sangamonian] was divided into four episodes or stages. Curiously, the oldest was marked by the "Kettle or Altamont" Moraine, the next by the Gary Moraine, then the Antelope Moraine, followed by still "later stages." It is interesting to note that they correlated the Altamont Moraine with that of Nantucket, and the Gary with the Cape Cod Moraine (p. 215). This was long-range correlation indeed!

By 1884, Newberry, who had held out longer than most for iceberg origin for much of the drift, finally understood that a great ice sheet "possessing a peculiar plasticity," and not icebergs floating in a great

---

[5]This paper also contains much of interest concerning superglacial erosion and deposition, some "odd" ideas on the secondary origin of the "blue color" of buried till, and references on the latter subject not easily found elsewhere.

[6]This is essentially the break within the Woodfordian at the front of the Valparaiso Moraine, and not the margin of Woodfordian drift as now understood (Frye and Willman, 1960).

sea, had deposited the till of the northern United States, and adduced evidence to disprove the iceberg theory. He now asserted that "the ice advanced with many arrests and retreats" (1884, p. 86), and that *"striated pavement"* boulder pavements were formed by readvances of the ice (p. 88). He cited contorted drift as evidence of ice readvance and stated that "two boulder clays separated by . . . beds of peat" were indications of *"interglacial warm periods"* (p. 90). He assumed many episodes of advance and retreat and disappearance of the ice, as he was now strongly influenced by Croll's (1875) explanation of glacial episodes caused by variations in the eccentricity of the earth's orbit.

In a closely reasoned paper, Salisbury (1893) discussed the criteria for the recognition for distinct glacial epochs. He listed 12 criteria, which included various kinds of organic and inorganic intertill beds, weathering zones, varying amounts of erosion, and different compositions of tills. His paper was a significant one and is still an important reference for study of multiple glacial deposits.

James Geikie invited Chamberlin, his long-time friend and corre- spondent, to contribute the section "Glacial phenomena of North Ameri- ca" to the third edition of *The Great Ice Age* (Geikie, 1894). It was thus that the first classification using geographic names, on which our present classification is based, came to be published outside the United States. The text and legend of the colored map shows "Kansan Forma- tion," "East Iowan Formation," and "East Wisconsin Formation." The East Wisconsin extended to the Kettle and Altamont Moraines in the west and the Shelbyville and "Main Morainic System" in the east. The East Iowan was shown in eastern Iowa and in an area in central northern Illinois, extending into southern Wisconsin (Chamberlin, 1894, p. 724–775; Pl. 15).

At Chamberlin's invitation, Geikie (1895) summarized at length the classification of European deposits as a basis for possible correlation with American deposits. In a following paper, Chamberlin (1895) made tentative suggestions about the correlation of American deposits with the European. He "cheerfully accepted Upham's suggested changes"[7] and introduced the term "Wisconsin Formation" instead of East Wiscon- sin, as he had proposed the previous year, and "Iowan" in place of East Iowan. He recognized that there were stages of halts and "possibly advances" in the latest stage of glaciation (p. 276).

FOREIGN RELATIONS

In the late 1870s and through the 1880s (and later), the glacial geologists of America, Great Britain, and the European continent, communicated with each other through personal visits, exchange of publications, and

[7]In a review of Geikie's book, Upham (1895, p. 56) suggested changing "East Iowan" to "Iowan" and "East Wisconsin" to "Wisconsin."

correspondence. The observations and reports of the great triumvirate of James Giekie of Scotland, T. C. Chamberlin of America, and Otto Torrell of Sweden, with conclusions based in part on the theories of James Croll, were received, studied, taught, and expanded by most American geologists. Those few who were more conservative and reluctant to accept multiple glaciation, and particularly to accept several interglacial periods, were in such minority that their views were hardly regarded as respectable. They tended to be ignored by the mainstream of workers, although from time to time they were graciously allowed to present their conflicting views.

The validity of multiple glaciation and interglacial episodes, and thus the theoretical basis for seeking evidence for these in the deposits were given impetus by the many papers of Croll, beginning in 1864 and culminating in his book *Climate and Time in Their Geological Relation; A Theory of Secular Changes of the Earth's Climate* (1875). The central part of Croll's theory was the change in eccentricity of the earth's orbit, which had produced (partly indirectly) changes in climate throughout geologic time and especially during the glacial period. Croll's book deserves attention today, not only for the theories presented, but also for the voluminous geological data adduced in support of the theories. His arguments were well received by many American geologists, and his ideas were as influential in the last quarter of the nineteenth century as those of Milankovich in the middle part of the twentieth century.

Otto M. Torrell, Swedish professor of geology and later head of the Swedish Geological Survey, whose work in glacial geology in Europe was highly regarded, visited America in 1877. His views were welcomed, inasmuch as they confirmed the findings of the Americans. He supported the concept of advance and retreat of the ice and in a paper summarizing his views states "the retrograde movement of a glacier during the period of melting is characterized by deposits formed . . . by local backward and forward movements of a glacier during successive intervals of time" (Torrell, 1877, p. 77).

James Geikie first came to the United States from Edinburgh in 1884. He was able to see and study glacial and interglacial deposits throughout the Middle West under the guidance of the men who were working in the various states (Newbiggin and Flett, 1917, p. 106–110). Always an enthusiastic proponent of multiple glaciation, he encouraged the Americans to recognize not only one interglacial time, but a multiplicity of them. Some of these are within the Wisconsinan Stage as we now know it. His ideas were brought to the notice of Americans not only in the three editions of his book, *The Great Ice Age,* but also in a paper in the newly established *Journal of Geology* (Geikie, 1895), which was followed by a paper of Chamberlin's (1895). It is interesting to note that Geikie chose for the frontispiece of the third edition of his book (1894) a magnificently reproduced photograph of an exposure "near Williamsport, Indiana," which can now be interpreted as showing five tills, the upper ones being of Wisconsinan age. The development of the concept of multiple glaciation and interglacial intervals in Europe and James Geikie's *avant garde* place in these are discussed at length by Newbiggin and Flett (1917, p. 149-210).

WORK OF LEVERETT

Frank Leverett began to work with Chamberlin in 1886 and was closely associated with him for many years. Leverett's work resulted in a series of papers which were published in the first of his U.S. Geological Survey monographs, *The Illinois Glacial Lobe*, in 1899. Anyone working in territory described by Leverett must begin by a detailed study of his various publications.[8]

In his 1899 monograph Leverett stated that he studied "several thousand well sections" and thus was the first to use subsurface records in detail, although it is to be noted that he relied mainly on "records" and not on sample study, as in present-day methods. McGee (1891) studied well records, but not as extensively as did Leverett.

In this 1899 monograph Leverett distinguished sharply between the "Iowan of Illinois" and the "Iowan of Iowa" (Pl. 6 and p. 131-152). He demonstrated that the Iowan of Illinois was younger than Illinoian, but older than Wisconsin (see also Leverett, 1897, 1898a, 1898b). The extreme northern part of his Iowan of Illinois is now called Altonian, the earliest division of the Wisconsinan. The Wisconsin was divided into "early" and "late"; the "early Wisconsin" was divided into Shelbyville, Champaign, Bloomington, and Marseilles till sheets (Pl. 6, p. 191-379). These are all within the Woodfordian as presently understood.

Leverett's *U.S. Geological Survey Monograph 41* on the Erie and Ohio Basins appeared in 1901. It covered Indiana, Ohio, Kentucky, much of Pennsylvania and western New York, and part of Ontario. In this monograph Leverett for the first time had the advantage of extensive topographic map coverage that brought a new dimension into the study of glacial geology. He extended the use of "geographic names" (p. 50), but mainly for the moraines and their tills. He retained the Iowan Stage and Peorian interval and classified the Wisconsin as "early," to include the Shelbyville, Champaign, Bloomington, and Marseilles Morainic Systems, followed by a "fifth interval of recession (unnamed), shown by shifting of ice lobes." This was followed by the "late Wisconsin drift sheet" with Valparaiso, St. John's or Salamonie, Mississinawa, Wabash, and Fort Wayne Morainic Systems and the lake beaches and their "correlative moraines." The Iowan, he stated "appears not to extend beyond the limits of the Wisconsin drift" (p. 51). He wrote "This outline

---

[8]Leverett appears to have made some changes in some of his monographs either during printing or after they had been approved for printing, so that in places conflicting views appear and modifications of earlier views are inserted.

is essentially the same as the writer's outline presented in Monograph XXXVIII, the only important modification being in the subdivisions of the late Wisconsin glacial stage. These are more complete in the Ohio District than in the Illinois, and require a corresponding elaboration of the outline."

Leverett studied well records mainly for drift thickness. Unless obvious organic layers were recorded, he did little interpretation of the stratigraphy. Following Chamberlin's earlier preference, he depended mainly on the surface character of the different drifts for his classification (Leverett, 1909). His first love was morphology and he described in great detail the morainic areas, their extent, appearance, and at times in a most detailed way, the rounded cone of a single two-acre kame.

Leverett classified the drift beyond the "Wisconsin moraine" on the east side of the Grand River Lobe in Pennsylvania as "Kansan," but later (1934) decided this outer drift was "Illinoian" and "pre-Illinoian." The more westerly part of the "Illinoian" is now known to be about 40,000 yrs old and is therefore Wisconsinan (Altonian) in age (White and others, 1969, p. 23–32). Apparently Leverett had second thoughts about this "fringe drift" and—obviously out of place—inserted a page (Leverett, 1901, p. 351) of exact and excellent description leading to the statement that a portion "nowhere more than 10 miles" is too eroded to be late Wisconsin and too fresh to be pre-Wisconsin. It is curious that he did not speculate that this might be his "Iowan." He sensed (p. 452) the difference in composition of the till between the coarse, sandy till (Kent) of the outer part of the Grand River Lobe and the clayey till (Lavery and Hiram) of the inner part, but did not carry the subject far enough to indicate a considerable time difference of retreat and then readvance of the ice to account for the difference.[9]

Leverett's "late Wisconsin" drift of the east side of the Scioto Lobe is, in present terms, Woodfordian as far north as Utica. North of Utica, in Licking and Knox Counties, Leverett (1901, Pl. 13) recognized a northeastern extension of the Wisconsinan drift to Danville and Jelloway. Because this drift is more eroded it was called "Illinoian" by White (1934) and by Forsyth (1961). Now that an Altonian Stage of the Wisconsinan is recognized, the possibility must be considered that Leverett was right in calling this Wisconsin, with a narrow and possibly discon-

---

[9]The history of the "fringe drift" (Wright, G. F., 1884, 1889) of the Grand River Lobe in Ohio outside of what was earlier regarded as the outermost Wisconsinan moraine has already been discussed by the author (White, 1951). It has been variously called "Wisconsin" (Leverett, 1901), "Illinoian" and "pre-Illinoian" (Leverett, 1934), "Illinoian" (White, 1951), and "Kansan" (Stout, 1943). Narrow as it is, it is now interpreted as being of two ages—the larger part, called "Inner Illinoian" by Shepps and others (1959, p. 21) is 40,000 years old and hence Wisconsinan-Altonian (White and others, 1969, p. 23–32). A narrow and discontinuous outer part (Shepps and others, 1959, p. 20) is pre-Wisconsinan and probably Illinoian. Thus, the Grand River Lobe has, at the surface and in the subsurface, Woodfordian and Altonian Wisconsinan drifts, which now have rock-stratigraphic names (White and others, 1969, p. 9).

tinuous, outer part of this tract, being of Illinoian age. [10] Leverett's "early
Wisconsin drift" of the Miami Lobe (1901, p. 304, Pl. 11) may possibly
be early Wisconsinan (Altonian) in the present sense. The "Illinoian
drift" beyond is Illinoian in the present sense and no part is Wisconsinan-
Altonian. He discussed the difference between the early Wisconsin drift
and that to the north in some detail (p. 352–353).

U.S. Geological Survey Monograph 53, The Pleistocene of Indiana
and Michigan and the History of the Great Lakes, by Leverett and
Taylor (1915) contains an extensive bibliography which is important for
historical purposes. The folding chart (p. 30), "Approximate time relations
of ice lobes, morainic systems, and moraines of the Wisconsin stage
of glaciation in Indiana and southern Michigan" also shows the Illinois,
Miami, Scioto, Lake Michigan, Saginaw, and Huron-Erie Lobes. The
map (Pl. 5), which was later reprinted without change (Leverett, 1931,
Pl. 18), shows the Shelbyville as the oldest and the Port Huron as
the youngest Wisconsinan drifts in the region. This is a key map to
show his concept of the divisions in the Wisconsinan.

It is interesting to note that "Illinoian and Iowan Drifts" are at one
place combined (Leverett and Taylor, 1915, p. 26), but on the following
page the "post-Sangamon and main loess (Iowan?)" indicates Leverett's
growing belief that the Iowan was more closely related to the Wisconsin
as he understood it than to the Illinoian. His uncertainty about the
age assignment of post-Illinoian, pre-Wisconsinan (Woodfordian) drift and
the realization that it is closely allied to the Wisconsinan Stage is well
illustrated by his later statement that "it may be more consistent to
regard the Illinoian, Iowan, and Wisconsin as a triple glaciation. . . .
The data at hand do not seem to be decisive in this matter" (Leverett,
1930, p. 194).

LEIGHTON AND THE EXTENSION AND
CLASSIFICATION OF THE WISCONSIN—1931-1956

In 1913, M. M. Leighton published his first paper on Pleistocene deposits
in the Mississippi Valley. His work was to continue for more than 50
years. [11] After more than 15 years of field investigation Leighton proposed

[10] Leverett actually spent less time in this "re-entrant angle" tract of Knox and southern
Ashland Counties than was usual in his studies (personal commun. in a conference with
Leverett at Ann Arbor, 1930). He was a master at dealing with questionable areas or interpretations. Note his statement that part of the fringe drift of the Grand River Lobe may be
Wisconsin (1901, p. 351). His detailed description and analysis as he proceeds from place
to place are convincing and masterful. When, after scores of pages of detailed descriptions
mile by mile, he jumps 20 or more miles, as from Utica to Mansfield, one must follow
a detailed map to see that for the intervening space there are no real data!

[11] It is interesting to note the long time span covered by the work of several other
geologists working on the glacial geology of the Mississippi Valley. A. C. Trowbridge
published his first glacial paper in 1912, and thus his publication record reaches to almost
60 years. W. C. Alden has a 54-year record. Frank Leverett just reached 50 years, and
G. F. Wright, F. T. Thwaites, Charles Whittlesey, James Walter Goldthwait, and Paul MacClintock closely approached 50-year records.

a classification that soon achieved wide usage. He concluded that "the Iowan drift is the first deposit of the last glacial stage and it must be known as the early Wisconsin or else the older terms, early Wisconsin, middle Wisconsin, and late Wisconsin must be dropped" (Leighton, 1931, p. 51). He proposed the terms "Manitoban Substage" for the Iowan, "Quebecan Substage" for the early and middle Wisconsin, and "Hudsonian Substage" for the late Wisconsin. However, two years later Leighton (1933, p. 168) proposed that the term "Iowan" be retained for the oldest subdivision of the Wisconsin, and that "Tazewell," "Cary," and "Mankato" be used for the early, middle, and late subdivisions, as previously differentiated by Leverett. These names came into general use for more than 30 years and still have some currency. When radiocarbon dating showed that the Mankato drift in Minnesota was older, not younger, than the Two Creeks forest bed, Leighton (1957) modified his classification by adopting the name "Valders" (Thwaites, 1943) for a substage younger than the Mankato.

Kay and Leighton (1933, p. 673) used the term "Eldoran Epoch," which had been introduced by Kay (1931) for the Wisconsin and Holocene Stages, and other new names for each of the three earliest ages and their following interglacial times. These names for combined glacial and interglacial stages never achieved general usage.

A major advance in elucidation of Wisconsinan history and in classification was the recognition by Leighton that a major glacial advance had taken place in post-Sangamonian time before the deposition of Shelbyville drift (earliest "classical Wisconsin"). Leighton (1926) observed that the loess which he had called "Late Sangamon Loess" rested on the Sangamon Soil and was only moderately weathered before a loess, then correlated with the Iowan glaciation, was deposited on it. He named the early Wisconsin loess "Farmdale" (*in* Wascher and others, 1948), and the Farmdale was made the first substage of the Wisconsin Stage (Leighton and Willman, 1950, p. 602). This Farmdale Loess was interpreted as silt blown from valley trains coming from Farmdale ice an unknown distance away. Later, Shaffer (1954, 1956) identified till in northwestern Illinois as the deposit of this ice and named the till "Farmdale."

The Farmdale Loess was later (Frye and Willman, 1960) subdivided into a lower unit, largely loess, named "Roxana Silt," and Farmdale was restricted to an upper organic silt and peat which had been widely dated. The name "Winnebago" was applied to the till in northern Illinois that was the source of the Roxana Silt, and the name "Altonian" was introduced for the earliest Wisconsinan glacial substage which included the Winnebago Till and the Roxana Silt.

Although the Wisconsinan classification of 1931 was modified by Frye and Willman (1960) in developing a multiple classification scheme, Leighton (1960, 1964) defended the previous classification and introduced the terms "Farm Creek," "Gardena," "St. Charles," "Bowmanville," and "Two Creeks" for intraglacial substages between the glacial substages.

IOWAN DRIFT

The Iowan drift, originally described as "East Iowan" (Chamberlin, 1894) and soon renamed "Iowan" (Chamberlin, 1895), was originally set off as a separate stage [12] older than the Wisconsinan ("classical Wisconsin"-Woodfordian). As its affinities in the type area of eastern Iowa were thought to be much closer to the Wisconsinan than to the older drifts, it was eventually classified as the earliest Wisconsinan substage (Leighton, 1933).

A problem with the Iowan, not present with any other stage, is that the term "Iowan" has been used for widely different features. This history has been summarized by Ruhe (1969, p. 93-95) and his map (Pl. 1) shows the "Iowan of Iowa" not as a till but as an "Iowan erosion surface" developed largely on the Kansas till. The till is overlain by "loam sediments," except on the southwest where it is covered by "Wisconsinan loess." At the extreme eastern edge the surface cuts across Illinoian till. Leverett (1909, p. 368) came to the same conclusion, but later was not so sure. Ruhe proposed the same explanation for the western part of "Iowan drift" in northwestern Iowa (Ruhe, 1969, p. 104).

The "Iowan of Illinois" (Leverett, 1899, Pl. 6) is a "real" till of Altonian age in extreme northern Illinois and southern Wisconsin, but farther south it is a till now interpreted as earliest Woodfordian (Frye and others, 1969, Figs. 1-1899, 1-1969; Fig. 2). Leverett also used the term "Iowan" in Illinois for the loess outside the Shelbyville (Woodfordian) drift and for the loess underlying the drift, but Alden and Leighton (1917) substituted the term "Peorian" for the loess, and Kay and Leighton (1933) restricted Peorian to the loess outside the Shelbyville and reintroduced Iowan for the loess beneath the Shelbyville. This was the Iowan of Illinois until Frye and Willman (1960) dropped Iowan as a glacial substage and introduced the term "Morton" for the loess previously called Iowan.

As the term "Iowan" was used in Iowa for a till, now believed to be the Kansas till, it is proper that it now be dropped. In northern Illinois this "real" till is now given the rock-stratigraphic name of Winnebago and is referred to the Altonian Substage of the Wisconsinan (Frye and others, 1969).

It may be noted that Carman (1917) explained the "Iowan drift" in northwestern Iowa as eroded Kansas till. Like Ruhe's explanation for eastern Iowa 50 years later, he suggested that erosion had truncated drift of two different ages. The idea was so revolutionary that it was not allowed to be printed by the Iowa State Geologist until it was considerably modified (Kay and Apfel, 1929, p. 114). Later, Carman's ideas were

---

[12] Much of Chamberlin's description was based on the detailed descriptions, diagrams, and illustrations of McGee, who, however, did not apply names to these units, but described them as "upper till" and "lower till" (McGee, 1891, p. 472-542).

more acceptable, particularly about the two ages of the drift, and were printed in a revision of his earlier bulletin (Carman, 1931).

INTRODUCTION OF ROCK-STRATIGRAPHIC TERMINOLOGY AND DISTINCTION BETWEEN ROCK AND TIME TERMS

Until about 1960 glacial deposits were referred to by names which carried both time and rock connotation.[13] For example, the term "Tazewell" for early "classical" Wisconsinan was used from the eastern United States to the western. By 1960 Pleistocene deposits began to be regarded as rock strata which not only could, but should, be described in rock-stratigraphic terms according to the Stratigraphic Code. Time and space does not permit a review of the exploding use of rock-stratigraphic terminology in the decade 1960 to 1970, but two examples will be used.

In Illinois the work of Frye and Willman (1960) on the drift of the Lake Michigan Lobe led to the time-stratigraphic classification of the Wisconsinan Stage into the Altonian Substage, in which were deposited the Winnebago Drift and Roxana Silt; the Farmdalian Substage, in which were deposited silt and peat; the Woodfordian Substage, in which were deposited tills and loesses; the Twocreekan Substage; and the Valderan Substage. Further elaboration of time-stratigraphic and rock-stratigraphic classifications was presented by Frye and others (1968, 1969), and by Willman and Frye (1970). The time terms "Altonian" and "Woodfordian" as Wisconsinan substages have now been used by some workers for deposits at least as far east as Pennsylvania (White and others, 1969, p. 9) and as far west as Saskatchewan (Christiansen, 1968, p. 335).

The author and his associates (Shepps and others, 1959) used rock-stratigraphic terms for Wisconsinan tills in northwestern Pennsylvania in 1959 and used "Cary" and "Tazewell" as time terms. In 1960 and later, Wisconsin tills in northeastern Ohio were defined as rock-stratigraphic units according to rigid Code specifications (White, 1960), again using "Tazewell" and "Cary" as time terms, but in 1966 (p. 20) "uncertain" was used in place of "Tazewell." A few years later, White (1969) and White and others (1969), were using the same rock-stratigraphic terms in Pennsylvania and Ohio, but the time terms now were "Wisconsinan Stage" with "Altonian," "Farmdalian," and "Woodfordian" Substages.

CONCLUSION—SUMMARY OF STAGES IN INVESTIGATION OF WISCONSINAN DEPOSITS

The surface material of the Central Interior is mainly late Pleistocene (Wisconsinan) in age. During the past 130 years the investigation of this

---

[13] Multiple classification, to the degree of time-stratigraphy and rock-stratigraphy, was introduced for the Pleistocene of Kansas by Frye and Leonard (1952).

"drift" and the classification of its parts were based on the materials themselves, the varying amount of surface weathering and dissection, the intertill materials, and other features. These investigations progressed through a series of stages, each roughly 20 to 30 years in duration.

Recognition of Drift and its Possible Iceberg Origin—1815-1840

Erratics and the matrix in which they were embedded were observed and an iceberg origin suggested by Drake as early as 1815, and soon thereafter by others, including Maclure in 1823. The observations increased rapidly in the 1830s as state surveys were organized. It was in this period that the concept of origin of the drift from icebergs floating in a continent-covering sea was increasingly common.

Recognition of Glacial Origin of the Drift—1841-1860

In this period the belief in iceberg origin of the drift slowly gave way to ice sheet origin. The activity of state surveys decreased from the level of the 1830s, but independent geologists continued field observations.

Recognition of Multiple Glaciation—1865-1894

In the early part of this period state surveys became more active. In the later part of the period the United States Geological Survey was established and soon supported work on glacial deposits under the leadership of T. C. Chamberlin. Interglacial deposits were recognized, and it was realized that more than one episode of glaciation was required to explain the sequence of deposits. By 1875 Chamberlin determined that the last glacial episode was complex, and in it there had been several advances and retreats of the ice sheets. In 1894 and 1895 he proposed the name "Wisconsin" and stated it had several divisions.

Beginning of Detailed Field Study—1895-1915

Topographic maps for parts of the Central Interior began to be available by 1895 and glacial studies took a quantum jump. Led by Frank Leverett, detailed field study, in some cases square mile by square mile was inaugurated. The four encyclopedic monographs of Leverett provided the basic knowledge of glacial geology of the Central Interior. Leverett's divisions of early, middle, and late Wisconsin, and the named moraines and morainal systems became a classification to be used for 40 years; many of his moraine names are still used. During this period he had varying views on the place of the Iowan.

Elaboration of Detailed Field Study—1915-1940

During this period topographic maps became available for most of the Central Interior. More and more detailed studies could be made, and moraine mapping was highly developed. Each moraine was believed to mark an important stand of the ice front and to record an ice advance to that position. Discordant moraines that cut off others at an angle were supposed to mark more important episodes of retreat, "regrouping," readvance, and a change of direction of advance of the continental glacier. It was assumed that each moraine was made of the till of the last ice advance to that position, and that this till extended northward from the moraine to the margin of the next younger moraine. It was in this period that Leighton proposed his widely used classification of the Wisconsinan substages—Iowan, Tazewell, Cary, and Mankato.

Air Photos Come into Use—1940-1950

From about 1940 air photos began to supplement maps as tools for observation and as bases for recording field data. By the use of these for observing topography, drainage, and soil types, much more refined mapping of surface drift, moraines, and other features became possible. Moraine mapping reached its highest degree of detail.

C-14 Dating—1950–

Soon after 1950[14] enough C-14 dates began to appear to make it possible for the first time to attach ages in years[15] to different drift units. At first the ages were regarded as shockingly low, but mental adjustment was soon made to the chronology they imposed (Horberg, 1955). The earliest Wisconsinan drift—"classical Wisconsin" or Tazewell—was only about 24,000 years old! More and more dates between 30,000 and 40,000 years appeared for "pre-Wisconsin" deposits. It was increasingly recognized that a drift, earlier than classical Wisconsin, but much younger than Illinoian, was widespread; and that the weathered or organic material, or both, in the subsurface was not Sangamonian. After stillborn proposals of "Ohioan," "Rockian," and other terms (but curiously not of "Iowan"), the name "Altonian" began to achieve wider and wider usage for this early Wisconsinan substage.

---

[14]The history of the discovery and use of C-14 dating in Pleistocene geology would require a lengthy paper in itself—parts have appeared in various places, but a connected study would produce a long thesis in geology or in history of geology.

[15]The term "absolute age" is sometimes used, but the term "radiocarbon age" is to be preferred, because radiocarbon years are not calendar or solar years, and are known to vary through time. Where the term "years," referring to age of deposit, occurs in this paper, it should be taken as "radiocarbon years."

Stratigraphic and Subsurface Study—1960-

The great increase in deep excavations that started in the late 1950s for superhighways, subways, strip mining, quarrying, and large building foundations was continuing at an accelerated rate, laying bare an almost embarrassment of riches in the form of drift anatomy that showed what was "really" below the surface over wide areas and not just in an occasional narrow ravine or in a stream cutbank.[16] Drill cores and samples from engineering investigations became available—indeed, more of them than there were Pleistocene geologists to study all of them. Pleistocene geology made a rapid right-angle turn from mainly study of surface morphology and surface drift (including soil study) to stratigraphic study of the whole drift column. It was discovered that—although there are some exceptions—the surface drift is usually only a few feet thick, and that the "real story" was below the surface. It was also learned that till sheet margins do not always coincide with moraines. Moraine mapping became less "interesting," for it became apparent in many cases that the surface drift did not make the bulk of a moraine, but was only a veneer over older material. While moraines remain important geomorphic features, each one must be critically re-examined to determine if indeed it does mark the edge of the last ice sheet in the region.[17]

Through extensive subsurface study of drift units, then tracing them to surface outcrop, it became more and more evident that these were really rock-stratigraphic units that could be treated according to the Stratigraphic Code. Although the distinctiveness of some Wisconsinan tills was known as early as 1930, or even before, it was only in this period that rock-stratigraphic names began to be applied to Wisconsinan drifts, and assigned to time-stratigraphic substages. These units began to be studied in the laboratory as rocks and it was discovered that over wide areas different tills maintained distinctive color, texture, lithology, and content of heavy, light, and clay minerals.

Exploratory Drilling—1960-

Drilling specifically for subsurface Pleistocene study has been done in several states, especially Illinois, Iowa, and Saskatchewan, and is now

---

[16] "There are extensive areas, for instance in southern Ontario, where 100–300 feet of Pleistocene stratigraphy are well visible in natural exposures, and therefore the 'real story' has been deciphered even without deep drilling . . . during the last 20 years" (Dreimanis, 1970, written commun.).

[17] The reader must be warned that the author's statements here are based on "conditioning" from working mainly on the drifts on the eastern sublobes of the Erie Lobe. Here the ice advanced upslope out of the basin and did not go far, so that the pulsating glacier piled thin sheet on thin sheet. In the Lake Michigan Lobe, where till sheets may indeed be thick, there was a rapid downslope flow of the ice, followed by a pulsing retreat with several major readvances, producing the situation described above under the heading "Elaboration of detailed field study—1915-1940."

being stepped up in intensity.[18] In many states samples from state highway and ground-water investigations have been available and have been studied in varying degrees of detail. Saskatchewan, where drift is more than 1,000 ft thick over wide areas, leads the way in exploratory drilling—each Pleistocene geologist is issued a powerful drilling rig, with two shifts of geologically skilled crews for collection of samples, cores, and electric logs, plus a field laboratory, and a headquarters chemical laboratory for more detailed mineral and chemical analyses. Investigations to a depth of 1,100 ft are routine.

Geophysical Methods—1970 (?)-

For more than 20 years it has been possible to distinguish some drift units by electrical and seismic methods. Electric logging of bore holes, routine in petroleum geology, has been almost unused in Pleistocene studies, except in Saskatchewan. The whole field of Pleistocene geophysics is still essentially uncultivated but will be explored and used widely in the future.

Paleontology

The history of study of Pleistocene pollen, grass and other seeds, insects, gastropods and other invertebrates, and vertebrates, almost entirely found in Holocene or interglacial and interstadial deposits, would make a multivolume monograph. The stratigraphic usefulness of these materials has not been great to this time, because most geologists are not comfortable in the biological field, and the biologists are uncertain in the geological field. More and more Pleistocene workers are becoming able to operate in the two worlds. Great results are already beginning to appear.

Another paleontological field is that of macro- and micro-fossils in till derived from fossiliferous pre-Pleistocene rocks. The fossil content of till—*Tasmanites*, bryozoa, conodonts, and other forms—may be characteristic of a given till. Studies in Holland and in Great Britain already give indication of this happy situation.

The Next Stage—1970-

Pleistocene geology in the Central Interior for the past 130 years has moved in 25- to 35-year cycles. The first period, to 1870, was of recognition of ice sheet origin, followed by the Chamberlin period to 1895, the Leverett period to 1930, the Leighton period to 1955, and the present

---

[18]The 1969 recommendation of the Quaternary Subcommittee of the National Advisory Committee [of Canada] on Research in the Geological Sciences has been accepted in the following wording: "The mapping should include stratigraphic investigations and test drilling to a greater extent than in the past" (Dreimanis, 1970, written commun.).

period dominated by those heretics, now become respectable, who treat Wisconsinan deposits as rock-stratigraphic units.

The next period, soon to be entered—for geological periods become exponentially shorter—will certainly be one in which subsurface geology, the study of the whole drift column, will predominate. Anything less will not be "respectable." Pleistocene units are rock units, and will be treated strictly as such. Geophysical methods, both in boreholes and from the surface, will be as routine as in petroleum geology, and will enable the geologist to look into the drift as he now can look into cores by radiography. Paleontology in its many branches will be extensively used. New isotopic dating methods will sooner or later be found to extend the record to 100,000 years and hopefully far beyond. The study of deep-sea Pleistocene sediments will be more and more correlated with terrestrial and continental shelf Pleistocene deposits. All this will require teamwork. The day of Leverett with his little trowel, or of Leighton (and many others!) with only a Pleistocene pick and an acid bottle, is over. The team will be composed of a geological drilling crew, mineralogist-petrographer, geochemist, geophysicist, geochronologist, and stratigrapher, led by a Pleistocene geologist. The cost, although high in dollars, will be moderate in comparison with the cost per day of marine geology and oceanographic ship-time or of North Slope petroleum exploration. After all, over half the people in the United States and all those of Canada live on, and more and more in, Pleistocene (mainly Wisconsinan) deposits that constitute a major part of their environment. These deposits are a key factor in that now magic term "environmental geology."

ACKNOWLEDGMENTS

An earlier draft of this paper was sent to other participants in this Symposium; their useful comments are much appreciated. I am particularly indebted to J. C. Frye and H. B. Willman for additional references, and especially to the latter for suggestions which saved me from some errors and which led to extensive revisions; and to R. P. Goldthwait for many suggestions and especially for gently tempering some of my heretical enthusiasms in the latter part of the paper. He has almost convinced me that "heresy by itself, however, is no token of truth."

BIBLIOGRAPHY

Agassiz, Louis, 1840, Etudes sur les glaciers: Ouvrage accompagne d'un atlas de 32 planches; aux frais l'auteur: Neuchatel, 347 p. and atlas of 32 pls.
Alden, W. C., and Leighton, M. M., 1917, The Iowan drift, a review of the evidences of the Iowan Stage of glaciation: Iowa Geol. Survey Ann. Rept. 1915, v. 26, p. 49–212.
Black, R. F., 1970, Glacial geology of Two Creeks Forest Bed, Valderan type

locality, and northern Kettle Moraine State Forest: Wisconsin Geol. and Nat. History Survey Inf. Circ. 13, 40 p.

Black, R. F., Bleuer, N. K., Hole, F. D., Lasca, N. P., Maher, L. J., 1970, Pleistocene geology of southern Wisconsin, a field trip guide: Wisconsin Geol. and Nat. History Survey Inf. Circ. 15, 175 p.

Carman, J. E., 1917, The Pleistocene geology of northwestern Iowa: Iowa Geol. Survey, v. 26, p. 233–445.

_____ 1931, Further studies on the Pleistocene geology of northwestern Iowa: Iowa Geol. Survey, v. 35, p. 15–193.

Chamberlin, T. C., 1877, Geology of eastern Wisconsin: Geology of Wisconsin, Survey of 1873–1877: Wisconsin Geol. Survey, v. 2, p. 91–405.

_____ 1878, On the extent and significance of the Wisconsin Kettle Moraine; n.p., n.d. [1878], 36 p.: a separate publication of a paper in Wisconsin Acad. Sci., Arts and Letters Trans., v. 4, p. 201–234.

_____ 1882, The bearing of some recent determinations on the correlation of the eastern and western terminal moraines: Am. Jour. Sci., v. 124, p. 93–97.

_____ 1883, Preliminary paper on the terminal moraine of the second glacial epoch: U.S. Geol. Survey, 3d Ann. Rept., p. 291–402.

_____ 1894, Glacial phenomena in North America, in Geikie, J., The Great Ice Age (3d ed.): London, p. 724–775.

_____ 1895, The classification of American glacial deposits: Jour. Geology, v. 3, p. 270–277.

_____ 1896, Nomenclature of glacial formations: Jour. Geology, v. 4, p. 872–876.

Chamberlin, T. C., and Salisbury, R. D., 1885, Preliminary paper on the Driftless Area of the upper Mississippi Valley: U.S. Geol. Survey, 6th Ann. Rept., p. 199–322.

Christiansen, E. A., 1968, A thin till in west-central Saskatchewan, Canada: Canadian Jour. Earth Sci., v. 5, p. 329–336.

Conrad, T. A., 1839, Notes on American geology; observations on characteristic fossils, and upon a fall of temperature in different geological epochs: Am. Jour. Sci., v. 35, p. 237–251.

Croll, J. A., 1875, Climate and time in their geological relations; a theory of secular changes of the earth's climate: London, Daldy, Isbister & Co., 577 p., 8 pls.

Drake, Daniel, 1815, Natural and statistical view or picture of Cincinnati and the Miami country: Cincinnati, Looker & Wallace, 251 p, 2 maps.

_____ 1825, Geological account of the Valley of the Ohio: Am. Philos. Soc. Trans., v. 2, p. 124–139.

Dreimanis, A., and Goldthwait, R. P., 1973, Wisconsin glaciations in the Huron, Erie, and Ontario Lobes: Geol. Soc. America Mem. 136, p. 71–105.

Flint, R. F., 1965, Introduction: Historical perspectives, in Wright, H. E., Jr., and Frey, D. G., eds., The Quaternary of the United States: Princeton, N. J., Princeton Univ. Press., p. 3–11.

Forsyth, J. L., 1961, Pleistocene geology, in Root, S. I., Rodriguez, Joaquin, and Forsyth, J. L., Geology of Knox County: Ohio Div. Geol. Survey Bull. 59, p. 107–138, pl. 5.

_____ 1965, Age of the buried soil in the Sidney, Ohio, area: Am. Jour. Sci., v. 263, p. 571–597.

Frye, J. C., and Leonard, A. B., 1952, Pleistocene geology of Kansas: Kansas Geol. Survey Bull. 99, 230 p.

Frye, J. C., and Willman, H. B., 1960, Classification of the Wisconsinan Stage in the Lake Michigan glacial lobe: Illinois Geol. Survey Circ. 285, 16 p.

_____ 1963, Development of Wisconsinan classification in Illinois related to radiocarbon chronology: Geol. Soc. America Bull., v. 74, p. 501-506.

Frye, J. C., Willman, H. B., and Black, R. F., 1965, Outline of glacial geology of Illinois and Wisconsin, in Wright, H. E., Jr., and Frey, D. G., eds., The Quaternary of the United States: Princeton, N. J., Princeton Univ. Press, p. 43-61.

Frye, J. C., Willman, H. B., Rubin, M., and Black, R. F., 1968, Definition of Wisconsinan Stage: U.S. Geol. Survey Bull. 1274-E, 22 p.

Frye, J. C., Glass, H. D., Kempton, J. P., and Willman, H. B., 1969, Glacial tills of northwestern Illinois: Illinois Geol. Survey Circ. 437, 45 p.

Geikie, James, 1894, The Great Ice Age (3d ed.): London, 850 p. (Also see review by Salisbury, R. D., 1894, Jour. Geology, v. 2, p. 730-747.)

_____ 1895, Classification of European glacial deposits: Jour. Geology, v. 3, p. 241-269.

Goldthwait, R. P., Dreimanis, Aleksis, Forsyth, J. L., Karrow, P. F., and White, G. W., 1965, Pleistocene deposits of the Erie Lobe, in Wright, H. E., Jr., and Frey, D. G., eds., The Quaternary of the United States: Princeton, N. J., Princeton Univ. Press, p. 85-97.

Gooding, Ansel, 1963, Illinoian and Wisconsin glaciations in the Whitewater Basin, southeastern Indiana, and adjacent areas: Jour. Geology, v. 71, p. 665-682.

Hitchcock, Edward, 1832, Report on the geology of Massachusetts, examined under the direction of the government of that state during the years 1830 and 1831: Am. Jour. Sci., v. 22, p. 1-70.

_____ 1841a, Final report on the geology of Massachusetts: Northampton, Mass., J. H. Butler, 831 p.

_____ 1841b, First anniversary address before the Association of American Geologists at their second annual meeting in Philadelphia, April 5, 1841: reprint New Haven, Conn., B. A. Hamlen, 48 p.; Am. Jour. Sci., v. 41, p. 232-275.

Horberg, C. L., 1955, Radiocarbon dates and Pleistocene chronological problems in the Mississippi Valley region: Jour. Geology, v. 63, p. 278-286.

Kay, G. F., 1931, Classification and duration of the Pleistocene period: Geol. Soc. America Bull., v. 42, p. 425-466.

Kay, G. F., and Apfel, E. T., 1929, The pre-Illinoian Pleistocene geology of Iowa: Iowa Geol. Survey, 304 p., 3 pls.

Kay, G. F., and Leighton, M. M., 1933, Eldoran epoch of the Pleistocene period: Geol. Soc. America Bull., v. 49, p. 669-674.

Keyes, C., 1928, Theory of multiple glaciations: Pan-Am. Geologist, v. 50, p. 131-144.

Leighton, M. M., 1913, An exposure showing post-Kansan glaciation near Iowa City, Iowa: Jour. Geology, v. 21, p. 431-435.

_____ 1926, A notable type Pleistocene section; the Farm Creek exposure near Peoria, Illinois: Jour. Geology, v. 34, p. 167-174.

_____ 1931, The Peorian loess and the classification of the glacial drift sheets of the Mississippi Valley: Jour. Geology, v. 39, p. 45-53.

_____ 1933, The naming of the subdivisions of the Wisconsin glacial age: Science, v. 77, p. 168.

_____ 1957, The Cary-Mankato-Valders problem: Jour. Geology, v. 65, p. 108-111.

____ 1960, The classification of the Wisconsin glacial stage of the north-central United States: Jour. Geology, v. 68, p. 529-552.

____ 1964, Elements in the classification of the late glacial Quaternary of midwestern North America: Advancing frontiers in geology and geophysics: Hyderabad, India, Osmania Univ. Press, p. 115-133.

Leighton, M. M., and Willman, H. B., 1950, Loess formations of the Mississippi Valley: Jour. Geology, v. 58, p. 599-623.

Lemke, R. W., Laird, W. M., Tipton, M. J., and Lindvall, R. M., 1965, Quaternary geology of the northern Great Plains, in Wright, H. E., Jr., and Frey, D. G., eds., The Quaternary of the United States: Princeton, N. J., Princeton Univ. Press, p. 15-27.

Leverett, F., 1897, The Pleistocene features and deposits of the Chicago area: Chicago Acad. Sci. Geol. and Nat. History Survey Bull. 2, 86 p.

____ 1898a, The weathered zone (Sangamon) between the Iowan loess and Illinoian till sheet: Jour. Geology, v. 6, p. 171-181.

____ 1898b, The Peorian soil and weathered zone (the Toronto Formation?): Jour. Geology, v. 6, p. 244-249.

____ 1899, The Illinois glacial lobe: U.S. Geol. Survey Mon. 38, 818 p., 24 pls.

____ 1901, Glacial formations and drainage features of the Erie and Ohio Basins: U.S. Geol. Survey Mon. 41, 802 p., 26 pls.

____ 1909, Weathering and erosion as time measures: Am. Jour. Sci., v. 177, p. 349-368.

____ 1929, Moraines and shore lines of the Lake Superior region: U.S. Geol. Survey Prof. Paper 154, 72 p.

____ 1930, Relative length of Pleistocene glacial and interglacial stages: Science, v. 72, p. 193-195.

____ 1931, Quaternary system, in Cushing, H. P., Leverett, Frank, and Van Horn, F. R., Geology and mineral resources of the Cleveland district, Ohio: U.S. Geol. Survey Bull. 818, p. 57-81.

____ (with contributions by F. W. Sardeson), 1932, Quaternary geology of Minnesota and parts of adjacent states: U.S. Geol. Survey Prof. Paper 161, 149 p., 5 pls.

____ 1934, Glacial deposits outside the Wisconsin terminal moraine in Pennsylvania: Pennsylvania Geol. Survey Bull., v. 7, 123 p.

Leverett, Frank, and Taylor, F. B., 1915, The Pleistocene of Indiana and Michigan and the history of the Great Lakes: U.S. Geol. Survey Mon. 53, 529 p., 32 pls.

Lyell, Charles, 1849, A second visit to the United States of North America: London, John Murray, 2 vols., 273, 287 p.

Maclure, William, 1823, Some speculative conjectures on the probable changes that may have taken place in the geology of the continent of North America: Am. Jour. Sci., v. 6, p. 98-102.

McGee, W. J., 1878, On the relative positions of the forest bed and associated drift formations in northeastern Iowa: Am. Jour. Sci., v. 15, p. 339-341.

____ 1891, The Pleistocene history of northeastern Iowa: U.S. Geol. Survey, 11th Ann. Rept., pt. 1, p. 189-577.

McGee, W. J., and Call, R. E., 1882, On the loess and associated deposits of Des Moines: Am. Jour. Sci., v. 124, p. 202-233.

Merrill, G. P., 1906, The development of the glacial hypothesis in America:

Popular Sci. Monthly, v. 68, p. 300-322.

Merrill, G. P., 1924, The first one hundred years of American geology: New Haven, Conn., Yale Univ. Press, 773 p. (Reprint, 1964, New York, Hafner Publishing Company, Inc., 773 p.)

Muller, E. H., 1965, Quaternary geology of New York, in Wright, H. E., Jr., and Frey, D. G., eds., The Quaternary of the United States: Princeton, N. J., Princeton Univ. Press, p. 99-112.

Newberry, J. S., 1870, On the surface geology of the basin of the Great Lakes, and the valley of the Mississippi: Annals Lyceum Nat. History New York, v. 9, p. 213-234.

_____ 1874, Geology of Ohio; Surface geology: Geol. Survey Ohio, Rept. 2, pt. 1, p. 1-80, maps.

_____ 1884, The drift deposits of Indiana: Indiana Geol. Survey, 14th Ann. Rept., p. 85-97.

Newbiggin, M. I., and Flett, J. S., 1917, James Geikie, the man and the geologist: Edinburgh, Oliver and Boyd, 227 p.

Orton, Edward, 1870, On the occurrence of a peat bed beneath deposits of drift in southwestern Ohio: Am. Jour. Sci., v. 50, p. 54-57, 293.

_____ 1873, Report on the third geological district; geology of the Cincinnati group; Hamilton, Clermont, Clarke Cos.: Ohio Geol. Survey, Rept. 1, pt. 1, p. 365-480.

Ruhe, R. H., 1969, Quaternary landscapes in Iowa: Ames, Iowa, State Univ. Press, 255 p., colored map.

St. John, S., 1851, Elements of geology, intended for the use of students: Hudson, Ohio, Pentagon Press, 334 p.

Salisbury, R. D., 1893, Distinct glacial epochs and the criteria for their recognition: Jour. Geology, v. 1, p. 61-84.

Shaffer, P. R., 1954, Farmdale drift: Science, v. 119, p. 693-694.

_____ 1956, Farmdale drift of northwestern Illinois: Illinois Geol. Survey Rept. Inv. 198, 25 p.

Shepps, V. C., White, G. W., Droste, J. B., and Sitler, R. F., 1959, Glacial geology of northwestern Pennsylvania: Pennsylvania Geol. Survey Bull. G 32, 54 p., colored map.

Stewart, D. P., and MacClintock, Paul, 1969, The surficial geology and Pleistocene history of Vermont: Vermont Geol. Survey Bull. 31, 251 p.

Stout, Wilber, 1943, Glacial geology, in Stout, Wilber, Ver Steeg, K., and Lamb, G. F., Geology of water in Ohio: Ohio Geol. Survey (4th ser.), Bull. 44, p. 17-50, pls. 1-3, 5, 7.

Thomas, David, 1819, Travels through the western country in the summer of 1816, including notices of the natural history . . .: Auburn, N.Y., David Rumsey, 320 p.

Thwaites, F. T., 1927, The development of the theory of multiple glaciation in North America: Wisconsin Acad. Sci. Trans., v. 23, p. 41-164.

_____ 1943, Pleistocene of part of northeastern Wisconsin: Geol. Soc. America Bull., v. 54, p. 87-144.

Torrell, O., 1877, On the glacial phenomena of North America: Am. Jour. Sci. (3d ser.), v. 13, p. 76-79.

Trowbridge, A. C., 1912, Geology and geography of the Wheaton quadrangle: Illinois Geol. Survey Bull. 19, 79 p.

Upham, W., 1883, The Minnesota Valley in the Ice Age: Am. Jour. Sci., v. 127, p. 34-42, 104-111.

_____ 1884, The geology of Faribault County (and other counties): Minnesota Geol. Survey, v. 1, p. 404-532.

_____ 1895, Review of the Great Ice Age and its relation to the antiquity of man, by James Geikie, 3d ed.: Am. Geologist, v. 15, p. 52-57.

Volney, C. F., 1803, Tableau du climat et du sol des États Unis d'Amérique. Suivi d'eclaircissements sur la Floride, sur la colonie Française au Scioto, sur quelques colonies Canadiennes et sur les sauvages. Enrichi de quatre planches gravées, dont deux cartes géographiques et une coupe figurée de la chûte de Niagara: Paris, Courcier, Dentu, An XII-1803, 2 v., xvi, 300 and 532 p., 2 folding maps, 2 folding plates.

_____ 1804, A view of the soil and climate of the United States of America: C. B. Brown, trans., Philadelphia, J. Conrad & Co. 446 p., maps (Reprinted, with introduction by G. W. White, New York, Hafner Publishing Company, Inc., 1968).

Wascher, H. L., Humbert, R. P., and Cady, J. G., 1948, Loess in the southern Mississippi Valley—Identification and distribution of loess sheets: Soil Sci. Soc. America Proc. 1947, v. 12, p. 389-399.

Wayne, W. J., 1963, Pleistocene formations of Indiana: Indiana Geol. Survey Bull. 25, 85 p.

Wayne, W. J., and Zumberge, J. H., 1965, Pleistocene geology of Indiana and Michigan, in Wright, H. E., Jr., and Frey, D. B., eds., The Quaternary of the United States: Princeton, N. J., Princeton Univ. Press, p. 63-84.

White, G. W., 1934, Illinoian drift region of northeast central Ohio: Ohio Jour. Sci., v. 37, p. 1-19.

_____ 1951, Illinoian and Wisconsin drift of the southern part of the Grand River Lobe in eastern Ohio: Geol Soc. America Bull., v. 62, p. 967-977.

_____ 1960, Classification of glacial deposits in northeastern Ohio: U.S. Geol. Survey Bull. 1121-A, 12 p.

_____ 1966, Glacial geology, in Winslow, J. D., and White, G. W., Geology and ground-water resources of Portage County, Ohio: U.S. Geol. Survey Prof. Paper 511, p. 17-38.

_____ 1967, The first appearance in Ohio of the theory of continental glaciation: Ohio Jour. Sci., v. 67, p. 210-217.

_____ 1969, Pleistocene deposits of the north-western Allegheny Plateau, U.S.A.: Geol. Soc. London Quart. Jour., v. 124, p. 131-151.

White, G. W., Totten, S. M., and Gross, D. L., 1969, Pleistocene stratigraphy of northwestern Pennsylvania: Pennsylvania Geol. Survey Bull. G-55, 88 p., 43 figs.

Whittlesey, Charles, 1866, On the fresh-water glacial drift of the northwestern states: Smithsonian Contr. Knowledge 197 (also in v. 15, 1867), 32 p.

Willman, H. B., and Frye, J. C., 1970, Pleistocene of Illinois: Illinois Geol. Survey Bull. 94, 204 p.

Winchell, N. H., 1873, The surface geology: Minnesota Geol. and Nat. History Survey, 1st Ann. Rept., p. 61-62.

_____ 1875, The geology of Mower County: Minnesota Geol. and Nat. History Survey, 3d Ann. Rept., p. 20-36.

_____ 1878, The geology of Rock and Pipestone Counties: Minnesota Geol. and Nat. History Survey, 6th Ann. Rept., p. 93-111.

Worthen, A. H., 1868, Geology and paleontology: Illinois Geol. Survey, v. 3, 574 p., 20 pls.

Wright, G. F., 1884, The glacial boundary in Ohio, Indiana and Kentucky: Cleveland, Ohio, Western Reserve Historical Co., Tract No. 60, p. 199–263.

_____ 1889, The ice age in North America and its bearing on the antiquity of man: New York, D. Appleton and Company, 622 p.

Wright, H. E., Jr., and Frey, D. G. (eds.), 1965, The Quaternary of the United States, a review volume for the VII Congress of the International Association for Quaternary Research: Princeton, N.J., Princeton Univ. Press, 922 p.

Wright, H. E., Jr., and Ruhe, R. V., 1965, Glaciation of Minnesota and Iowa, *in* Wright, H. E., Jr., and Frey, D. G., eds., The Quaternary of the United States: Princeton, N.J., Princeton Univ. Press, p. 29–41.

MANUSCRIPT RECEIVED BY THE SOCIETY DECEMBER 27, 1971

Printed in the United States of America

REGIONAL

GEOLOGICAL SOCIETY OF AMERICA
MEMOIR 136
© 1973

Late Wisconsin Fluctuations of the Laurentide Ice Sheet in Southern and Eastern New England

HAROLD W. BORNS, JR.

Department of Geological Sciences, University of Maine, Orono, Maine 04473

ABSTRACT

The age of the late Wisconsin maximum of the Laurentide ice sheet off the coast of New England and on Long Island, New York, is not closely designated. Radiocarbon and stratigraphic evidence from Martha's Vineyard suggests that the glacier margin may have been close to its maximum position as late as 15,300 yrs ago; indirect evidence from Long Island infers that the ice sheet had reached a maximum and had begun to recede prior to 17,000 yrs ago. In any case, by at least 14,200 yrs ago the glacier margin had retreated from its maximum late Wisconsin position at the Ronkonkoma Moraine on Long Island, had constructed recessional frontal deposits, and had retreated north of Rogers Lake on the southern Connecticut coast. Subsequent readvances culminated near Middletown, Connecticut, some time after 15,000 yrs ago, and in Cambridge, Massachusetts, after 14,000 yrs ago. Whether these readvances were synchronous is unknown because of the absence of close limiting dates and because of the lack of evidence for readvance in the intervening area. The northwestward recession of the glacier margin from the present coast in eastern Maine was accompanied by a marine transgression and deposition of hundreds of submarine moraines between 13,500 and 12,500 yrs ago. This general recession was interrupted by a readvance which culminated at the Pineo Ridge Moraine approximately 12,700 yrs ago. Although it may have resulted from general climatic change, the Pineo Ridge readvance just as likely may have been caused by a vastly decreased calving rate associated with isostatic uplift and marine recession from coastal Maine. This is well documented as having occurred simultaneously with the Pineo Ridge readvance. Thereafter, the ice sheet thinned and separated over the highlands of northwestern Maine leaving residual ice to the southeast. Active ice, receding into the St. Lawrence Valley of southeastern Quebec deposited the Highland Front Moraine approximately 12,700 to 12,600 yrs ago. In summary, (1) a major amelioration of climate that began prior to 14,200 yrs ago resulted in very rapid dissipation of the ice sheet in New England at least by 12,500 yrs ago, with the exception of small glaciers that possibly persisted in the highlands; (2) no conclusive evidence has been recognized for any climatic reversals during the dissipation of the ice sheet in New England; and (3) although major events in New England compare with those of the Great Lakes region, no minor events have proven correlation with the possible exception of the Pineo Ridge and Port Huron readvances.

INTRODUCTION

This paper reviews the recession of the Laurentide ice sheet from eastern and southern New England. Emphasis is placed on the chronology of ice recession, on the possible significance of end moraines, and on the comparison of glacial chronologies in New England and in the Great Lakes region.

When making comparisons of glacial events between New England and the Great Lakes region, for example, consideration has to be given to modern glaciological theory which raises the question of whether synchroneity should be expected among fluctuations of different glacier systems or among lobes of the same system (Meier, 1965). Regional climatic changes obviously affect the mass balance of an ice sheet. However, due to the time required for dynamic response of ice sheets to mass balance changes, the complete terminal response may be delayed considerably and may not occur at the same time in each lobe.

The length of time required for complete dynamic response of large ice sheets is on the order of several hundreds or thousands of years (Meier, 1965, p. 800). Moreover, the dynamic response of the glacier may induce additional climatic events resulting in a complex feedback relationship that further complicates interpretation of terminal fluctuations. In addition, local perturbations along the margin of an ice sheet can influence the behavior of individual lobes. These, superimposed upon the effects of climatic events affecting the ice sheet as a whole, add to the lack of confidence in expecting synchroneity in fluctuations along the full length of the margin of the ice sheet. Interpretation of glacier behavior is complicated even further by the possibility that terminal fluctuations can be caused by surges, sea-level changes, and isostatic land-level changes, as well as by climatic events.

LATE WISCONSIN MAXIMUM

The late Wisconsin terminal position of the Laurentide ice sheet east of the Hudson River is marked by the Ronkonkoma-Vineyard-Nantucket Moraine line (Schafer, 1961; Kay, 1964; Schafer and Hartshorn, 1965), and perhaps by the distribution of coarse gravel on the continental shelf (Schlee and Pratt, 1970, p. 28; Fig. 1). The time when the ice sheet reached that maximum is not closely designated, although several radiocarbon dates afford bracketing ages. These dates suggest that Laurentide ice was advancing through southern New England approximately 20,000 yrs ago (Schafer and Hartshorn, 1965, p. 119-120), and thus must have attained a maximum after this time. Sirkin (1967, p. 271) inferred, on the basis of pollen stratigraphy, that the ice sheet had reached a maximum position on western Long Island and had begun to recede prior to 17,000 yrs ago. Kaye (1964, p. 138) presented evidence that the ice margin was still at or near its terminal position on Martha's Vineyard as late as approximately 15,300 yrs ago (W-1187; Levin and others, 1965, p. 363).

Long Island, New York, and Southern New England

Marginal positions of the receding ice sheet on Long Island, New York, and in southern New England are marked by Harbor Hill, Charlestown, Buzzard's Bay, and Sandwich Moraines (Fig. 1). The Harbor Hill, Charlestown, and probably the Buzzard's Bay and Sandwich Moraines were deposited prior to 14,240 yrs ago (Y950-51; Stuiver and others, 1963, p. 320), the age of the earliest organic sediments in Rogers Lake 30 km north of the Harbor Hill Moraine on the Connecticut coast (Fig. 1). Whether any of these moraines represent a significant readvance of the glacier margin is unknown.

Flint (1953, p. 899) reported that subsequent retreat was interrupted by the Middletown readvance in the Connecticut Valley. This readvance, which amounted to about 26 km, culminated in the Middletown area of central Connecticut (Fig. 1) shortly before about 13,000 yrs ago (Flint, 1956, p. 275).

In the northwestern part of the Boston Basin the receding ice readvanced at least 5 km and constructed the Fresh Pond Moraine in Cambridge (Fig. 1; Chute, 1959, p. 197). This moraine and its outwash are underlain, in part, by the Boston marine clay which is about 14,000 yrs old (Kaye and Barghoorn, 1964, p. 75). Based upon this fact, and the ages of freshwater peats overlying the Boston marine clay, and also on an interpretation of a sea-level and crustal-rise curve for Boston, Kaye and Barghoorn (1964, p. 77) have suggested that the Cambridge readvance occurred 13,800 to 12,500 yrs ago.

Flint (1953, p. 901) suggested the possible correlation of the Middletown and Cambridge readvances. This suggested correlation cannot yet be

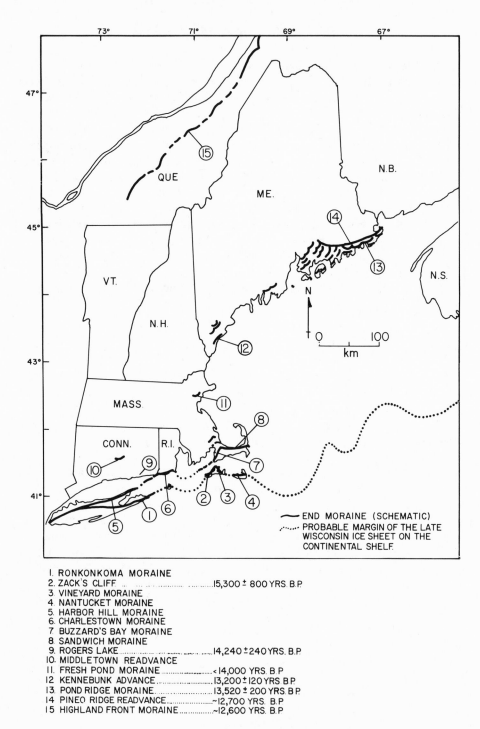

I. RONKONKOMA MORAINE
2. ZACK'S CLIFF ....15,300 ± 800 YRS. B.P.
3. VINEYARD MORAINE
4. NANTUCKET MORAINE
5. HARBOR HILL MORAINE
6. CHARLESTOWN MORAINE
7. BUZZARD'S BAY MORAINE
8. SANDWICH MORAINE
9. ROGERS LAKE...14,240 ±240 YRS. B.P.
10. MIDDLETOWN READVANCE
II. FRESH POND MORAINE........................< 14,000 YRS. B.P.
12. KENNEBUNK ADVANCE........................13,200 ± 120 YRS. B.P.
13. POND RIDGE MORAINE........................13,520 ± 200 YRS. B.P.
14. PINEO RIDGE READVANCE...................~12,700 YRS. B.P.
15. HIGHLAND FRONT MORAINE...............~12,600 YRS. B.P.

Figure 1. A schematic representation of end moraines and ice margin positions of the late Wisconsin Laurentide ice sheet in southern and eastern New England and on the continental shelf.

proved or disproved convincingly because of the following considerations:

1. The ice sheet may have started to recede from its maximum position prior to 17,000 yrs ago or as late as 15,300 yrs ago, but certainly prior to 14,200 yrs ago. It is not known how much time elapsed between deglaciation and the accumulation of the oldest organic sediments in Rogers Lake. The Middletown readvance culminated approximately 25 km north of Rogers Lake. Therefore, the Middletown readvance may be as old as 15,000 to 16,000 yrs, and the possibility that the advance may be younger than 13,000 yrs cannot be excluded.

2. Although the Cambridge readvance occurred after 14,000 yrs ago, its age is not closely bracketed.

3. No recognizable marginal features link deposits of the Middletown and Cambridge readvances. White (1947, p. 757) reported two tills from the intervening area. However, it has not been demonstrated that either of these tills is related physically or chronologically to either advance.

In summary, the data offer no compelling reason to correlate the Middletown and Cambridge readvances, nor do they demand that these readvances represent regional events.

Maine and Southeastern Quebec

In Maine the fluctuating margin of the ice sheet retreated approximately parallel to the coast leaving a belt of submarine moraines (Borns, 1966, p. 13). This recession was accompanied by a marine invasion of the coastal region, extending into the river valleys of central Maine (Goldthwaite, 1949, p. 65). Glaciomarine sediments up to 50 m thick were deposited in the coastal region. Central Maine is characterized by ground moraine, eskers up to 160 km long extending to the southeast-facing slopes of the northeast-trending highlands of the state (Leavitt and Perkins, 1935, p. 74), and glaciomarine sediments in the major river valleys. The deposits on the northwest-facing slopes of the highlands are poorly known. However, reconnaissance (Borns and Calkin, unpub. data) suggests that the region may be characterized by end moraines left by a fluctuating glacier margin retreating downslope into the St. Lawrence Lowland.

In eastern Maine, Borns (1966, p. 13; 1967) reported a northeast-trending 30-km-wide belt of hundreds of moraines (Fig. 1) deposited along a fluctuating glacier margin as it retreated northwest from a position on the continental shelf. Most, and perhaps all, of these moraines were deposited below sea level that prevailed at that time. Within the moraine complex as many as 30 local marginal fluctuations are recognized, many of which were synchronous among the many former ice lobes of the coastal region.

Pond Ridge Moraine (Leavitt and Perkins, 1935, p. 46; Borns, 1967) in Cutler, Maine, was formed by one of these local readvances. Exposures in this moraine at North Cutler (Fig. 1) demonstrate that the moraine is underlain by glaciomarine silt and that the moraine is composed of an interbedded complex of coarse and fine glacial and glaciomarine sediments. The interbedded sediments contain a sparse assemblage of marine mollusks and pelecypods with shallow water affinities, as well as seaweed. C-14 dates for these materials indicate that the moraine was formed approximately 13,500 yrs ago (Stuiver and Borns, unpub. data).

Bloom (1960, p. 28) reported washboard moraines in southwestern coastal Maine that define recessional positions of the ice margin. However, these moraines are not as numerous nor as geographically ubiquitous as those in eastern Maine. General recession through southwestern Maine was interrupted by the Kennebunk glacial advance which reached the sea along a line at least 20 km and perhaps 45 km long (Bloom, 1960, p. 136; Fig. 1). Evidence for this advance is not continuously traceable in southwestern Maine, suggesting that the evidence may have resulted from one or perhaps several local marginal fluctuations at close, but slightly different, times. Clearly this was the situation in eastern Maine. Recent exposures in the Kennebunk town gravel pit have yielded marine shells that provide a C-14 age of approximately 13,200 yrs for this segment of the Kennebunk advance (Stuiver and Borns, unpub. data).

Approximately 25 C-14 dates on marine organisms contained in the glaciomarine sediments closely bracket the time of the marine submergence and the formation of the moraines of the coastal region between 13,500 and 12,500 yrs ago (Borns, 1967; Stuiver and Borns, 1967, p. 59–60; Stuiver and Borns, unpub. data).

Generally, the moraines of coastal Maine were formed at or below sea level. In several areas the distribution and sedimentology of the moraines and associated marginal deposits, such as deltas, indicate that the position of the glacier margin was controlled at sea level at various times. This control was presumably accomplished through a balance between the budgetary and mechanical aspects of the glacier and the melting and calving activities of the sea. Hollin (1962) noted such relationships in his study of the fluctuations of the Antarctic ice sheet.

The glacier recession that produced the coastal moraine complex was interrupted by a major readvance in eastern Maine that culminated in the sea at Pineo Ridge (Fig. 1) approximately 12,700 yrs ago (Borns, 1967).

Although the Pineo Ridge readvance probably correlates with the Port Huron readvance (Farrand and others, 1969) of the Great Lakes region and may have resulted from a significant climatic change, the field evidence clearly demonstrates that the readvance was accompanied by a lowering of the shoreline.

End moraines have not been recognized between the proximal side of the coastal moraine belt along the line of recession in central Maine or on the southeastward-facing slopes of the Blue and Border Mountains which comprise the northeastward-trending highlands of Maine. However, moraines are present on the northwestward-facing slopes of these highlands in Maine and Quebec (McDonald, 1968; Shilts, 1969, 1970; Borns and Calkin, 1970, p. 12; McDonald and Shilts, 1971; Borns, unpub. data). Just prior to deposition of the moraines closest to the divide, the ice sheet thinned and separated over the highlands leaving a considerable, but as yet undetermined, mass of wasting ice to the southeast (Borns, 1963, p. 704; Borns and Calkin, 1970, p. 12).

The Highland Front Moraine (Fig. 1), the most extensive of the moraines formed as the glacier margin receded down the northwest-facing slope of the St. Lawrence Valley in Quebec, is approximately 12,700 to 12,600 yrs old (Gadd, 1964). The Pineo Ridge Moraine in Maine may have formed very shortly before the ice sheet separated over the highlands of northwestern Maine, while the Highland Front and associated moraines on the northwest-facing slope of the St. Lawrence Valley in Quebec could have formed shortly thereafter at the retreating margin of the still-active ice sheet.

Although this model possibly could account for the similarity in ages of the Pineo Ridge and Highland Front Moraines, and for the lack of moraines in central Maine, the data are insufficient to prove or disprove it at present.

CONCLUSIONS

An evaluation of the available data relevant to chronology and events of the recession of the late Wisconsin Laurentide ice sheet in southern and eastern New England leads to the following conclusions:

1. A major amelioration of climate began prior to 14,200 yrs ago which resulted in a rapid dissipation of the ice sheet in New England at least by 12,500 yrs ago. The possibility that residual ice masses and reconstituted small glaciers persisted in the highlands after this time cannot be excluded.

2. Presently, no conclusive evidence is recognized for any climatic reversals during the dissipation of the ice sheet in New England.

3. Major events associated with the late Wisconsin ice sheet in New England compare with those of the Great Lakes region. However, no minor events have proven correlation, with the possible exception of the Pineo Ridge and Port Huron readvances.

ACKNOWLEDGMENTS

The writer is indebted to M. Stuiver and C. Kaye for the contributions cited in the text, to many colleagues in New England and Canada for ideas gleaned from discussions in the field, and to G. H. Denton and W. R. Farrand for critically reading the manuscript. The writer also expresses thanks to the National Science Foundation as many of the ideas expressed herein were, in part, developed through research supported by the foundation through grants GA-2823, GA-404, and GA-1563 to the University of Maine, Orono.

REFERENCES CITED

Bloom, A. L., 1960, Late Pleistocene changes of sea level in southwestern Maine: Maine Geol. Survey, 143 p.

Borns, H. W., Jr., 1963, Preliminary report on the age and distribution of the late Pleistocene ice in north-central Maine: Am. Jour. Sci., v. 261, p. 738–740.

—— 1966, An end-moraine complex in southeastern Maine [abs.]: Geol. Soc. America, Abs. for 1966, Spec. Paper 101, 485.

—— 1967, Guidebook, Friends of the Pleistocene, 30th Ann. Reunion, Machias, Maine: Orono, Maine, Maine Univ. Press, 19 p.

Borns, H. W., Jr., and Calkin, P. E., 1970, Multiple glaciation and dissipation of the last ice sheet in northwestern Maine [abs.]: Geol. Soc. America, Abs. with Programs (Ann. Mtg.), v. 2, no. 1, p. 12.

Chute, N. E., 1959, Glacial geology of the Mystic Lakes–Fresh Pond area, Massachusetts: U.S. Geol. Survey Bull. 1061-F, p. 187–216.

Farrand, W. R., Zahner, R., and Benninghoff, W. J., 1969, Cary–Port Huron interstade: Evidence from a buried bryophyte bed, Cheboygan County, Michigan: Geol. Soc. America Spec. Paper 123, p. 249–262.

Flint, R. F., 1953, Probable Wisconsin substages and late Wisconsin events in northeastern United States and southeastern Canada: Geol. Soc. America Bull., v. 64, p. 897–919.

—— 1956, New radiocarbon dates and late-Pleistocene stratigraphy: Am. Jour. Sci., v. 254, p. 265–287.

Gadd, N. R., 1964, Moraines in the Appalachian region of Quebec: Geol. Soc. America Bull., v. 76, p. 1249–1254.

Goldthwait, L., 1949, Clay survey, 1948: Maine Development Commission, Rept. State Geologist 1947–1948, p. 63–69.

Hollin, J. T., 1962, On the glacial history of Antarctica: Jour. Glaciology, v. 4, p. 173–195.

Kaye, C. A., 1964, Outline of Pleistocene geology of Martha's Vineyard, Massachusetts: U.S. Geol. Survey Prof. Paper 501-C, p. 134–139.

Kaye, C. A., and Barghoorn, E. S., 1964, Late Quaternary sea-level change and crustal rise at Boston, Massachusetts, with notes on the autocompaction of peat: Geol. Soc. America Bull., v. 75, p. 63–80.

Leavitt, H. W., and Perkins, E. H., 1935, Glacial geology of Maine, v. 2: Maine Technology Expt. Sta. Bull. 30, 232 p.

Levin, B., Ives, P. C., Oman, C. L., and Rubin, M., 1965, U.S. Geological Survey radiocarbon dates VIII: Radiocarbon, v. 7, p. 372, 398.

McDonald, B. C., 1968, Deglaciation and differential rebound in the Appalachian region of southeastern Quebec: Jour. Geology, v. 76, p. 664–677.

McDonald, B. C., and Shilts, W. W., 1971, Quaternary stratigraphy and events in southeastern Quebec: Geol. Soc. America Bull., v. 82, p. 683–698.

Meier, M. F., 1965, Glaciers and climate, in Wright, H. E., Jr., and Frey, D. G., eds., The Quaternary of the United States: Princeton, N.J., Princeton Univ. Press, p. 795–805.

Schafer, J. P., 1961, Correlation of the end moraines in southern Rhode Island: U.S. Geol. Survey Prof. Paper 424-D, p. 68–70.

Schafer, J. P., and Hartshorn, J. H., 1965, The Quaternary of New England, in Wright, H. E., Jr., and Frey, D. G., eds., The Quaternary of the United States: Princeton, N.J., Princeton Univ. Press, p. 113–127.

Schlee, J., and Pratt, R. M., 1970, Atlantic continental shelf and slope of the United States: U.S. Geol. Survey Prof. Paper 529-H, 29 p.

Shilts, W. W., 1969, Pleistocene geology of the Lac-Mégantic region, southeastern Quebec, Canada (Ph.D. thesis): Syracuse, N. Y.: Syracuse Univ., 154 p.

____ 1970, Introduction to the Quaternary history of the highlands region of western Maine, southeastern Quebec, and northern New Hampshire, in Borns, H. W., Jr., Calkin, P. E., Koteff, C., Pessl, F., and Shilts, W. W., New England Intercollegiate Geol. Conf. Guidebook: p. 25-28.

Sirkin, L. A., 1967, Late Pleistocene pollen stratigraphy of western Long Island and eastern Staten Island, New York, in Cushing, E. J., and Wright, H. E., Jr., eds., Quaternary paleoecology: New Haven, Conn., Yale Univ. Press, p. 249-274.

Stuiver, M., and Borns, H. W., Jr., 1967, Deglaciation and early postglacial submergence in Maine [abs.]: Geol. Soc. America, Abs. for 1967, Spec. Paper 115, p. 59-60.

Stuiver, M., Deevey, E. S., Jr., and Rouse, I., 1963, Yale natural radiocarbon measurements VIII: Radiocarbon, v. 5, p. 312-341.

White, S. E., 1947, Two tills and the development of glacial drainage in the vicinity of Stafford Springs, Connecticut: Am. Jour. Sci., v. 245, p. 754-778.

MANUSCRIPT RECEIVED BY THE SOCIETY DECEMBER 27, 1971

Printed in the United States of America

GEOLOGICAL SOCIETY OF AMERICA
MEMOIR 136
© 1973

Wisconsinan History of the Hudson-Champlain Lobe

G. Gordon Connally

Department of Geological Sciences, State University of New York at Buffalo, Buffalo, New York 14207

Leslie A. Sirkin

Department of Earth Science, Adelphi University, Garden City, New York 11530

ABSTRACT

The Hudson-Champlain Valley is the only continuous lowland between the classic glacial areas of the Midwest and coastal New England, and presumably it contains the most complete Wisconsinan record east of the Erie-Ontario Lobe. A date of 26,800 yrs B.P. on intraglacial peat in New Jersey establishes a maximum age for the Woodfordian advance of the Hudson-Champlain Lobe.

On western Long Island, deposition of the Ronkonkoma and Harbor Hill Moraines was followed by readvance and deposition of the Roslyn Till, and finally, by a stillstand on the north shore. Deglaciation from the Ronkonkoma Moraine began about 17,000 yrs B.P. In the Wallkill Valley, the southwestern physiographic continuation of the Hudson Valley, the terminal Woodfordian position is the Culvers Gap Moraine. Recessional positions are recorded at the Augusta, Sussex, Pellets Island, and Wallkill Moraines. The age of the Wallkill Moraine is established at 15,000 yrs B.P. The Woodfordian terminus of the Hudson-Champlain Lobe is traced northward from the Denville re-entrant in the Terminal Moraine in New Jersey, rather than westward, connecting the Ronkonkoma and Culvers Gap Moraines.

47

As the ice margin retreated north of the Hudson Highlands, Lake Albany was initiated in the mid-Hudson Valley. Readvances are recorded near Kingston, New York (the Rosendale readvance), and Glens Falls, New York (the Luzerne readvance). The Luzerne readvance has been inferred at 13,200 yrs B.P. When the ice margin retreated into the Champlain Valley, water levels dropped, forming Lake Quaker Springs and then Lake Coveville. The Bridport readvance is recorded in Lake Coveville sediments. Following the Bridport readvance the ice margin retreated to the vicinity of Burlington, Vermont, the maximum extent of Lake Coveville. When the ice retreated to the Highland Front Moraine, 12,600 yrs ago, water levels dropped again to form Lake Fort Ann, a lake that was restricted to the Champlain Valley.

Time-distance relationships are used to show migration of the herb and spruce pollen zones following deglaciation and to interpolate ages for undated deglacial events. Correlations are proposed for the Ronkonkoma-Culvers Gap Moraine, the Wallkill Moraine, and the Rosendale, Luzerne, and Bridport readvances.

INTRODUCTION

The purposes of this paper are to summarize the Wisconsinan history of the Hudson-Champlain Glacial Lobe and to emphasize the deglacial history during the Woodfordian Substage. The Hudson-Champlain Lowland is the only continuous, north-south lowland between the classic areas of the Midwest and coastal New England. As such, the record of this lobe is presumably the most complete east of the Great Lakes. However, the glacial geology of the area covered by the lobe received little attention by modern geologists prior to the 1960s. Recent studies have revealed mainly the results of Woodfordian glaciation, although some evidence for pre-Woodfordian glaciation has been encountered.

Geographic Extent

The Hudson-Champlain Lobe advanced southward through the Champlain Lowland into the Hudson Valley Lowland and then into the Hudson and Wallkill Valleys. The lobe crossed over the Hudson Highlands and traversed northern New Jersey as far as the Terminal Moraine. During early phases of deglaciation the lobe was confined by the New England Upland on the east and the Catskill Plateau on the west. During later phases, the ice was confined by the Green Mountains on the east and the Adirondack Mountains on the west. During its maximum extent the Hudson-Champlain Lobe coalesced with the Connecticut Valley Lobe on the east and the Ontario Lobe on the west. The deepest axial portion of the lobe followed the outcrop belt of Cambrian and Ordovician carbonates and shales, except where it crossed the crystalline rocks of the Hudson Highlands and flowed onto the Triassic Lowland in New Jersey.

Previous Work

The most pertinent work on western Long Island was performed in the late nineteenth century by Nathanial L. Britton, John Bryson, and Charles L. Hollick, and by J. B. Woodworth (1901). The early work culminated in the treatise of Myron L. Fuller (1914). Fleming (1935), MacClintock and Richards (1936), Perlmutter (1949), and Upson (1955) made the only modern contributions, and these were contained within regional studies.

Work in the Hudson Highlands is confined to the comprehensive report of Salisbury (1902) on the glacial geology of New Jersey and to the discussion by Thompson (1936) of the Hudson gorge.

After the papers by Peet (1904) and Woodworth (1905), the mid-Hudson Valley was subject mainly to quadrangle analysis by Gordon (1911), Stoller (1911, 1916, 1920), Holzwasser (1926), and Cook (1930, 1942, 1943). Notable exceptions are papers by Fairchild (1916) and Chadwick (1928) on the northern Hudson Valley and by Happ (1938) on the Rondout (Minisink) Valley. Regional correlations by Flint (1953, 1956), MacClintock (1954), and Denny (1956) suggested possible ice margins in the Hudson Valley, but were not based on field work there. Many other papers were published on the proglacial lakes in the Hudson and Champlain Valleys, culminating in the classic work of Chapman (1942). However, the glacial geology of the Champlain Valley was largely ignored except for quadrangle reports by Woodworth (1908), Kemp (1910), Barker (1916), and Kemp and Alling (1925), although early reports by the State Geologists of Vermont mention isolated features.

Since 1960, sufficient work has been reported to warrant this general history of the region. However, this paper should be considered as a progress report since many areas remain unstudied. Sirkin (1967a, 1967b, 1968, 1971) has examined the geomorphology, stratigraphy, and palynology of western and central Long Island. Foord and others (1970) have looked at the subsurface relationships in north-central Long Island, and Donner (1964) has examined pollen stratigraphic relationships in eastern Long Island. Connally (1967a, 1968c) and Connally and Sirkin (1967, 1970) have described the geomorphology, palynology, and late-glacial history of the Wallkill Valley. LaFleur (1965a, 1965b) has described the glacial geology of the Troy area, and Connally and Sirkin (1969b, 1971) have described the events in the Glens Falls region.

Modern work in the Champlain Valley began with a statewide, quadrangle-by-quadrangle project by the Vermont Geological Survey that resulted in the Surficial Geologic Map of Vermont (Stewart and MacClintock, 1970) and the accompanying report (Stewart and MacClintock, 1969). Simultaneously, work began in the northeast corner of the Adirondacks (Denny, 1967, 1970) and an environmental survey was later initiated by the New York Geological Survey (Connally, 1967b). Connally (1968a, 1970) reported individually on his Vermont work and Connally and Sirkin (1969b) summarized the glacial history of the Lake Champlain-Lake George region.

Woodworth (1905), Flint (1953), and Connally and Sirkin (1969a) have previously attempted summaries of the entire Hudson-Champlain Lobe.

GLACIAL ADVANCE

Presumably, the striae of the Adirondack Mountains and Catskill Plateau, the Green Mountains and western Massachusetts, and of the Hudson and Champlain Valleys resulted from the most recent advance, the Woodfordian. Striae west of the lowland have been recorded in many quadrangle studies and show a uniform flow pattern west of south. Striae east of the lowland in Vermont and in a few quadrangles in New York show a uniform flow pattern east of south. Striae in the lowland record a general southerly flow direction. Thus, the Woodfordian glacier is inferred to have flowed southward as a lobe, down the axis of the lowland and to have spread radially westward and eastward. Studies of the Newburgh quadrangle west of the Hudson River (Holzwasser, 1926) and the Poughkeepsie quadrangle east of the river (Gordon, 1911) support the hypothesis of radial lobate flow, as does the study of Connally (1967b) in the Champlain Valley.

Drumlin patterns in the Hudson Valley support the southerly trend indicated by the striae. However, the drumlins in the southern Hudson Valley and in the Wallkill Valley appear to be composed of a different till than is present in other exposures of ground moraine. The "drumlin till" is very compact and low in carbonate. It is oxidized for 10 to 18 ft and is leached for 5 to 10 ft. The usual ground moraine is open textured, locally stratified, and always oxidized throughout exposures. Unfortunately, the two types of drift have never been observed in superposition. Except that the Hudson Valley tills have a silty-clay matrix derived from the shaly bedrock, they present a "two till problem" similar to that reported by Schafer and Hartshorn (1965) in New England. It is possible that the drumlin till of the Hudson Valley represents a pre-Woodfordian deposit and that the thick oxidized and leached zone at the surface represents a Sangamonian or Farmdalian geosol.

LONG ISLAND

Geomorphology

The Ronkonkoma and Harbor Hill Moraines (Fig. 1) have been well-known morphologic features since they were first formally described by Fuller (1914, p. 163). It has been largely ignored that these moraines are the deposits of three glacial lobes: the Hudson-Champlain Lobe in western Long Island, the Connecticut Valley Lobe in central Long Island, and the Narragansett Valley Lobe in eastern Long Island. To the east the Harbor Hill Moraine may be traced into southern New England, while to the west it has been presumed to override the Ronkonkoma

Figure 1. Ice margins on Long Island. From south to north, the Ronkonkoma Moraine, the Harbor Hill Moraine, and the "ice stand on the necks" are shown by solid hachured lines

Moraine. Because there is no evidence that these moraines differ significantly in age, they are here assumed to be deposits of the same glaciation. However, the possibility exists that the Ronkonkoma Moraine may be the product of an earlier glaciation.

Stratigraphy

The stratigraphy of the Long Island drift is less well known and understood than the morphology. Fuller (1914) defined the Manhasset Formation as containing a lower outwash (the Herod Gravel), the medial Montauk Till, and an upper outwash (the Hempstead Gravel). An upper till unit has since been traced by Sirkin (1968) in western and central Long Island and designated as the Roslyn Till (Sirkin, 1971). He also demonstrated a lower till unit in the Ronkonkoma Moraine in western Long Island which was observed grading southward into outwash. However, this unit is not the Montauk Till, as described by Fuller or demonstrated in the type section by Newman and others (1968). Although the relationship between the Montauk Till and the moraines is uncertain in the Hudson-Champlain Lobe, Foord and others (1970) have demonstrated that a possible Montauk equivalent is associated with the Ronkonkoma Moraine in the Connecticut Valley Lobe. We suggest that the Herod Gravel is a pro-Montauk facies while the Hempstead Gravel, of which the Ronkonkoma Moraine is composed almost entirely, was deposited during the stillstand that caused the Ronkonkoma Moraine. This interpretation is in opposition to that of Fuller (1914) and MacClintock and Richards (1936). The Harbor Hill Moraine in the Connecticut Valley Lobe is composed of outwash similar to the Hempstead Gravel and is thus thought to represent a recessional stillstand. Foord and others (1970) report an upper drift, which may include the Roslyn Till, associated with the Harbor Hill Moraine and infer a readvance to that position.

In the Hudson-Champlain Lobe the stratigraphy is much more complicated. Both the Ronkonkoma and Harbor Hill Moraines are composed of thick outwash, presumably the Hempstead Gravel, that is capped by the thin Roslyn Till. In the sections north of the end moraine the outwash is generally folded or contorted and suggests overriding. Sirkin (1968) demonstrated continuity of the Roslyn Till across western Long Island. Thus, it is possible either that the Ronkonkoma and Harbor Hill Moraines are palimpset, as are the moraines of Ohio (Totten, 1969), and represent pre-Woodfordian glaciation(s), or that the Roslyn Till represents readvance over earlier deposited Woodfordian moraines. Interbedded lacustrine, but unfossiliferous sands, silts, and clays have been observed beneath Roslyn Till and as probable facies of the outwash in the end moraine in both central and western Long Island. However, as weathering zones have never been observed, the latter hypothesis is accepted by the authors.

Sirkin (1967a, Fig. 1) has demonstrated features that he interprets as evidence of a recessional stillstand along the north shore of Long Island—his "ice stand on the necks." The event is presumed to be sequentially younger than that which deposited the Roslyn Till. It left stagnant ice deposits and drumlinoid topography along the sides of the bays that indent the coast, and it may have been accompanied by minor readvance of ice tongues into these bays.

Chronology and Environments

Pollen analysis of bog sediments in this region reveals the presence of the herb pollen zone and the extension of that zone into late Wisconsinan time when ice apparently occupied the northern margin of Long Island. Although radiocarbon ages for this interval have not been successfully obtained for either basal bog sediments or glacial sediments in the area, correlation of the pollen stratigraphy of the Long Island region with radiocarbon-dated, late Pleistocene sequences in nearby regions has provided probable ages for the onset of glacial recession from the moraines and for the establishment of vegetation in this area (Fig. 6).

Evidence for intraglacial deposits in the Delaware Valley of New Jersey is based on peat samples that yielded an age of 26,800 yrs B.P. (Sirkin and others, 1970). The peat contains a mixed deciduous pollen spectrum, and the event is correlated with the Farmdalian Substage of the Lake Michigan Basin. This provides a maximum age for the advance of the Hudson-Champlain Lobe.

It is inferred that glacial recession had begun by 17,000 yrs B.P. and that tundra vegetation was established at that time with ice still occupying the northern margin of western Long Island. Pollen stratigraphy in central and northern New Jersey indicates the presence of tundra vegetation (that is, the herb pollen zone) about 17,000 yrs ago, and of spruce dominance (the spruce pollen zone) about 14,000 yrs ago (Sirkin and others, 1970). Tundra vegetation is recorded between 14,300 and 12,150 yrs ago in southern Connecticut (Davis, 1969).

Correlations of the herb and spruce pollen zones in Martha's Vineyard with those zones in New England (Ogden, 1963) and with the upper part of the herb pollen zone and the spruce pollen zone in western Long Island (Sirkin, 1967a, 1967b) point to decreasing ages for the onset of tundra (the base of the herb pollen zone) and for glacial recession, both northeastward and northwestward from western Long Island. This inference is supported by the absence of "older" radiocarbon-dated pollen stratigraphy in central Long Island (Sirkin, 1971) or in eastern Long Island (Donner, 1964). It may also be inferred that recession of the Hudson-Champlain Lobe left most or all of western Long Island ice free, and that this area was colonized first by tundra and next by spruce forests when the Connecticut Valley Lobe occupied or was receding from eastern Long Island, southern coastal New England, and the offshore islands.

In summary, we infer a Woodfordian advance and deposition of the Ronkonkoma Moraine, followed by deposition of the Harbor Hill Moraine, and then readvance to the Ronkonkoma Moraine. In western and central Long Island the readvance reached the Ronkonkoma Moraine and draped both it and the Harbor Hill Moraine with the Roslyn Till. Following the readvance, the ice receded to the north shore and left ice-contact deposits and outwash in the bays and inlets. In southeastern Long Island ice-contact kames north of the end moraine may mark the extent of that readvance. The Harbor Hill Moraine of eastern Long Island is apparently younger than its classic counterpart in western Long Island and even possibly younger than the last glacial stillstand in the west.

WALLKILL VALLEY

Geomorphology

Five moraines have been described in the Wallkill Valley: the Culvers Gap, Augusta, Sussex, Pellets Island, and Wallkill Moraines (Fig. 2). The Culvers Gap was originally described by Salisbury (1902) and then redefined by Herpers (1961) as the Ogdensburg-Culvers Gap recessional moraine. Minard (1961) traced this moraine over Kittatinny Mountain to the Delaware Valley, and Connally (in prep.) has remapped the entire unit as the terminal Woodfordian moraine of the Hudson-Champlain Lobe.

South of the Culvers Gap Moraine, Salisbury (1902, Pl. 8) shows flow lines oriented north-south across Kittatinny Mountain, but north of the moraine they are oriented northeast-southwest, parallel to the Wallkill Valley. In addition, tillstones representing Ridge and Valley lithologies are present south of the moraine and crystalline rocks of the

Hudson Highlands (Reading Prong) are absent. North of the moraine, Ridge and Valley lithologies are subordinate to crystallines. Thus, the Wisconsinan drift between the Culvers Gap Moraine and the Terminal Moraine is either pre-Woodfordian or was deposited by Woodfordian ice slightly older than the Hudson-Champlain Lobe.

The Augusta Moraine was mentioned by Salisbury (1902, p. 374) and by Connally and Sirkin (1967). It has been mapped by Connally as a recessional moraine about 2 mi north of the Culvers Gap Moraine. The Sussex Moraine (new name) has also been mapped as a recessional moraine at Sussex, New Jersey, about 9 mi north of the Culvers Gap Moraine.

The Pellets Island Moraine occurs about 24 mi northeast of the Culvers Gap Moraine and was first described by Connally and Sirkin (1967, p. A7). When the ice stood at the Pellets Island Moraine, it dammed a lake to the south that drained southwestward into the Delaware River drainage system, with an outlet at an elevation of about 500 ft. The Pellets Island Moraine includes deposits that were originally mapped as the separate New Hampton Moraine by Connally and Sirkin (1967). This moraine is probably equivalent to morainal segments that are banked against the Hudson Highlands east of the Hudson River.

The Wallkill Moraine occurs 15 mi north of the Pellets Island Moraine and was first described by Connally and Sirkin (1967, p. A9). When the ice stood at the Wallkill Moraine, it dammed a second lake that drained eastward to Lake Albany, in the Hudson Valley, via Moodna Creek. This lake had an outlet at about 400 ft. Both the Wallkill and Pellets Island Moraines are the product of an active ice front as shown by the multiple till sections reported from wells in their vicinities by Frimpter (1970).

Chronology and Environments

Pollen stratigraphy in the mid-Wallkill Valley demonstrates the presence of tundra vegetation (the herb pollen zone) as early as 15,000 yrs B.P., the age of emplacement of the Wallkill Moraine (Connally and Sirkin, 1970). This age is approximately correlative with, or somewhat older than, the inception of tundra vegetation in southern New England as determined by Deevey (1958) and Davis (1969). Similarly, the inception of the spruce record (the spruce pollen zone) prior to 12,850 yrs B.P. predates that event in southern Connecticut reported by Davis (1969) as 12,150 yrs B.P. Both low deposition rates and a 2,000-yr duration for the tundra indicate that the ice probably receded from the southern

Figure 2. Ice margins in the Wallkill Valley. From south to north, the Culvers Gap, Augusta, Sussex, Pellets Island, and Wallkill Moraines, and the Rosendale readvance are shown by solid hachured lines.

and middle Wallkill Valley prior to 15,000 yrs B.P., as discussed in Connally and Sirkin (1970).

In summary, we infer that the Woodfordian ice advanced to the Culvers Gap Moraine and then receded northeastward with stillstands of active ice at the Augusta, Sussex, and Pellets Island positions, while damming a southward-draining lake. When the ice receded north of the Hudson Highlands, the lake drained eastward toward Lake Albany and lowered its outlet. A final stillstand is recorded by the Wallkill Moraine about 15,000 yrs B.P., before the ice evacuated the Wallkill Valley completely. In mild contrast to Long Island, there does not seem to be a history of readvance during earliest Woodfordian time.

TERMINAL MORAINE

The classic tracing of the Terminal Moraine by Chamberlin (1883), reinforced by Salisbury (1902), would seem to preclude the correlation of events between the Wallkill Valley and Long Island as proposed here. However, careful examination of Salisbury's work (1902, p. 234) shows that he was uncertain about the moraine in the vicinity of Denville, New Jersey. It must be remembered that both Salisbury and Chamberlin were tracing the *limit of glaciation* and not stage boundaries. Salisbury followed Chamberlin in extending the moraine westward, forming the Denville re-entrant.

Examination of the modern 7-1/2′ topographic maps revealed another possible tracing of the terminal moraine, northwest from the Denville re-entrant (Fig. 3). Field reconnaissance has confirmed this trend. The massive moraine actually continues northwestward as far as the base of the Highlands, presumably overriding the westerly trending moraine. Within the Highlands, Salisbury (1902, p. 466–478, Pl. 8) illustrates a difference in drift lithology and flow direction from Picatinny Arsenal, at the base of the Highlands, north and then west toward the Culvers Gap Moraine. These differences coincide with our projected ice margin. The lithologic differences are explained by north-south flow lines southwest of the moraine and southwesterly flow lines, parallel to the Highlands, northeast of the moraine.

Thus, we propose morphostratigraphic correlation of the Woodfordian terminal moraine as shown in Figure 3. The ice margin here outlined is consistent with lobate flow emanating from the Hudson Valley Lowland. Although much field work is needed, it is probable that recessional positions will also ultimately be identified in the Highlands.

ROSENDALE AND LUZERNE READVANCES AND LAKE ALBANY

As the Woodfordian ice margin retreated north of the Hudson Highlands, Lake Albany began to form between the ice front and a dam

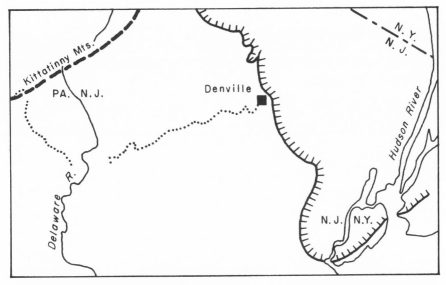

Figure 3. The Terminal Moraine in New Jersey. The Woodfordian terminal moraine of the Hudson-Champlain Lobe (hachured line) is shown overriding the older terminal moraine (dotted line) at the Denville re-entrant.

to the south. Lake Albany expanded northward with the retreating ice front but was interrupted by two readvances (Fig. 4). The expansion of Lake Albany was temporarily interrupted by a southward readvance of the ice into the Wallkill Valley, described by Connally (1968b) as the Rosendale readvance. Following this readvance the ice front continued to recede northward. Cook (1942, 1943) documented the Lake Albany water plane at the confluence of Catskill Creek and the Hudson Valley, and LaFleur (1965b) presented a sequence of diagrams illustrating an expanding Lake Albany in the Capitol District. Connally and Sirkin (1971) have shown that the ice front continued to retreat 20 to 35 mi into the Lake George Basin and Champlain Valley before the Luzerne readvance into Lake Albany near Glens Falls, New York.

Rosendale Readvance

As the ice front readvanced into the Wallkill Valley, it redammed the valley and caused the formation of glacial Lake Tillson at 230 ft. A delta built at the south end of Lake Tillson is composed of more than 30 ft of topset beds, which show that the lake was a feature of readvance and rising waters rather than merely recession and a falling water level. Frimpter (1970, wells 159-357-15, 158-400-7, and 156-402-d) reports multiple tills as much as 13 mi north of Rosendale, which gives a minimum figure for the magnitude of the readvance. When the ice front receded from the Rosendale readvance, Lake Albany expanded northward once again at an elevation of about 200 ft at the confluence

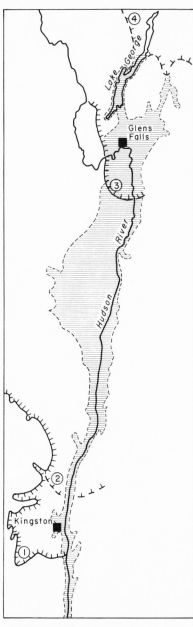

Figure 4. Lake Albany. The maximum extent of Lake Albany (dashed line) is shown in relation to (1) the Rosendale readvance (solid hachured line), (2) the minimum known withdrawal of ice (dashed hachured line) prior to that readvance, (3) the Luzerne readvance (solid hachured line), and (4) the minimum known withdrawal of ice prior to that readvance (dashed hachured line).

of the Wallkill and Hudson Valleys. The position of the Rosendale readvance cannot be determined precisely but can be bracketed within a distance of 2 mi (Fig. 4).

Luzerne Readvance

The Luzerne readvance extended south of Glens Falls to the vicinity of Willton, New York, and dammed waters of the upper Hudson River in the Adirondacks, causing glacial Lake Warrensburg (Miller, 1925; Connally and Sirkin, 1971). Craft (1970) reported that local glaciers in the Adirondacks were actively discharging sediment into Lake Warrensburg. Connally and Sirkin (1971) have suggested an age of 13,200 yrs B.P. for the Luzerne readvance.

Chronology and Environments

In the Glens Falls region of the northern Hudson Valley, tundra vegetation (the herb pollen zone) was established from sediments that contain pollen of tundra plants (Connally and Sirkin, 1971). The maximum possible age for the spruce pollen zone in this region has been set at 12,400 ± 200 yrs B.P., based on samples from the same site. The duration of the herb pollen zone in this region was probably about 750 yrs. The duration of the spruce zone is approximately 3,400 yrs, using the 12,400 yr age as the maximum age and the age of 9,000 yrs for the end of the zone in southern New England reported by Deevey (1958) and Davis (1969).

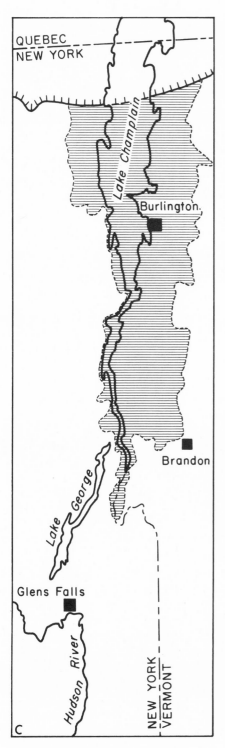

a threshold near Whitehall, New York. Lake Fort Ann drained southward into the Hudson River via Wood Creek and was dammed on the north by the ice front, probably as it stood at the Highland Front Moraine of Gadd (1964) (Fig. 5C).

Wagner (1969) has suggested a complex history for the shoaling of Lake Fort Ann and the later invasion of marine waters to form the Champlain Sea. However, the history of the Hudson-Champlain Lobe essentially ends with the establishment of the Highland Front Moraine and Lake Fort Ann. Control for later events rests with the topography of the St. Lawrence Valley.

RATES OF
GLACIAL RECESSION

The rate of glacial recession is indicated by the ages of basal bog sediments on the glacial depositional surfaces. It is inferred that deglaciation began about 17,000 yrs B.P. in the end moraine region of western Long Island, and 15,000 yrs ago in the mid-Wallkill Valley and in southern New England (Fig. 6A). In the northern Hudson Valley, deglaciation began over 13,200 yrs ago. In the Champlain Valley, deglaciation was over by about 12,600 yrs ago, based on a bog-bottom date of the herb pollen

Figure 5. Champlain-Hudson Valley lakes. A. Lake Quaker Springs (dashed line) and the probable ice margin (hachured line) bordering that lake. B. Lake Coveville (dashed line) and the approximate ice border during the maximum phase (solid hachured line) and the Bridport readvance (dashed hachured line). C. Lake Fort Ann (dashed line) and the approximate position of the Highland Front Moraine (hachured line).

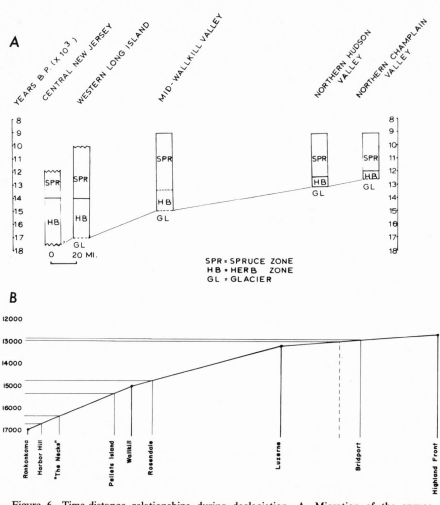

Figure 6. Time-distance relationships during deglaciation. A. Migration of the spruce and herb pollen zones following deglaciation. B. Interpolated ages for deglacial events based on assumed ages for the Ronkonkoma, Wallkill, and Highland Front Moraines and the Luzerne readvance.

zone in the southern St. Lawrence Lowland obtained by Terasmae and LaSalle (1968). Thus, about 4,000 yrs are necessary for glacial recession from Long Island to the northern Hudson Valley, and 600 yrs for deglaciation of the Champlain region.

The persistence of glacially controlled climates is evident at least as far north as the northern Hudson Valley, as attested to by the persistence of the herb pollen zone in that region. Influx of spruce and other boreal elements may have been transitional, but spruce forests (that is, the spruce pollen zone) were not established until about 12,400 yrs B.P. The ages for the late-glacial pollen zones provide a time scale for the northward migrations of the tundra and the spruce forests following deglaciation. A minimum of 3,800 yrs (17,000 to 13,200 yrs B.P.) is inferred for the transgression of tundra (near the ice front) to the northern Hudson Valley, while only about 1,600 yrs (14,000 to 12,400 yrs B.P.) were required for spruce forests to reach that region. This forest migration, once started, is shown to be much more rapid than deglaciation. The lag in the migration of the spruce forests may be related to climate, which may have persisted in severity until about 14,000 yrs B.P., but warmed considerably after that time. This warming may also be interpreted from the rate of recession curve (Fig. 6B), which demonstrates that the ice margin receded from western Long Island to the mid-Hudson region, a distance of about 65 mi, in about 2,000 yrs (17,000 to 15,000 yrs B.P.), and from the mid-Hudson to northern Hudson Valley, a distance of 125 mi, in about 1,800 yrs (15,000 to 13,200 yrs B.P.). Final recession from the northern Hudson Valley to the northern Champlain Valley, a distance of 110 mi, took place in only 600 yrs (13,200 to 12,600 yrs B.P.). These observations may provide a clue to the mode of deglaciation. It may have been predominantly downwasting with gradual recession and minor fluctuations of the ice front until 15,000 yrs ago, followed by more rapid downwasting accompanied by backwasting and two major readvances until 13,200 yrs ago. Final deglaciation then accelerated as a result of downwasting and backwasting of Champlain Valley ice accompanied by stagnation of upland ice.

If a simple time × distance relationship is assumed, as in Figure 6B, the ages of events may be interpolated between the Ronkonkoma Moraine and the Wallkill Moraine, between the Wallkill Moraine and the Luzerne readvance, and between the Luzerne readvance and the Highland Front Moraine. The age of the Augusta Moraine–Harbor Hill Moraine is about 16,700 yrs, the Sussex Moraine and "ice stand on the necks" date from about 16,400 yrs B.P., and the Pellets Island Moraine dates from about 15,600 yrs B.P. The Rosendale readvance is no older than 14,800 yrs. The Bridport readvance and Lake Coveville date from about 12,900 yrs B.P., and Lake Quaker Springs from perhaps 13,000 yrs B.P.

CORRELATIONS

Pre-Woodfordian drift has not been identified in either the Hudson or Champlain Valleys. The change in flow direction south of the Culvers Gap Moraine suggests that the drift between the Culvers Gap Moraine and the Terminal Moraine may be pre-Woodfordian but stratigraphy is lacking. Lacustrine deposits are present beneath Woodfordian till in the Luzerne Mountain section of Connally and Sirkin (1971) and have been reported throughout the Hudson Valley by James F. Davis (oral commun.). These deposits, along with the drumlin till of the Wallkill Valley, may date from pre-Woodfordian time.

Table 1 shows the proposed correlations for Woodfordian units from the Ronkonkoma Moraine to the Highland Front Moraine. Where no obvious correlation exists, a column is left blank.

TABLE 1. CORRELATION OF THE HUDSON-CHAMPLAIN VALLEY ICE MARGINS WITH WESTERN NEW YORK, CENTRAL NEW YORK, THE MOHAWK VALLEY, AND WESTERN NEW ENGLAND

Age B.P.	Time	Western New York	Central New York*	Mohawk Valley	Hudson-Champlain	Western New England
12,600	Upper Woodfordian (Port Huron) Substage	—	Pinnacle Hills? Moraine	?	Highland Front Moraine	Highland Front Moraine
12,800		Hamburg Moraine	Waterloo-Auburn Moraine	Valley Heads Moraine	Bridport readvance	Cherry River Moraine
13,150	Cary-Port Huron interstade	Gowanda Moraine	Valley Heads Moraine	Valley Heads Moraine	Luzerne readvance	Burlington drift
		?	Valley Heads Moraine	Valley Heads Moraine	Rosendale readvance	Middletown readvance
		Lake Escarpment Moraine	Valley Heads Moraine	Valley Heads Moraine	Wallkill Moraine	?
15,000	Main Woodfordian Substage	?	?	?	Pellets Island Moraine	Coastal Connecticut Moraines
		?	Arkport? Moraine	—	Sussex Moraine	"Ice stand on the necks"
17,000		Kent Moraine	Almond Moraine	—	Augusta Moraine	Harbor Hill Moraine
24,000		Kent Moraine	Almond Moraine	—	Culvers Gap Moraine	Ronkonkoma Moraine

*The Pinnacle Hills and Waterloo-Auburn Moraines are defined by Fairchild (1932), and the Arkport and Almond Moraines by Connally (1964).

The Culvers Gap and Ronkonkoma Moraines are considered the earliest Woodfordian deposits of the Hudson-Champlain Lobe. As such, they correlate with the Kent drift border of western New York and Ohio. The Montauk Till in eastern Long Island is lithologically similar to the till exposed in the southern cliffs of Block Island and may correlate with it. The age of emplacement of these moraines is not known but the interglacial peat in the Delaware Valley dated at 26,800 yrs B.P. is probably much older. As recession had commenced by 17,000 yrs B.P., this is a minimum age for the moraines.

Both the stratigraphic relationships and the interpolated ages in Figure 6B suggest that the Harbor Hill and Augusta Moraines and the Sussex Moraine and "ice stand on the necks" are closely associated in time with the maximum Ronkonkoma-Culvers Gap position.

No correlation of the Pellets Island Moraine is readily apparent, but it may be equivalent to ice marginal positions in coastal Connecticut.

The 15,000 yr age of the Wallkill Moraine compares favorably with that of the Valley Heads Moraine of western New York. Calkin (1970) reported a date of 14,900 yrs B.P. for the cessation of outwash deposition from the Lake Escarpment Moraine, which suggests correlation of these two units. This further implies correlation with the Valley Heads Moraine of the Finger Lakes and Mohawk Valley, although the ice apparently remained at the Valley Heads position while recession was underway to the west.

The maximum position of the Rosendale readvance may be traced into the Rondout Valley, west of the Wallkill Valley, where it is equivalent to a position shown by Happ (1938). Rich (1935, Pl. 2) extended his lowest (and youngest) ice margin from the Catskills into the Rondout Valley, where it appears to equate with the Rosendale readvance. This, in turn, suggests that the older ice marginal positions of Rich may be equivalent to the Wallkill and Pellets Island Moraines, south of the Rosendale readvance. As the Rosendale readvance and the Middletown readvance in the Connecticut Valley were defined on the basis of similar evidence, and as both occur at about the same latitude, correlation of these events is also proposed. An age is suggested between 14,800 and 14,000 yrs B.P.

Connally and Sirkin (1971) have suggested correlation of the Luzerne readvance with the Gowanda Moraine in western New York. The ice in the Finger Lakes and Mohawk Valley probably still remained at the Valley Heads Moraine through both the Rosendale and Luzerne readvances. The Burlington drift margin near Bennington, Vermont, has been proposed as the New England equivalent of the Luzerne readvance. If this correlation is accepted, then the Rosendale readvance, and perhaps all the moraines in the Wallkill Valley, correlate with the Shelburne drift of Vermont.

Because the 13,150 yr age of recession from the Luzerne readvance is pre-Port Huron, the Bridport readvance was correlated by Connally and Sirkin (1971) with the Port Huron Moraine in the Michigan and Huron Basins, with the Hamburg Moraine in western New York, and with the Waterloo-Auburn Moraine in the Finger Lakes region. The ice margin probably still remained at the Valley Heads position in the Mohawk Valley. Connally and Sirkin (1969a, 1969b) have tentatively suggested correlation of the Bridport readvance and the Cherry River Moraine of McDonald (1968) in Quebec.

ACKNOWLEDGMENTS

We thank James A. Bier for permission to use portions of his *Landforms of New York* as a base for Figures 1 and 2.

REFERENCES CITED

Barker, E. E., 1916, Ancient water levels of the Crown Point embayment: New York State Mus. and Sci. Service Bull. 187, p. 165–190.

Calkin, P. E., 1970, Strand lines and chronology of the glacial Great Lakes in northwestern New York: Ohio Jour. Sci., v. 70, p. 78–96.

Chadwick, G. H., 1928, Ice evacuation stages at Glens Falls, New York: Geol. Soc. America Bull., v. 39, p. 901–922.

Chamberlin, T. C., 1883, Preliminary paper on the terminal moraine of the second glacial epoch: U.S. Geol. Survey Ann. Rept. 3, p. 291–402.

Chapman, D. H., 1942, Late-glacial and postglacial history of the Champlain Valley: Vermont Geol. Survey, State Geol. 23d Rept., p. 48–83.

Connally, G. G., 1964, The Almond Moraine in western New York (Ph.D. thesis): East Lansing, Michigan, Michigan State Univ.

—— 1966, Surficial geology of the Mount Mansfield 15-minute quadrangle, Vermont: Open-file report to State Geologist, 33 p.

—— 1967a, The glacial history of the mid-Hudson region, New York [abs.]: Geol. Soc. America Spec. Paper 101, p. 254–255.

—— 1967b, Surficial resources of the Champlain Basin, New York: Maps and report to New York State Office of Planning Coordination, 111 p.

—— 1968a, Glacial geology of the Brandon-Ticonderoga region, Vermont [abs.]: Geol. Soc. America Spec. Paper 115, p. 255–256.

—— 1968b, Glacial geology of the Mount Mansfield Quadrangle, Vermont [abs.]: Geol. Soc. America Spec. Paper 115, p. 257.

—— 1968c, The Rosendale readvance in the lower Wallkill Valley, New York, *in* National Assoc. Geol. Teachers Guidebook, Eastern Section, New Paltz, New York, p. 22–28.

—— 1970, Surficial geology of the Brandon-Ticonderoga 15-minute quadrangle, Vermont: Vermont Geol. Survey, Studies in Vermont Geology No. 2, 45 p.

Connally, G. G., and Sirkin, L. A., 1967, The Pleistocene geology of the Wallkill Valley, *in* Waines, R. H., ed., New York State Geol. Assoc. Guidebook, 39th ann. mtg., p. A1–A16.

—— 1969a, Deglacial events in the Hudson-Champlain Valley and their possible equivalents in New England [abs.]: Geol. Soc. America Abstracts with Programs for 1969, Pt. 1, Northeastern Section, p. 9.

_____ 1969b, Deglacial history of the Lake Champlain-Lake George lowland, *in* Barnett, S. G., ed., New York State Geol. Assoc. Guidebook, 41st ann. mtg., 20 p.

_____ 1970, Late glacial history of the upper Wallkill Valley, New York: Geol. Soc. America Bull., v. 81, p. 3297-3306.

_____ 1971, The Luzerne readvance near Glens Falls, New York: Geol. Soc. America Bull., v. 82, p. 989-1008.

Cook, J. H., 1930, The glacial geology of the Capitol District: New York State Mus. and Sci. Service Bull. 285, p. 181-190.

_____ 1942, The glacial geology of the Catskill quadrangle: New York State Mus. and Sci. Service Bull. 331, p. 189-237.

_____ 1943, Glacial geology of the Coxsackie quadrangle: New York State Mus. and Sci. Service Bull. 332, p. 321-357.

Craft, J. L., 1970, Late Pleistocene glacial climate of the Adirondack Mountains, northeast New York, U.S.A [abs.]: Am. Assoc. Quaternary Res., 1st ann. mtg., Bozeman, Montana.

Davis, M. B., 1969, Climatic changes in southern Connecticut recorded by pollen deposition at Rogers Lake: Ecology, v. 50, p. 409-422.

Deevey, E. S., 1958, Radiocarbon-dated pollen sequences in eastern North America: Zurich Geobot. Inst. Veruff., v. 34, p. 30-37.

Denny, C. S., 1956, Wisconsin drifts in the Elmira region, New York, and their possible equivalents in New England: Am. Jour. Sci., v. 254, p. 82-95.

_____ 1967, Surficial geologic map of the Dannemora quadrangle and part of the Plattsburgh quadrangle, New York: U.S. Geol. Survey, GQ 635.

_____ 1970, Surficial geologic map of the Mooers quadrangle and part of the Rouses Point quadrangle, Clinton County, New York: U.S. Geol. Survey Map I-630.

Donner, J. J., 1964, Pleistocene geology of eastern Long Island, New York: Am. Jour. Sci., v. 262, p. 355-376.

Fairchild, H. L., 1916, Pleistocene features in the Schenectady-Saratoga-Glens Falls section of the Hudson Valley: Geol. Soc. America Bull., v. 27, p. 65-66.

_____ 1932, New York moraines: Geol. Soc. America Bull., v. 43, p. 627-662.

Fleming, R. L. S., 1935, Glacial geology of central Long Island: Am. Jour. Sci., v. 36, p. 216-238.

Flint, R. F., 1953, Probable Wisconsin substages and late-Wisconsin events in northeastern United States and southeastern Canada: Geol. Soc. America Bull., v. 64, p. 897-904.

_____ 1956, New radiocarbon dates and late Pleistocene stratigraphy: Am. Jour. Sci., v. 254, p. 265-287.

Foord, E. E., Parrott, W. R., and Ritter, D. F., 1970, Definition of possible stratigraphic units in north-central Long Island, New York, based on detailed examination of selected well cores: Jour. Sed. Petrology, v. 40, p. 194-204.

Frimpter, M. H., 1970, Ground-water basic data, Orange and Ulster Counties, New York: New York Water Resources Comm. Bull. 65, 93 p.

Fuller, M. L., 1914, The geology of Long Island, New York: U.S. Geol. Survey Prof. Paper 83, 223 p.

Gadd, N. R., 1964, Moraines in the Appalachian region of Quebec: Geol. Soc. America Bull., v. 75, p. 1249-1254.

Gordon, C. E., 1911, Geology of the Poughkeepsie quadrangle, New York: New

York State Mus. and Sci. Service Bull. 148, 121 p.

Happ, S. C., 1938, Significance of Pleistocene deltas in the Minisink Valley: Am. Jour. Sci., v. 236, p. 417–439.

Herpers, H., 1961, The Ogdensburg-Culvers Gap recessional moraine and glacial stagnation in New Jersey: New Jersey Geol. Survey Geol. Rept. Ser., no. 6, 16 p.

Holzwasser, F., 1926, Geology of Newburgh and vicinity: New York State Mus. and Sci. Service Bull. 270, 95 p.

Kemp, J. F., 1910, Geology of the Elizabethtown and Port Henry quadrangles, New York: New York State Mus. and Sci. Service Bull. 138, 173 p.

Kemp, J. F., and Alling, H. L., 1925, Geology of the Ausable quadrangle: New York State Mus. and Sci. Service Bull. 261, 126 p.

LaFleur, R. G., 1965a, Glacial lake sequences in the eastern Mohawk–northern Hudson region, in Hewitt, P. C., and Hall, L. M., eds., New York State Geol. Assoc. Guidebook, 37th ann. mtg., p. C1–C23.

_____ 1965b, Glacial geology of the Troy, N.Y., quadrangle: New York State Mus. and Sci. Service Map and Chart Ser., no. 7, 22 p. and map.

MacClintock, P., 1954, Leaching of Wisconsin glacial gravels in eastern North America: Geol. Soc. America Bull., v. 55, p. 1143–1164.

MacClintock, P., and Richards, H. G., 1936, Correlations of Pleistocene marine and glacial deposits of New Jersey and New York: Geol. Soc. America Bull., v. 47, p. 289–338, 1982–1994.

McDonald, B. C., 1968, Deglaciation and differential postglacial rebound in the Appalachian region of southeastern Quebec: Jour. Geology, v. 76, p. 664–677.

Miller, W. J., 1925, Remarkable Adirondack glacial lake: Geol. Soc. America Bull., v. 36, p. 513–520.

Minard, J. P., 1961, End moraines on Kittatinny Mountain, Sussex County, New Jersey: U.S. Geol. Survey Prof. Paper 424-C, p. 68–70.

Newman, W. S., Thurber, D. L., Krinsley, D. H., and Sirkin, L. A., 1968, The Pleistocene geology of the Montauk Peninsula, in Finks, R. M., ed., New York State Geol. Assoc. Guidebook, 40th ann. mtg., p. 155–173.

Ogden, J. G., III, 1963, The Squibnocket cliff peat; radiocarbon dates and pollen stratigraphy: Am. Jour. Sci., v. 261, p. 344–353.

Peet, C. E., 1904, Glacial and postglacial history of the Hudson and Champlain Valleys: Jour. Geology, v. 12, p. 415–469.

Perlmutter, N. M., 1949, Geological correlation of logs of wells in Long Island, New York: New York Water Power and Control Comm. Bull. GW-18.

Rich, J. L., 1935, Glacial geology of the Catskills: New York State Mus. and Sci. Service Bull. 299, 180 p.

Salisbury, R. D., 1902, The glacial geology of New Jersey: New Jersey Geol. Survey Final Rept. Ser., v. 5, 802 p.

Schafer, J. P., and Hartshorn, J. H., 1965, The Quaternary of New England, in Wright, H. E., and Frey, D. G., eds., The Quaternary of the United States: Princeton, N.J., Princeton Univ. Press, p. 113–128.

Sirkin, L. A., 1967a, Late Pleistocene pollen stratigraphy of western Long Island and eastern Staten Island, New York, in Cushing, E. J., and Wright, H. E., eds., Quaternary paleoecology: New Haven, Conn., Yale Univ. Press, p. 249–274.

_____ 1967b, Correlation of late glacial pollen stratigraphy and environments in the northeastern U.S.A.: Rev. Paleobotany and Palynology, v. 2, p. 205–218.

_____ 1968, Geology, geomorphology, and late-glacial environments of western

Long Island, New York, *in* Finks, R. M., ed., New York State Geol. Assoc. Guidebook, 40th ann. mtg., p. 233-253.

____ 1971, Surficial glacial deposits and postglacial pollen stratigraphy in central Long Island, New York: Pollen et Spores, v. 13, p. 93-100.

Sirkin, L. A., Owens, J. P., Minard, J. P., and Rubin, M., 1970, Palynology of some Pleistocene peat samples from the coastal plain of New Jersey: U.S. Geol. Survey Prof. Paper 700-D, p. D77-D87.

Stewart, D. P., and MacClintock, P., 1969, The surficial geology and Pleistocene history of Vermont: Vermont Geol. Survey Bull. 31, 251 p.

____ 1970, Surficial geologic map of Vermont, *in* Doll, C. G., ed., Vermont Geol. Survey map.

Stoller, J. H., 1911, Glacial geology of the Schenectady quadrangle: New York State Mus. and Sci. Service Bull. 154, 44 p.

____ 1916, Glacial geology of the Saratoga quadrangle: New York State Mus. and Sci. Service Bull. 183, 50 p.

____ 1920, Glacial geology of the Cohoes quadrangle: New York State Mus. and Sci. Service Bull. 215-216, 49 p.

Terasmae, J., and LaSalle, P., 1968, Notes on late-glacial palynology and chronology at St. Hilaire, Quebec: Canadian Jour. Earth Sci., v. 5, p. 249-257.

Thompson, H. D., 1936, Hudson Gorge in the Highlands [abs.]: Geol. Soc. America Proc. 1935, p. 111.

Totten, S. M., 1969, Overridden recessional moraines of north-central Ohio: Geol. Soc. America Bull., v. 80, p. 1931-1946.

Upson, J. E., 1955, Ground water resources on Long Island: Am. Water Works Assoc. Jour., v. 47, p. 341-347.

Wagner, W. P., 1969, The late Pleistocene of the Champlain Valley, Vermont: New York State Geol. Assoc. Guidebook, 41st ann. mtg., p. 65-76.

Woodworth, J. B., 1901, Pleistocene geology of portions of Nassau County and the Borough of Queens: New York State Mus. and Sci. Service Bull. 48, p. 618-670.

____ 1905, Ancient water levels of the Champlain and Hudson Valleys: New York State Mus. and Sci. Service Bull. 84, p. 65-265.

____ 1908, Pleistocene geology of the Mooers quadrangle: New York State Mus. and Sci. Service Bull. 83, p. 3-60.

MANUSCRIPT RECEIVED BY THE SOCIETY DECEMBER 27, 1971

GEOLOGICAL SOCIETY OF AMERICA
MEMOIR 136
© 1973

Wisconsin Glaciation in the Huron, Erie, and Ontario Lobes

A. DREIMANIS

*Department of Geology, University of Western Ontario,
London, Ontario, Canada*

R. P. GOLDTHWAIT

*Department of Geology, The Ohio State University,
Columbus, Ohio 43210*

ABSTRACT

The names of the Huron, Erie, and Ontario Lobes imply that glaciers followed these lake depressions, but the flow patterns of these lobes were complex and changed several times during the Wisconsin glaciation. The sublobes of the southwestern part of the so-called Erie Lobe were more often an extension of the ice coming down Huron Basin, and ice from both basins participated in these sublobes. A review of the studies of Wisconsin-age deposits in the area of the three lobes indicates the emphasis that has been placed on investigations of tills by multiple methods, on paleontological studies of the interstadial and late-glacial deposits, and on radiocarbon dating. A threefold time-stratigraphic division of the Wisconsin Stage in this area is based upon synchronous fluctuations by several glacial lobes, and upon climatologic inferences from paleontologic studies. Early and late Wisconsin experienced maximum glacial advances; middle Wisconsin was dominated by interstadial retreats.

The first Wisconsin glacial advance reached into the St. Lawrence Lowland only, and was followed by a glacial retreat during the St. Pierre Interstade about 65,000 radiocarbon yrs B.P. The second major glacial advance (by several lobes) went farther, but did not reach as far south as the late Wisconsin glaciation in Indiana and Ohio. In Pennsylvania and New York it was more extensive than the late Wisconsin, if the Olean Drift is indeed of early Wisconsin age. The source of ice was centered in the eastern Canadian Laurentide area.

Mid-Wisconsin glacier margins retreated several times far into the Huron and Ontario Basins, or even north of them. Three main retreats were probably interrupted by two readvances which reached into the Erie Basin but not south of the Lake Erie watershed. Mid-Wisconsin time began more than 50,000 yrs B.P. and ended about 23,000 yrs B.P.

During late Wisconsin time the major glacial advance in the western part of the region investigated reached its farthest extent south in at least three pulses: 21,000, 19,500, and 18,000 yrs ago. The oscillating retreats were interrupted by three documented readvances: about 17,000, 15,000, and 13,000 yrs B.P. The source of ice was primarily in the western Laurentide center. No evidence has been found here for the Valders readvance which took place in the Lake Michigan Lobe.

INTRODUCTION

The Huron, Erie, and Ontario Lobes occupied the eastern Great Lakes region, splitting into several sublobes down valleys along the periphery of the ice sheets (Fig. 1). The names Huron, Erie, and Ontario Lobes imply correctly that the streams followed these existing lake depressions; however, their flow patterns have been more complex than their names may indicate. For instance, the southwestern portion of the so-called Erie Lobe was composed of ice from the Erie, Georgian Bay, and Huron Lobes, as indicated by the lithology and fabric of till at places like Sidney, Ohio (Forsyth, 1965), from landforms, and striae on bedrock (Flint and others, 1959). Harrison (1960), from lithologic investigations of tills of the White River Sublobe in central Indiana, concluded that the tills were deposited by the Erie Lobe without any contribution from the two other lobes, but we question this. First, Harrison's determination of minerals indicates glacial paths could have been different from those that he had ascertained. Second, Gunn (1967) has found jasper pebbles originating from the area north of Lake Huron or northern Michigan, and in the area to the south (Brown and Morgan Counties). Tillite cobbles from the area north of Georgian Bay have been noticed by the authors in many places from east-central Indiana to the Scioto Sublobe areas. The jasper and tillite pebbles suggest a contribution by the Huron and Georgian Bay Lobes.

The most westerly of the sublobes, Decatur, which has recently been considered as a part of the Erie Lobe (Willman and Frye, 1970), is probably an extension of the Huron and the Saginaw Lobes. Because of the absence of long distance provenance studies of tills in the Decatur Sublobe, the related Peoria Loess from the Wabash River valley can be used for testing this tentative reclassification. This loess contains consid-

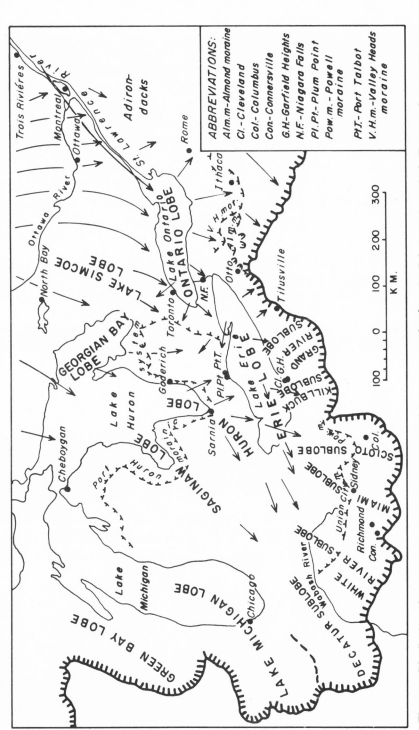

Figure 1. Distribution of glacial lobes and those sublobes which have been related to the Erie Lobe in geologic literature for the Wisconsin glaciation. The arrows indicate glacial movements concluded from striae and drumlins. Note: the Almond and Valley Heads Moraines mark the approximate southern boundary of the Ontario Lobe. The Olean Drift area south of it was traversed by an older glacial flow coming from the northeast.

erably more epidote (Frye and others, 1962) than has been found in the tills of the Erie Basin (Dreimanis and others, 1957), or even in the tills of mixed Huron and Erie lithology from an interlobate area (Westgate and Dreimanis, 1967). Thus, the assignment of the Decatur and White River Sublobes (or even the Miami, Scioto, and Kilbuck Sublobes) to the Erie Lobe should be re-examined.

The upglacier sources of the Erie and other lobes also changed during the glacial stages, especially in the Wisconsin, as the areas of glacial outflow shifted. For instance, till minerals and pebbles show that the Erie Lobe had been fed at different times by glacial flows originating on the Canadian Shield either as far east as south-central Quebec, or as far west as the Ontario-Quebec boundary (Dreimanis and others, 1957). At another time the center of glacial outflow was as far south as the Lake Ontario Basin (Holmes, 1952). As for the Huron Lobe, southern Lake Huron was a center of local glacial outflow during Cary time, whereas the center shifted toward the Canadian Shield during the following Port Huron advance (Dreimanis, 1967).

All these factors have been considered when interpreting the lithology, texture, fabric, and other characteristics of the tills deposited in these lobal areas. The assignments to distant upglacier source areas have been done mainly in southern Ontario and northern New York, where heavy mineral and carbonate investigations of till matrix were used to decipher the regional movements of the glacial lobes or flows (Dreimanis and others, 1957; Dreimanis, 1961; Connally, 1964). South of the Great Lakes, in Pennsylvania, Ohio, and Indiana, more attention has been paid to the empirical differentiation of the sublobes than to their upglacier provenances. Therefore, in Figures 3 and 4, the lobal provenance of the sublobe till is given with various degrees of uncertainty.

TILL STUDIES FOR STRATIGRAPHIC CORRELATIONS

During the past 30 yrs this region has become one of the main testing grounds for various criteria of till investigation and correlation. For instance, Holmes (1941) began till fabric investigation, and various procedures were tested by Dreimanis and Reavely (1953), Sitler and Chapman (1955), Harrison (1957), Dreimanis (1959a), and Ostry and Deane (1963). MacClintock and Dreimanis (1964) recognized reorientation of till fabric by a subsequent glacial overriding. Pebble lithology has been identified by nearly every till investigator of this region, the most comprehensive regional studies being those of Holmes (1952) and Anderson (1957). Holmes (1960) discussed the shapes of the pebbles south of Lake Ontario. Dreimanis and Reavely (1953) and Anderson (1957) studied rock and mineral fragments of coarse sand. Regional heavy mineral investigations, aimed mainly at deciphering the provenance of tills, were begun by Dreimanis and others (1957) and were continued in other areas by Dreimanis (1960), Moss and Ritter (1962), Frye and others (1962), Sitler

(1963), Connally (1964), Forsyth (1965), and Rosengreen (1970). Light minerals of the sand grade, particularly feldspars and quartz, have been studied by Totten (1960), White (1967), Gross (1967), Vagners (1970), and Gross and Moran (1971). New techniques for rapid carbonate determinations were developed by Skinner and Halstead (1958) and Dreimanis (1962). Carbonate percentages in till matrix were found to be a useful criterion for regional differentiation and correlation of tills (Dreimanis, 1960, 1961, 1969b; Dreimanis and Karrow, 1965; Karrow, 1967; White, 1967; Wilding and others, 1971). Sitler (1963, 1968) has investigated till in thin sections. As illite and chlorite are the predominant clay minerals in nonweathered tills and loesses of the Huron, Erie, and Ontario Lobes (Droste, 1956; Droste and Doehler, 1957; Forman and Brydon, 1961; Frye and others, 1962), clay minerals have been useful only empirically for differentiation or correlation of tills in this region (Rosengreen, 1970; Teller, 1970; Gooding, 1971). Trace elements in tills have been used only recently for regional differentiation of tills (Wilding and others, 1971; May and Dreimanis, this volume). The relationship of till composition to bedrock lithology has been investigated by various authors, leading to diverse theoretical conclusions (Harrison, 1960; Vagners, 1966; Dreimanis and Vagners, 1969; Gross and Moran, 1971). Granulometric compositions have been used widely for description and differentiation of tills, particularly by Shepps (1953), Shepps and others (1959), Dreimanis (1961), and Karrow (1963). A new method to characterize till by "size factors" was developed by Shepps (1958). Dreimanis and Vagners (1969) have concluded that the lithologic and granulometric composition of tills depends upon the bimodal distribution of rocks and their constituent minerals in tills. They introduced the term and defined the concept of "terminal grade," which is useful for selecting the appropriate particle size grades for investigation of minerals in tills (Dreimanis, 1969b). During the past few years, quantitative statistical methods have been applied to regional studies; for instance, in trend-surface analyses (Gross, 1967; White, 1969; Gross and Moran, 1971) and in multivariate statistical techniques (May and Dreimanis, this volume, p. 221–228).

Stratified drift has been investigated by techniques similar to those applied to tills. Attention has been focused on their texture and lithology; for instance, by Kempton and Goldthwait (1959), Dell (1959), Frye and others (1962), Moss and Ritter (1962), Chapman and Dell (1963), Denny and Postel (1964), Dreimanis and others (1966), Goldthwait (1968), Lajtai (1969), and Steiger and Holowaychuk (1971). One of the main purposes of these investigations has been to decipher the provenance of stratified drifts. Engineering tests have seldom been used as stratigraphic criteria.

The use of multiple analytical data for description, correlation, and differentiation of glacial and related deposits has become a common practice since the middle of this century, and most stratigraphic or regional reports on the last ice-age deposits now contain such data. See, for example, Dreimanis and Reavely (1953), Dreimanis (1958, 1961), Shepps and others (1959), Day (1960), Denny and Lyford (1963), Karrow (1963, 1967), Muller (1964), Forsyth (1965), Dreimanis and others (1966), Van Wyckhouse (1966), Westgate and Dreimanis (1967), White (1967), Goldthwait and Rosengreen (1969), White and others (1969), and Thomas (1970).

OTHER STRATIGRAPHIC CRITERIA

Morphology

For a long time, systematic investigations of the last ice-age deposits in the eastern Great Lakes region were based mainly upon landforms. Occasionally, natural or artificial sections and well logs were also studied by applying field criteria (color, description of texture, and the like). Because of the large areas investigated, a wealth of information was gathered, and presented in regional reports and maps (Winchell, 1872; Gilbert, 1871, 1873; Chamberlin, 1883, 1894; Wright, 1889; Chamberlin and Leverett, 1894; Dryer, 1894; Leverett, 1895, 1902, 1931; Coleman, 1901, 1933, 1937, 1941; Sherzer, 1902; Wilson, 1905; Taylor, 1913; Fairchild, 1909, 1932; Goldthwait, 1910; Leverett and Taylor, 1915; Johnston, 1916; Malott, 1922; Thornbury, 1937; Deane, 1950; Chapman and Putnam, 1966; Hough, 1958, 1963; Shepps and others, 1959; Zumberge, 1960; Von Engeln, 1961; Goldthwait and others, 1961, 1965; Muller, 1965; and Wayne and Zumberge, 1965). The county reports and master's theses dealing with smaller areas, in Ohio and Indiana especially, are too numerous to be listed here. In the Ontario, Erie, and Huron Basins the relation of the end moraines to the beaches and outlets of the proglacial lakes provided useful criteria for deciphering the retreat of the Wisconsin ice sheet; the differences in the amount of uplift and tilt of the beaches helped to determine their relative ages. South of the Erie–Ohio River divide, the nature of Wisconsin stratigraphy is different from that of the Great Lakes basins, because subparallel streams carried all outwash southward away from the ice. Here the chronology of the Wisconsin is demonstrated by a series of valley trains with ice-contact margins and by a few till sheets, distinct from each other in lithology and composition, which end near one or another of the many moraines.

Sections, Test Holes, Geophysics

Along the north shore of Lake Ontario, wave-cut sections up to 100 m high provided an excellent opportunity for an early start of stratigraphic investigations. Several reports may be considered quite modern by present standards; for instance, those by Hinde (1877) and Coleman (1894, 1901). Even prior to these studies, multiple drifts separated by "forest beds" were recognized by Orton and Newberry in Ohio and Indiana (White, this volume, p. 3–34). The number of sections suitable for stratigraphic investigations has been increasing during the last two decades, partly because of continually stepped-up urban construction in places such as Toronto, Hamilton, Cleveland, Toledo, and Columbus, and especially because of new highway cuts. The shortage of Quaternary geologists, rather than lack of sections, leads to the sporadic nature of stratigraphically described exposures.

In areas where the exposures are not sufficiently deep, continuous stratigraphic records and samples for analyses have been obtained by drilling (Dreimanis and others, 1966; Lewis and others, 1966; Hobson and Terasmae, 1969), and stratigraphic boundaries have been traced by geophysical methods (Janssens, 1963; Hobson and others, 1969; Wall, 1968). These are the approaches which must be more widely used in the resolution of remaining questions.

Soils as Chronologic Criteria

In Indiana, Leverett and Taylor (1915) noted the difference in carbonate leaching between the Wisconsin and Illinoian drifts, and Thornbury (1940) used the depth of leaching of carbonates in surface soils for accurate mapping of the boundary between Wisconsin and Illinoian drift areas. MacClintock (1954), Goldthwait and Forsyth (1962), and Forsyth (1967) attempted to differentiate substages of the last glacial stage by this method, and Dreimanis (1959b) demonstrated its feasibility for correlations between distant areas of similar soil development. Dreimanis (1957b), Meritt and Muller (1959), and Rosengreen (1970) drew attention to the dependence of the depth of leaching upon the carbonate content in the material that had been leached. Thus, the unusually deep Russell soils in southern Ohio, which were considered to be of "early" Wisconsin age by Goldthwait and Forsyth (1962), are now shown (Rosengreen, 1970) to be a function of a low carbonate content in the unique Boston Till.

Paleosols of Wisconsin interstades, buried underneath glacial drift, have been found mainly in the south half of this region (White, 1953, 1967; Goldthwait, 1958; Gooding, 1963; Forsyth, 1965; Wayne, 1966; White and others, 1969; Dreimanis, 1971). Gooding and others (1959) interpreted some of the clay-enriched leached tops of calcareous gravels, which have been found underneath calcareous till, as the result of recent weathering extending from the present soils downward along the joints in till. However, Goldthwait (1959a) has shown that in some cases the continuous clay-enriched zones are remnants of paleosols that developed on the gravel.

Paleontologic Studies

Organic sediments between tills of Wisconsin age, or organic remains incorporated in tills, are more widely preserved than paleosols from the south boundary of the Wisconsin glaciation to the St. Lawrence Valley and James Bay Lowlands. Recently, the palynology of the early Wisconsin deposits has been studied in Quebec, Ontario, New York, and Indiana (Terasmae, 1958, 1960; Kapp and Gooding, 1964; Muller, 1964), as have the mid-Wisconsin deposits in Ontario, Ohio, and Michigan (Dreimanis, 1958; Dreimanis and others, 1966; Forsyth, 1965; Hobson and Terasmae, 1969; Karrow and Terasmae, 1970; Zumberge and Benninghoff, 1969; Berti, 1971).

Larger fossils have been studied from the intertill deposits in these lobes since before the turn of the century (Coleman, 1894). The main types of fossils studied have been mollusks, wood, plant seeds, mammal bones, insects, diatoms, and ostracods (Hinde, 1877; Coleman, 1894, 1933, 1941; Scudder, 1895; Leonard, 1953; LaRocque and Forsyth, 1957; Burns, 1958; Wayne, 1959, 1965; Dreimanis and others, 1966; Karrow, 1967; Coope, 1968; Churcher, 1968). The fossils have provided data about the paleoecology and climate of the interstadial intervals during the last ice age.

Similar investigations have been made of some late Wisconsin organic deposits lying on top of the uppermost till. Many of them are at localities where proboscidian fossils have been found (Thomas and others, 1952; Skeels, 1962; Drumm, 1963; Forsyth, 1963; Dreimanis, 1968; Churcher and Karrow, 1963; Cleland, 1966). Most bog-lake studies deal mainly with palynology; for instance, the studies by Terasmae (1958, 1960), Terasmae and Mirynech (1964), Kapp (1964), Gooding and Ogden (1965), Ogden (1965, 1966, 1967, 1969), Cushing (1965), Lewis and others (1966), and Davis (1967). Mollusks are considered by Terasmae (1965) and LaRocque (1966, 1967, 1968, 1970).

Chronology

In order to establish a chronology for the late-glacial time, the evolution of the Niagara Falls was studied by Spencer (1907) and Johnston (1926), and varved clays were studied by Antevs (1925, 1928) and DeGeer (1926). Unfortunately, several assumptions in both interpretations were incorrect.

The introduction of radiocarbon dating provided a more accurate chronology of the Wisconsin deposits. Absolute dates may still vary a little as refinement and correction of carbon-14 scales continue. Wood, plant detritus, peat, and bones from more than 100 upper Pleistocene sites have been radiocarbon dated in the region discussed in this paper, thus providing a framework for the chronology and a means of correlation with other regions (Flint, 1963).

Two stratigraphic units, the middle Wisconsin of the entire Erie-Ontario Lobe and the beginning of the late Wisconsin ice advance over Ohio, have been dated by the largest number of radiocarbon age determinations (Fig. 2) ever published from any one glaciated area. When Flint and Rubin (1955) began to discuss radiocarbon dates older than 18,000 yrs B.P., half of the determinations compiled from all over the world were from this region. Many dates, some still unpublished, are now accumulating for the proglacial lake phases, which were contemporaneous with the oscillating retreat of the late Wisconsin ice.

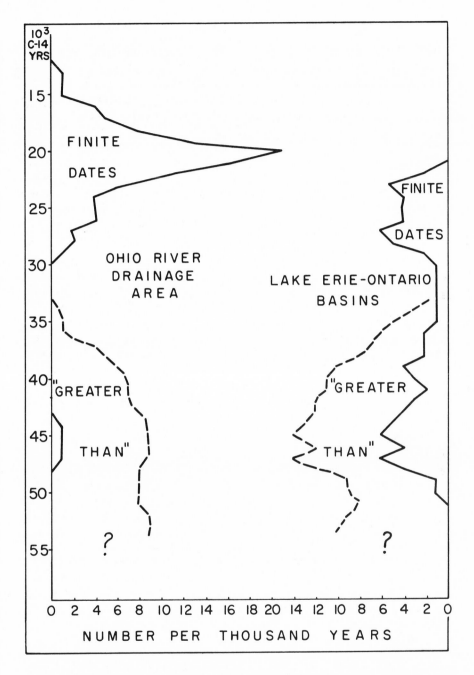

Figure 2. Possible radiocarbon date ranges determined on wood, plant detritus, or peat found underneath or in till, in numbers per thousand years, considering also one standard deviation. For instance, a date 40,300 ± 1,000 yrs B.P. is plotted once as the 39-40, once as the 40-41, and once as the 41-42 thousand.

MAJOR STRATIGRAPHIC DIVISIONS

Multiple criteria have been used for studies of the Wisconsin stratigraphy in the region of the Ontario, Erie, and Huron Lobes. Evidence of more or less synchronous glacial fluctuations in all three lobal areas, supported by palynologic investigations, and dated by radiocarbon age determinations, have gradually led to development of a threefold time-stratigraphic division of the last glacial stage: early, middle, and late Wisconsin. These now divide at 53,000 and 23,000 radiocarbon yrs. The terms have been used informally in Ohio and Ontario (Forsyth and LaRocque, 1956; Dreimanis, 1957a; Goldthwait, 1958), and formally since 1959 (Dreimanis, 1959c, 1960). Early Wisconsin used here should not to be confused with early Wisconsin as used long ago by Leverett (1902) for drift now exclusively late Wisconsin or Woodfordian.

This three-fold time division fits the changing activities of ice in these eastern lobes and sublobes (Fig. 3) far more directly and effectively than the three-fold Altonian-Farmdalian-Woodfordian divisions breaking at 28,000 and 23,000 radiocarbon yrs in the Michigan and Green Bay Lobes (Willman and Frye, 1970). The early and the late Wisconsin witnessed extensive glacial advances reaching into southern Ohio and Indiana. The middle Wisconsin, about 23,000 to 53,000 yrs B.P., was dominated by lengthy interstadial retreats, such as the Sidney Interstadial in Ohio (Forsyth and LaRocque, 1956; Forsyth, 1965; Goldthwait and others, 1965), the Port Talbot and Plum Point Interstadials in Ontario (Dreimanis, 1957a, 1958, 1971; Dreimanis and others, 1966; Karrow, 1969) and in Pennsylvania (White and others, 1969), and the New Paris Interstadial in Indiana (Gooding, 1963; Kapp and Gooding, 1964). These overlap in time.

In each of the three major time-stratigraphic divisions, several short, but synchronous and therefore stratigraphically significant, glacial fluctuations can be distinguished—for instance, the glacial retreat in the Decatur, White River, Miami, and Scioto Sublobes during the Connersville Interstadial (Gooding, 1963), the even greater retreat in all the local areas discussed in this paper and by Mörner and Dreimanis (this volume) during the Erie Interstade (Dreimanis, 1958), and another retreat in the Huron, Erie, and Ontario Basins during the Lake Arkona Phase (Leverett and Taylor, 1915; Dreimanis, 1967). Final decision on the time-stratigraphic ranks of all the major and minor divisions depends upon their correlations with other areas. At this time we judge the early, middle, and late divisions to correlate with many other places around the world as well.

EARLY WISCONSIN

First Glacial Advance

The first recorded glacial advance of the Labradorean ice sheet during Wisconsin time deposited the Bécancour Till in the St. Lawrence Lowland

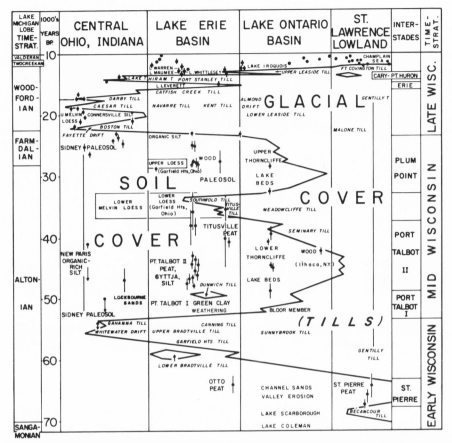

Figure 3. Diagrammatic stratigraphy of the Wisconsin deposits from the St. Lawrence Lowland (right side) to central Ohio and Indiana (left side). The vertical scale represents time in thousands of radiocarbon years B.P.; the horizontal scale is the approximate distance along the glacial path of the Ontario-Erie Lobe. The jagged line represents the advances and retreats of the Ontario-Erie Lobe, including also participation of the Georgian Bay and Huron Lobes. Areas where the glacial advance from the north did not merge with the Erie Lobe are marked as a lens. Most of the litho-stratigraphic names are the same as in Figures 4 and 7. Heavy dots are radiocarbon dates, with one standard deviation shown as a line; indefinite dates are indicated by a downward-pointing arrow.

and, after crossing the lowland for another 100 km, laid down the Johnsville Till over deeply weathered gravels, probably of Sangamon age, in the Quebec part of the Appalachians (McDonald and Shilts, 1971). Goldthwait (1970) cites evidence for local glaciers and cirque cutting at this time in the White (and Adirondack?) Mountains to the south when mean summer temperatures were 9.3° C lower than at present. However, there is no evidence that this first advance of the ice sheet reached the Great Lakes. Instead, by blocking the outlet of the Ontario Basin, it caused a rise of water level and formed Lake Scarborough. This lake was as high as the late Wisconsin Lake Iroquois, and its outlet was probably also in the Rome, New York, area. A large delta

(Scarborough Formation), rich in fossil plant and animal remains, was deposited at Toronto, Ontario; it has been the object of paleontological and sedimentological investigations for nearly 100 yrs (Karrow, 1967). Terasmae (1960) concluded from palynologic study that the Toronto area climate was cool at that time, with mean annual temperatures about 6° C lower than at present.

Farther southwest the climatic conditions were more like those of the present time but cooler in the summer. It is quite possible that the top portion of some deposits and soils, considered to be of Sangamon age, belong to the early Wisconsin. Thus, in the pollen diagram from the Smith Farms "Sangamon" soil (Kapp and Gooding, 1964) southwest of Richmond, Indiana, the upper portion of unit 8 (silt and fine sand) contains three palynologically different zones. The lower zone (51 to 86 cm below the top) is rich in pollen of *Quercus* and *Carya*, the middle zone (33 to 51 cm below the top) is completely disturbed by involutions, and the upper zone (5 to 33 cm below the top) shows dominance of *Pinus* and *Picea* pollen. We propose that the involuted middle layer represents the first cold episode of the Wisconsin glaciation, and the upper (the pine and spruce zone) belongs to the following St. Pierre Interstade. Kapp and Gooding (1964) also list mosses from the upper surface of two buried Sangamon soils, and from two Connersville Interstadial silts of late Wisconsin age. All these moss assemblages are similar and suggest climates colder than the present Indiana climate; that is, boreal conditions. It is suggested here that the mosses from the Sangamon soil surfaces at Cummin's Farm (unit 3) and Bergendorfer Farm (unit 4), both west of Connersville, Indiana, belong to the early Wisconsin. Kapp and Gooding already imply this correlation (1964, p. 321): "As the Wisconsin continental glacier approached this area, closing the Sangamon interglacial age, elements of the boreal flora dispersed southward."

St. Pierre Interstade

The St. Pierre Interstade is the only nonglacial early Wisconsin interval in the Laurentide ice area that has several finite as well as infinite radiocarbon age determinations: Gro 1711, 67,000 ± 1,000 yrs B.P.; Gro 1766, 64,000 ± 2,000 yrs B.P.; and GrN 1799, 65,300 ± 1,400 yrs B.P. (type area near Trois Rivières, Quebec); and GrN 3213, 63,900 ± 1,700 yrs B.P. (Otto Peat Bed, New York). These dates are so close to the possible maximum range of the radiocarbon method that they may be considered as minimum dates and are subject to large adjustments to correct calendar years.

Terasmae (1958) concludes from palynologic data that the mean annual temperature in the St. Lawrence Lowland was 2° to 4° C lower than at present, and Muller (1964) concludes that the Otto Peat Bed was deposited under conditions similar to those existing today north of North Bay, Ontario. Frenzel (1967) notes that this means a lowering of the mean annual temperature at Otto by 5° C during the deposition of the peat bed. Temperature depression in the upper pine-spruce zone (unit 8) at Smith Farm, Indiana, would be comparable. Frenzel (1967) thinks that the St. Pierre type deposits correspond to the European Brørup Interstadial, and the Otto Peat Bed corresponds to the slightly older Ammersfoort Interstadial. Both belong to the European early glacial, or Würm I of the last ice age (van der Hammen and others, 1967).

During the St. Pierre Interstade the St. Lawrence Lowland was ice free, as were the Appalachians in Quebec (McDonald and Shilts, 1971). White Mountain–Adirondack local glaciers probably retreated but did not disappear. How far the Laurentide ice sheet retreated in Labrador is still unknown.

In the Lake Ontario Basin the St. Pierre Interstade is represented by an erosional interval during which valleys up to 60 m deep were cut into the Scarborough and the underlying Don Formations (Coleman, 1933, 1941; Karrow, 1967, 1969). These valleys are filled by the so-called channel sands, which contain a variety of fossil remains, particularly mollusks and mammals, probably reworked from the Don Formation. As one of the buried valleys extends at least 20 ft below the present Lake Ontario level, the outlet through the St. Lawrence Lowland must have been lower than now, probably due to depression by the preceding first Wisconsin glacial advance.

Main Glacial Advance

The farthest reach of ice in early Wisconsin time was south of the Ontario and Erie Basins into central Ohio and the adjoining part of Indiana (Karrow, 1963, 1967; Gooding, 1963; Wayne, 1963; Goldthwait and Forsyth, 1965; Dreimanis and Karrow, 1965; Forsyth, 1965; Dreimanis and others, 1966; Goldthwait and Rosengreen, 1969). Although the opinions on the age of the Olean Drift south of Lake Ontario vary (Denny and Lyford, 1963; Muller, 1965), the deep leaching of its gravels (MacClintock, 1954), combined with the differences in lithology and ice-flow directions between the Olean and the younger drifts (Dreimanis, 1960; Denny and Lyford, 1963; Connally, 1964), support a correlation of the Olean Drift with the other tills and drifts of the major early Wisconsin glacial advance (Fig. 4).

Lithology, fabric, and other criteria indicate that several glacial lobes participated in this main early Wisconsin advance: (1) an unnamed glacial lobe from the Adirondacks or east of them (Olean Drift); (2) the Ontario-Erie Lobe (Sunnybrook Till, Canning Till, Upper Bradtville Till); and (3) the Huron Lobe or a combination of the Huron–Georgian Bay and Erie Lobes (Rocky Fork Till of the Gahanna Drift; tills below soil at Sidney, Ohio). Because of the red-brown till inclusions, Gooding (1963) and Thomas (1970) interpreted the Whitewater Till as having been deposited by a Patrician Superior–Michigan Basin ice lobe (Gooding, 1963,

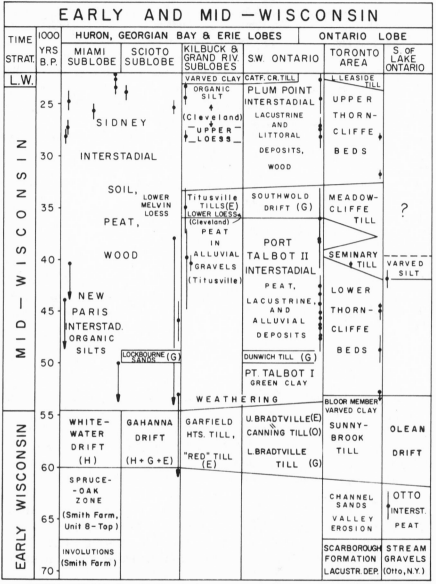

Figure 4. Proposed correlation of litho- and bio-stratigraphic units of early and middle Wisconsin age. References for the litho- and bio-stratigraphic names: Miami Sublobe (Gooding, 1963); Scioto Sublobe (Forsyth, 1965; Goldthwait, 1968; Goldthwait and Rosengreen, 1969); Kilbuck and Grand River Sublobes (White, 1968; White and others, 1969); southwestern Ontario (Dreimanis and others, 1966; Karrow, 1963); Toronto area (Karrow, 1969); south of Lake Ontario (Muller, 1965). Radiocarbon dates are plotted the same way as in Figure 3. The time-stratigraphic assignments of these names in some cases differ from those proposed in the references.

p. 675). However, recent findings of red-brown till in central Michigan (Eschman, oral commun.) would permit also an assignment of the White-water Drift to the Saginaw Lobe. The recently discovered Garfield Heights Till (Dreimanis, 1971; Dreimanis and Berti, in prep.) at Cleveland is another Erie-Ontario Lobe till. The "red till" inclusions in Titusville Till, described by White and others (1969) from northwestern Pennsylvania and Ohio, resemble the Upper Bradtville Till. We believe it possible that the Millbrook Till or some of the pre-Millbrook tills in northern Ohio, described by Totten (1965, p. 355) and White (1967, p. 12-16), are of early Wisconsin age. The main ice thrust of the early Wisconsin came from the northeastern (Labradorean) center, because in northeastern Pennsylvania and western New York the maximum Wisconsin advance was during the early Wisconsin (if the Olean Drift is of this age). Farther west, in Ohio and Indiana, the major early Wisconsin advance was not the most extensive of the last ice age. This is even more evident in Illinois where the later Woodfordian maximum glacial advance extended farther than any of the Altonian advances (Willman and Frye, 1970).

In central Ohio the Scioto Sublobe deposited the Gahanna Till up to a line south of Columbus, but it was overlapped by late Wisconsin drift (Boston, Darby, and Caesar Tills) that extends 30 to 80 km farther south (Figs. 3 and 4). This is indicated by the radiocarbon dates in Gahanna Till and by the deep soil on the intermediate level (Lancaster) outwash in Hocking Valley that is related to Gahanna Till (Kempton and Goldthwait, 1959). Some of the buried clay-enriched zones in gravel under the till appear to be too deep and continuous to be the result of leaching along joints through the till (Gooding and others, 1959; Gold-thwait, 1959a). Consequently, the 500 km^2 of buried outwash and ice-contact gravels (Lockbourne) mark the end of that early glaciation which terminated with widespread wasting of ice, rather than by systematic deposition of moraines. Previous publication (Goldthwait and Forsyth, 1962; Goldthwait and Rosengreen, 1969), indicating that an early Wisconsin drift lies south of the late Wisconsin drift in Highland County, were based on stratigraphic relations, but radiocarbon dates proved it to be "early late" Wisconsin, 21,500 yrs old, in this marginal area. The deeper Russell soils are now shown (Rosengreen, 1970) to be a function of half as much initial carbonate content in the unique Boston Till as in the next younger till. Totten (1965) suggests that some of the well-known moraines in north-central Ohio were made in early Wisconsin time, because they have such till cores (White, 1971).

In the Miami Sublobe the extent of early Wisconsin ice is not so clear. That it extends under the Brush Creek Interstadial peat, which is 22,500 to 50,000 yrs old (Forsyth, 1965) at Sidney, Ohio, is certain. Was this the same ice that produced Whitewater Till with red-brown inclusions farther south near Richmond, Indiana (Gooding, 1963), and in Preble County, Ohio (Thomas, 1970)? No absolute dates are available there (> 41,000 yrs B.P.). The correlation (Fig. 3) is conjectural and is based upon some till similarities and these limiting dates. In any case, Gooding (1963) also shows that the younger late Wisconsin drifts overlap the Whitewater Drift and extend about 35 km farther south in Indiana.

MIDDLE WISCONSIN

Middle Wisconsin time is the equivalent of the latter part of Altonian plus Farmdalian in the Lake Michigan Lobe—53,000 to 23,000 radiocarbon yrs B.P. Middle Wisconsin deposits have been investigated and identified at widely scattered localities throughout the area. The radiocarbon-dated organic sites shown in Figure 5 represent only about half of them. The main subunits of the middle Wisconsin in these eastern lake basins are the Port Talbot and the Plum Point Interstades, separated by one or more glacial readvances into the Lake Ontario and Lake Erie Basins. In the Ohio River Basin the very long Sidney Interstade corresponds to the entire middle Wisconsin, indicating that the above glacial readvances did not extend that far southwest. In central Ohio all direct evidence of middle Wisconsin glacial deposits is lacking. Apparently the glacial lobes that deposited the Dunwich, Titusville, and other tills in the Erie Basin stopped north of the Lake Erie–Ohio River divide.

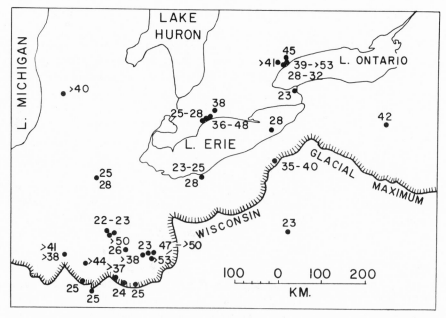

Figure 5. Distribution of middle Wisconsin radiocarbon dates (in thousands of years B.P.). Where several dates have been determined from one site or a small area, the range only is given, because of lack of space.

Lake Basin Tills and Organic Sediments

An important section of middle Wisconsin sediments is along the Scarborough Bluffs at Toronto, Ontario (Karrow, 1967; Mörner, 1971). Its main unit is the Thorncliffe Formation of sands, silts, and clays up to 50 m thick. These were deposited in a lake which had to be dammed by ice in the St. Lawrence Valley (Dreimanis, 1969a). The lake level was first well above Lake Iroquois but later fell to the level of that lake. In the upper part of the Scarborough Bluffs, two tills (Seminary and Meadowcliffe) are interbedded with the Thorncliffe Formation sands and silts. Karrow (1967) interprets them as deposits of the Labradorean Ontario Lobe which just reached the Toronto area. However, the Titusville Till in Pennsylvania, which is of middle Wisconsin age, according to radiocarbon dates (White and others, 1969), has a heavy mineral content typical of the Ontario-Erie Lobe, so one of the middle Wisconsin Ontario Lobe advances must have extended into the Erie Basin. Another indication is the 41,900 ± 900 yrs B.P. date (Y-1401) on driftwood from lacustrine sand near Ithaca, New York, which requires a glacial margin about 50 km south of Lake Ontario (Stuiver, 1969).

The most detailed stratigraphic investigations of middle Wisconsin deposits have been made along the north shore of Lake Erie in the Port Talbot-Plum Point area. This is the type area of many Wisconsin stratigraphic units of the Erie Basin, and it has been studied by a variety of field and laboratory methods, including lithologic, textural and fabric determinations, and pollen analysis and other paleontological investigations. Over 20 radiocarbon dates have been obtained for these deposits (Dreimanis, 1958, 1970, 1971; Dreimanis and others, 1966; Berti, 1971). The sediments are up to 40 m thick and contain the type sections of the Port Talbot I and Port Talbot II organic deposits, Plum Point Interstadial deposits, Dunwich Till, and Southwold Till (Fig. 4). Both tills were deposited by Huron–Georgian Bay Lobes entering the Erie Basin from the north.

Several significant exposures, some of them filling the gaps in the Port Talbot sequences, others providing additional information, have been studied south of Lake Erie. Thus, the Garfield Heights section at Cleveland, Ohio (White, 1953, 1968; Leonard, 1953; Coope, 1968; Dreimanis and Berti, in prep.), adds information about loesses, their fauna and weathering. The Titusville section in northwestern Pennsylvania (White and Totten, 1965; White, 1969; White and others, 1969) sheds light upon the events of about 40,000 yrs B.P. that are not well represented at Port Talbot. White (1969) describes several tills (Titusville, Mogadore, Millbrook) in northwestern Pennsylvania and northern Ohio, and considers them to be of late Altonian age (that is, middle Wisconsin). Where exposed at the surface, these tills are leached deeper than any of the late Wisconsin tills in the nearby area. Buried paleosol also has been noted on the Millbrook Till (White, 1967), which is why the authors suggest it may belong to the early Wisconsin, although White insists it is in later Altonian.

Silts and Paleosols in Ohio River Drainage Area

One very thoroughly studied middle Wisconsin paleosol is 17 km south-west of Lake Erie at Sidney, Ohio (LaRocque and Forsyth, 1957; Forsyth, 1965). At nearby Brush Creek, a peat bed, from which radiocarbon ages of 22,000 to greater than 50,000 yrs B.P. were obtained, contains mainly pine pollen. Farther south, in central Ohio, remnants of middle Wisconsin buried soils on gravels have been found in several places (Goldthwait, 1958; Goldthwait and others, 1965). From the Whitewater Basin in eastern Indiana, Gooding (1963) describes the New Paris Interstadial organic silts, which are dated as being older than 40,500 yrs B.P. (L-478B). Mollusks found in this silt at the American Aggregates section suggest climate considerably colder than at present. Still farther west, Wayne (1966) mentions that weathering separates a probable early Wisconsin loess from the main loess deposit along the Ohio and the lower Wabash Valleys. Goldthwait (1968) also suggests a break in the Melvin Loess of southwestern Ohio (Fig. 3), but believes this was late in middle Wisconsin time.

Climatic Interpretations

The climate of the warmest interstadial episodes of the middle Wisconsin was cool and boreal, judging from the palynological investigations in the Erie–Ontario Lobe area (Dreimanis, 1958; Gooding, 1963; Forsyth, 1965; Dreimanis and others, 1966; Karrow, 1967, 1969; Berti, 1971), and also in the interlobate Lake Michigan–Saginaw Lobe area (Zumberge and Benninghoff, 1969). Frenzel (1967) interprets the Port Talbot II data of Dreimanis (1958) to mean that winters were about 12° C colder and summers were about as warm as at present, but shorter. While the principal character of the middle Wisconsin climate was that of a long, cool interstadial, slight climatic differences between individual lobal areas could have caused either retreats or advances that were not necessarily synchronous along the entire margin of the Laurentide ice sheet. One characteristic feature is common to all lobes: the middle Wisconsin glacial advances were less extensive than those of the early or late Wisconsin. The strongest advance was between the Port Talbot and the Plum Point Interstades (Fig. 3).

LATE WISCONSIN

The late Wisconsin time is known in this region also as "Main" Wisconsin (Dreimanis, prior to 1966), "classical" Wisconsin (Prest, 1970), or by the Michigan Lobe time-stratigraphic terms: Woodfordian, Twocreekan, and most of Valderan. It lasted from 23,000 to 10,000 radiocarbon yrs ago.

First Advances near the Maximum Limit

During the late Wisconsin, the Laurentide ice sheet reached its maximum Wisconsin extent in Ohio and Indiana. Not all the glacial sublobes advanced at the same rate nor reached their outer limits at the same time. The Huron Lobe advanced first, entering Ohio prior to 23,000 yrs B.P. (Figs. 6 and 7). Ice moved upgrade, blocking one proglacial lake outlet after another, and about 23,000 yrs B.P. raised the lake waters in the Erie Basin at least 35 m above the highest Lake Maumee level (Dreimanis, 1971). At that time the Ontario Lobe was still at the Niagaran escarpment (Hobson and Terasmae, 1969; Karrow and Terasmae, 1970). The insect remains found in the mats of organic debris inundated at Cleveland, Ohio (Coope, 1968), and palynologic data from similarly flooded sediments in the St. David's buried gorge at Niagara Falls (Hobson and Terasmae, 1969; Karrow and Terasmae, 1970), indicate that the climate in the Erie Basin was cold, and was probably similar to that which is found now in open woodlands close to the timber line. In view of the numerous discoveries of ice-wedge casts and involutions in Wisconsin age materials in northern Indiana and Illinois (Wayne, 1967; Flemal and others, this volume, p. 229–250), it is suggested that the

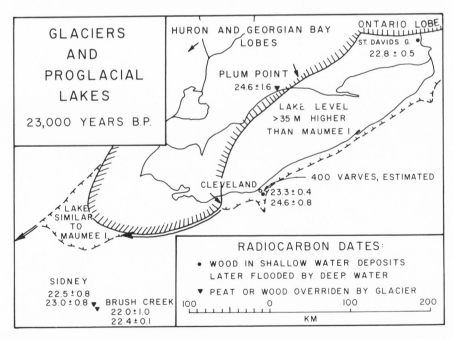

Figure 6. The possible configuration of glacial lobes (bounded by a solid hachured line), proglacial lakes (dashed hachured lines), and their outlets (double arrows), 23,000 yrs B.P.

1000 YRS B.P.	SAGINAW & HURON LOBES		HURON, GEORG BAY & ERIE LOBES			GEORG.B. & ERIE LOBES		ONTARIO LOBE	
	DECATUR SUBLOBE	E. MICHIGAN SW. ONTARIO	WHITE RIVER SUBLOBE	MIAMI SUBLOBE	SCIOTO SUBLOBE	KILBUCK AND GRAND R. SBL	SW. ONTARIO	TORONTO AREA	S. OF LAKE ONTARIO
12	?	TILLS OF PT. HURON MOR. SYSTEM	LAKE WHITTLESEY AND WARREN IN HURON AND ERIE BASINS				HALTON T / WENTWORTH T.	UPPER LEASIDE TILL	POST-VALLEY HDS TILLS
13	CARY — PORT HURON INTERSTADE							LACUSTR. DEP	?
	LAKE ARKONA IN HURON AND ERIE BASINS								
14	CARY TILL	LAGRO FM: NEW HOLLAND TILL MEMB.(H)	CLAYEY TILL (H)	LAKE TILL TYMCHT. T HIRAM TILL	ASHTABULA T HIRAM TILL HAYESVILLE T.	PT. STANLEY DRIFT (E)		VALLEY HEADS DRIFT
15		UNION CITY MORAINE	POWELL MOR	LAVERY TILL INGERSOLL MOR	LACUSTR. SILT & CLAY, BEACHES, DELTAS	LOWER		
	ERIE INTERSTADE:				WEATHERING			
	ALLUV. GRAV.							
16	"LOWER"	CARTERS-BURG TILL	BLOOMINGTON DRIFT	DARBY TILL	KILB. SBL.	GR. SBL.	CATFISH	
17	SANDY AND SILTY TILLS		FARMERSVILLE MOR.	REESVILLE MOR.			CREEK	LEASIDE
				ORGANIC SILT				
18		MEMBER	CHAMPAIGN DRIFT	CAESAR TILL				
19			SHELBYVILLE DRIFT HARTWELL M.				DRIFT	
20	CERRO GORDO MORAINE WEATHERING	CRAWFRDV. MORAINE VERTIGO ALP. OUGHT. BED	CONNERS-VILLE INTERSTAD.	FOREST BED OR	NAVARRE TILL	KENT TILL	TILL	KENT TILL ALMOND DRIFT
21	SHELBYVILLE MOR SYSTEM	CENTER GR. TILL MEMB. SHELBYV. MOR.	FAYETTE DRIFT	SILT BOSTON DRIFT	(G+E)			
22				MT. OLIVE MOR				

Figure 7. Proposed correlation of litho- and bio-stratigraphic units, end moraines, and proglacial lakes of late Wisconsin age. References for the terms used: Decatur Sublobe (Willman and Frye, 1970); eastern Michigan and southwestern Ontario (Dreimanis, 1961); East White Sublobe (Wayne, 1963, 1965); Miami Sublobe (Gooding, 1963); Scioto Sublobe (Goldthwait and Rosengreen, 1969; Goldthwait, this paper); Kilbuck and Grand River Sublobes (White, 1960, 1969); southwestern Ontario (Dreimanis, 1961; Karrow, 1963; Mörner and Dreimanis, this volume); Toronto area (Karrow, 1967); the area south of Lake Ontario (Connally, 1964; Muller, 1965). The time-stratigraphic assignments in some cases differ from those proposed in the above references.

tree limit was forced southward into north-central Ohio-Indiana-Illinois by discontinuous permafrost just prior to glaciation. Some 50 radiocarbon datings in western Ohio and eastern Indiana (Fig. 2, left column) show that numerous trees were buried by the advance of ice between 23,000 and 17,000 yrs ago.

Several deductions are assumed in the interlobate correlations of Figure 2 (left column):

1. Pairs or groups of carbon-14 dates, which are internally consistent, are reliable for correlation even if they are not exactly the same.

2. Nearly all dates represent ice advance, not retreat, because the logs and branches were green and fresh in till. They are bent in U-shapes and retained their bark (Goldthwait, 1958; Burns, 1958).

3. The abundant wood did not travel far from its growth point because 85 to 98 percent of pebbles and fine debris can be derived from within 8 km.

4. Forests were young, generally immature (98 percent of *Picea* under 100 yrs old) and spotty, perhaps mostly in valleys.

In western Ohio distinction of till sheets by minor differences in compositions has been difficult or conjectural. Where detailed studies have been made, as in Whitewater Valley, Indiana, they have produced a multiple till stratigraphy (Gooding, 1963). Only now in thesis form is this stratigraphy extended tentatively by petrologic, clay mineral, and mechanical criteria into southwestern Ohio (Brace, 1968; Thomas, 1970; and other theses at Miami University). Recent Ohio Geological Survey studies in Clinton and Highland Counties (Teller, 1967; Rosengreen, 1970) lend stratigraphic detail to the broader, early reconnaissance studies based on geographic distribution of pebble lithologies, moraines, soils, loess, and some stratigraphy. The complicated crisscrossing and overlapping of end moraine crests (Fig. 8) left confusion, especially in Warren, Clinton, and Highland Counties of southwestern Ohio. Recent critical carbon-14 dates (Teller and others, in prep.) and identification of till sheets by stone-counts and by heavy and clay minerals (Rosengreen, 1970) leave little doubt that the ice margin near the outer Wisconsin drift limit of the Scioto and Miami Sublobes oscillated for 4,000 yrs.

An early Scioto Lobe advance, 21,500 ± 500 radiocarbon yrs ago produced (1) the northeast-southwest striae that occur on the Brassfield Limestone in northwest Greene County under later northwest-southeast striae and Miami drift; (2) the Mt. Olive Moraines (Rosengreen, 1970) and the Boston Till, which is now the southernmost Wisconsin drift at VanderVoort (Teller, 1967); and (3) the intermediate outwash levels with acid soils in Clinton, Warren, and Ross Counties, and perhaps also the Fayette and Center Grove Tills in sublobes farther west (Gooding, 1963; Wayne, 1965; Wright, 1970; Thomas, 1970).

In the Miami Sublobe the interstadial called Connersville (Gooding, 1963) in the Whitewater Basin, Indiana, is marked by buried organic silts from which five dates of 20,100 to 18,750 yrs B.P. can be cited (I-610, OWU-490, L-653C, W-724, W-738). Overlapping in time in the Scioto Sublobe is an unnamed interstadial that is represented by a "forest bed" common in wells of Clinton and Highland Counties. Six radiocarbon dates of 19,800 to 20,910 yrs B.P. (I-4795, W-2465, W-2459, I-3715, ISGS 42, ISGS 44) have been obtained from the "forest bed."

A slightly later (19,000 ± 500 radiocarbon yrs B.P.) maximum extension of the Miami Sublobe ice deposited the "Shelbyville drift" (Gooding, 1961; Wright, 1970), which is correlated to the Shelbyville of Illinois by radiocarbon dates only. Farther east it ended in the complex multiple tills and gravels of the Hartwell end moraine and possibly a buried Scioto Lobe moraine, the Wilmington Moraine of Teller (1967).

An irregular later push, 18,000 to 18,500 yrs B.P. (W-91, W-331, Y-448, OWU-331), of the Scioto Sublobe over the northeastern end of the Hartwell Moraine overran all the earlier Wisconsin moraines in places in Clinton, Highland, and Ross Counties, and formed the outer Cuba Moraine. The Cuba Moraine may tie northward into the Xenia Moraine (and east of the Scioto River it is the outer moraine of Wisconsin age), but it involves a distinct Caesar Till sheet. Probably this was the time

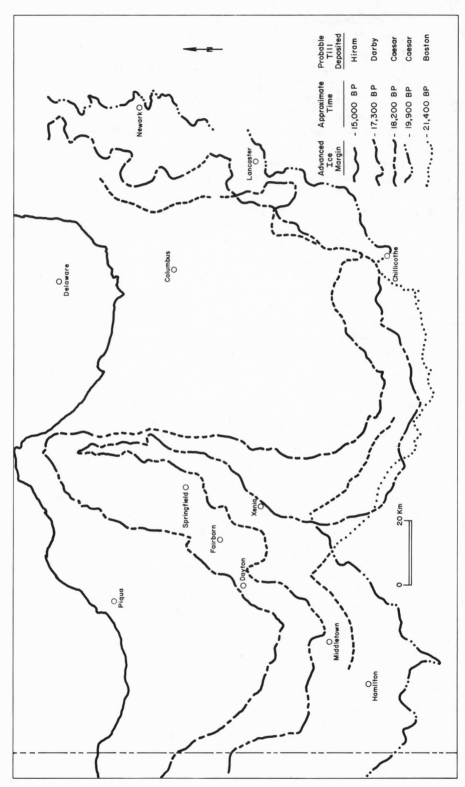

Figure 8. Hypothetical ice limit positions in southwestern Ohio based upon analysis of carbon-14 dates on wood in till and lithologic composition of the tills.

when the huge Kennard Outwash of west-central Ohio was built down Little Miami River, extending 55 km south into Clifton Gorge. This whole area, including the lower Scioto River (Kingston Outwash) was given a final significant coating of Upper Melvin Loess (Goldthwait, 1968).

Gooding (1971) recognizes an intermediate till sheet and outwash (Champaign Drift) indicating a readvance between "Shelbyville" and "Bloomington" time. This has been found in the Camden Moraine of the Miami Sublobe in Ohio, but it is not a widespread nor far-reaching readvance.

A later readvance at 17,200 ± 400 yrs B.P. of both sublobes is known from mechanical and lithological differences between tills in the Reesville Moraine (Scioto Sublobe) and the Farmersville Moraine and boulder belt (Miami Sublobe). It probably accounts for a date of 17,292 ± 436 yrs B.P. (OWU-76) in Ross County outer drift. This readvance produced the Darby Till, which can be distinguished on top of Caesar Till as far north as Columbus, Ohio (Forsyth, 1965; Goldthwait and Rosengreen, 1969). Nowhere did it reach quite as far as earlier advances, as indicated by the loess cover present only to the south (Goldthwait, 1968). The readvance and retreat did produce the almost loess-free outwashes in every valley from Whitewater (Bloomington Outwash) and Miami (Mad River Valley Train), to Hocking (Carroll Outwash) and Licking (Utica Valley Train). Farther west, in the White River Sublobe, this readvance is probably represented by the Cartersburg Till Member (Wayne, 1963).

In the area east of the Scioto Sublobe (right of the Scioto Sublobe in Fig. 7), only one silty and sandy till, called by various local names, was deposited during the early part of the late Wisconsin. In the Huron Lobe area close to Lake Huron, sandy and silty tills similar to the Catfish Creek Till have been found (Leverett and Taylor, 1915; Dreimanis, 1961), but some of the coarse-textured tills may be older than the Catfish Creek Till.

Erie Interstade and the Following Readvance

About 15,500 ± 500 yrs ago, the margin of the Erie Lobe, wasting rapidly by calving along a proglacial lake in the Erie Basin, had retreated as far eastward as the Niagara Peninsula, and the Huron Lobe was at least as far northward as Goderich, Ontario (Dreimanis, 1967; Mörner and Dreimanis, this volume p. 107–134). In many scattered locations in southwestern Ontario, lacustrine silts and clays, nearshore sands, and deltaic and beach gravels, probably of Erie Interstadial age, separate the Catfish Creek Till or its equivalents from younger, finer textured tills. The till-covered laminated clays in Seneca County, Ohio, probably relate to the same interval. During the maximum retreat, the lake level in the Erie Basin was lower than at present, indicating a low eastward outlet toward the Hudson River Valley. Lake-level changes from high to low

and back to high again must have been rapid, because no paleosols have
been found on the buried Erie Interstade beaches in Ontario. On higher
ground in northwestern Pennsylvania, weathering during this interstade
has been noticed by White and others (1969) on some surfaces of the
Kent Till where it is covered by calcareous Lavery Till. Carbon-14 dating
of Erie Interstade deposits is lacking, but the interstade is bracketed by
a date of 14,300 ± 450 yrs B.P. on top of the Wabash Moraine at Edon,
Ohio (W-198), and by dates of 16,500 yrs B.P. near the Reesville and
Farmersville Moraines. Wood buried in till inside the Powell Moraine at
Liberty, Ohio, dates 14,780 ± 192 yrs B.P. (OWU-83).

The glacial readvance that followed the Erie Interstade incorporated
the lacustrine deposits in the Huron and the Erie Basins and produced
tills rich in clay and silt. In the Huron Lobe area of Ontario these
tills are still unnamed, but they are related to the Mitchell, Lucan,
and Seaforth Moraines (Dreimanis, 1961). The Port Stanley Till of the
Ontario part of the Erie Lobe corresponds to the fine-textured tills south
of Lake Erie listed in Figure 7 (Dreimanis and Karrow, 1965). Certainly
the amount and nature of the clay content of Hiram Till of northeastern
Ohio and northwestern Pennsylvania indicate ice passage over lake beds
(White, 1960). As the Hiram Till is traced southwestward, its clay content
drops near each end-moraine group in west-central Ohio (Smith, 1949)
and eastern Indiana (Gooding, 1971). The most abrupt drop takes
place in the south near the Broadway and Powell Moraines (Scioto Sublobe) or the Union City–Mississinawa Moraines (Miami Sublobe), and
along the southern edge of the New Holland Till Member in northeastern
Indiana (Wayne, 1963, 1968). An increase in the calcium/magnesium
ratio in parent till and a change in soil type from Miami catena (south)
to Morley-Blount (north) is especially evident (Steiger and Holowaychuck,
1971). It is concluded that the ice that deposited the Hiram Till probably readvanced southward far enough to just reach the Ohio River watershed.

The remaining time is very short (1,000 yrs?) for the production of
the prominent recessional moraines (Salamonie–St. Johns, Wabash,
Wayne, Defiance) north of the Ohio River watershed in Indiana and
Ohio. However, the Wabash Moraine shows a prominent change in stone-
counts, in direction of ice movement as judged by moraine orientation
(Goldthwait, 1956b), and in clay content (Gooding, 1971). Hence,
it is shown as a time-break on the glacial map of Ohio (Goldthwait
and others, 1961). Certainly this halt and slight readvance erased the
last vestiges of the Kilbuck, Scioto, Miami, and White River Sublobes.

In Pennsylvania and New York, the Lake Escarpment Morainic Complex
of the Erie Lobe (White, 1960) was formed after the Erie Interstade,
as indicated by a recent radiocarbon date of 14,900 ± 450 yrs B.P.
(I-4216: Calkin, 1970). This morainic complex consists of the silty or
clayey Lavery, Hiram, and Ashtabula Tills. It has been correlated in
turn with the prominent Valley Heads Moraine of the Ontario Lobe

in northern New York State (MacClintock and Apfel, 1944; Muller, 1965). Although both of these moraines have been correlated with the Port Huron maximum (Muller, 1965), recent studies by Wall (1968) and Calkin (1970), and the presence of Lake Maumee beaches along the north side of the Lake Escarpment Moraine (Leverett, 1902; Evans, 1970), indicate that the Port Huron Moraine was constructed later.

Cary-Port Huron Interstade

After the Powell-Union City and related recessional moraines were deposited, much of the retreat occurred by calving in proglacial Lake Maumee, early Lake Saginaw, and Lake Arkona (Hough, 1958, 1963). Therefore, retreat was quite rapid—up to 500 m per year. This is even faster than the average recession during the final dissipation of the Laurentide ice sheet 10,000 to 7,000 yrs B.P. At the culmination of this Cary-Port Huron Interstade, the Ontario-Erie Lobe had retreated into the Ontario Basin east of Toronto (Karrow, 1969; Dreimanis, 1969a) and the Huron Lobe into northern Michigan (Farrand and others, 1969).

The Cary-Port Huron retreat coincides with the Lake Arkona phase in the Huron and Erie Basins, and some very low-level lake phases in Erie Basin (Dreimanis, 1969a). Evidence of a desiccated till surface buried underneath the till of the Port Huron glacial advance south of Sarnia, Ontario (Adams, 1970), indicates that a similar low-water phase existed also in the Huron Basin. Nonarboreal vegetation dominated by *Cyperaceae*, and by spruce trees that appeared some distance farther south, followed the retreating glacier and concomitant lowering lake levels, as shown by the bryophyte bed at Cheboygan on the northern tip of the Michigan lower peninsula (Miller and Benninghoff, 1969). Because of its alkali-acid pretreatment, the most reliable date of the bryophyte bed is 13,300 ± 400 yrs B.P. (Farrand and others, 1969), which agrees with other Arkona-Whittlesey dates (Dreimanis, 1966; Calkin, 1970).

A glacial readvance of at least 50 to 150 km reached the Port Huron Morainic System and is contemporaneous with Lakes Whittlesey and Saginaw. Lake Whittlesey is radiocarbon dated at 13,000 ± 500 yrs B.P. The Port Huron readvance deposited tills different in composition from the Cary Tills in the Huron Lobe (Wyoming Till), in the Erie-Ontario Lobe (Wentworth, Halton, and Upper Leaside Tills), and the upper St. Lawrence Valley (Fort Covington Till). The lithologic, textural, and fabric differences (Dreimanis, 1961; MacClintock and Stewart, 1963; Karrow, 1963; Dreimanis and Karrow, 1965) suggest a rearrangement of the pattern of glacial movement following the Cary-Port Huron withdrawal. Some confusion in correlations has been created by the reddish-brown color of the Port Huron Till in Michigan (Farrand and others, 1969), because the red color has been formerly considered to be typical of Valders Till. As indicated by the radiocarbon dates of the bryophyte bed underneath this till, the color alone is not a reliable stratigraphic criterion.

The Port Huron glacial advance to the Paris and Galt Moraines in the Ontario-Erie Basin must have been followed by a rapid retreat, inasmuch as Lake Whittlesey beaches surround the drumlins east of these moraines (Karrow, 1963). Lake Whittlesey was followed by several lake phases in the Erie and Huron Basins, each lower than the preceding one due to lower outlets opened by glacial retreat (Hough, 1958, 1963; Lewis, 1969; Calkin, 1970). Beginning with Lake Iroquois, development of the proglacial lakes in the Ontario Basin was similar to those in the Lake Erie Basin (Coleman, 1937; Karrow and others, 1961; Mirynech, 1967; Prest, 1970). Most of the lake phases lasted for relatively short periods of time—only a century or two. Strong erosion and heavy deposition along the shores of Lake Iroquois and Lake Algonquin indicate a longer period of time for those two lakes.

Boreal woodlands, dominated by spruce, followed the retreating glacial lobes; among their inhabitants the greatest attention has been paid to mastodons and mammoths (Thomas and others, 1952; Skeels, 1962; Drumm, 1963; Forsyth, 1963; Dreimanis, 1968). About 10,000 to 11,000 yrs B.P., pine began to replace spruce forests, particularly on better drained terranes (Ogden, 1966, 1967).

Up to now, no strong evidence has been found for Twocreekan and Valderan equivalents in the eastern Great Lakes region, so these are omitted as time-stratigraphic units in this discussion.

ACKNOWLEDGMENTS

This study has been supported by the National Research Council of Canada Grant A4215 and Grant-in-Aid from Ohio State University.

REFERENCES CITED

Adams, J. I., 1970, Effect of ground water levels on stress history of the St. Clair clay till deposit [discussion]: Canadian Geotech. Jour., v. 7, p. 190–193.

Anderson, R. C., 1957, Pebble and sand lithology of the major Wisconsin glacial lobes of the central lowland: Geol. Soc. America Bull., v. 68, p. 1415–1450.

Antevs, E., 1925, Retreat of the last ice sheet in eastern Canada: Canada Geol. Survey Mem. 146, 142 p.

—— 1928, The last glaciation with special reference to the ice retreat in northeastern North America: Am. Geog. Soc. Research Ser. 17, 202 p.

Berti, A. A., 1971, Palynology and stratigraphy of the mid-Wisconsin in the eastern Great Lakes region, North America (Ph.D. thesis): London, Ont., Univ. Western Ontario, 178 p.

Brace, B. R., 1968, Pleistocene stratigraphy of the Hamilton quadrangle (M.Sc. thesis): Oxford, Ohio, Miami Univ., 138 p.

Burns, G. W., 1958, Wisconsin age forests in western Ohio: II. Vegetation and burial conditions: Ohio Jour. Sci., v. 58, p. 220–230.

Calkin, P. E., 1970, Strand lines and chronology of the glacial Great Lakes in northwestern New York: Ohio Jour. Sci., v. 70, p. 78-96.

Chamberlin, T. C., 1883, Terminal moraine of the second glacial epoch: U.S. Geol. Survey Ann. Rept. 3, p. 291-402.

‗‗‗ 1894, Glacial phenomena of North America, in Geikie, James, The Great Ice Age (3d ed.): New York, D. Appleton & Co., p. 724-774.

Chamberlin, T. C., and Leverett, F., 1894, Further studies of the drainage features of the Upper Ohio basin: Am. Jour. Sci., v. 47, p. 247-283.

Chapman, L. J., and Dell, C. I., 1963, Revision of the early history of the retreat of the Wisconsin glacier in Ontario based on the calcite content of sands: Geol. Assoc. Canada Proc. 15, p. 103-108.

Chapman, L. J., and Putnam, D. F., 1966, The physiography of southern Ontario (2d ed.): Toronto, Ont., Univ. Toronto Press, 386 p.

Churcher, C. S., 1968, Mammoth from the middle Wisconsin of Woodbridge, Ontario: Canadian Jour. Zoology, v. 46, p. 219-221.

Churcher, C. S., and Karrow, P. F., 1963, Mammals of Lake Iroquois age: Canadian Jour. Zoology, v. 41, p. 153-158.

Cleland, C. D., 1966, The prehistoric animal ecology and ethnozoology of the upper Great Lakes region: Michigan Univ. Mus. Anthropol. Paper 29, 294 p.

Coleman, A. P., 1894, Interglacial fossils from the Don Valley, Toronto: Am. Geologist, v. 13, p. 85-95.

‗‗‗ 1901, Glacial and interglacial beds near Toronto: Jour. Geology, v. 9, p. 285-310.

‗‗‗ 1933, The Pleistocene of the Toronto region, accompanied by Map 41g: Ontario Dept. Mines Bull., v. 41, pt. 7, 69 p.

‗‗‗ 1937, Lake Iroquois: Ontario Dept. Mines Bull., v. 45, pt. 7, p. 1-36.

‗‗‗ 1941, The last million years: Toronto, Ont., Toronto Univ. Press, 216 p.

Connally, G. G., 1964, Garnet ratios and provenance in the glacial drift of western New York: Science, v. 144, p. 1452-1453.

Coope, G. R., 1968, Insect remains from silts below till at Garfield Heights, Ontario: Geol. Soc. America Bull., v. 79, p. 749-756.

Cushing, E. J., 1965, Problems in the Quaternary phytogeography of the Great Lakes region, in Wright, H. E., Jr., and Frey, D. G., eds., The Quaternary of the United States: Princeton, N.J., Princeton Univ. Press, p. 403-416.

Davis, M. B., 1967, Late-glacial climate in northern United States: a comparison of New England and the Great Lakes region, in Cushing, E. F., and Wright, H. E., Jr., Quaternary paleoecology: New Haven, Conn., Yale Univ. Press, p. 11-43.

Day, C., 1960, Stratigraphy of the lower till in the Catfish Creek area (B.Sc. thesis): London, Ont., Univ. Western Ontario, 77 p.

Deane, R. E., 1950, Pleistocene geology of the Lake Simcoe district, Ontario: Canada Geol. Survey Mem. 256, 108 p.

DeGeer, G., 1926, On the solar curve as dating the ice age, the New York moraine and Niagara Falls through the Swedish timescale: Geog. Annaler, v. 4, p. 253-284.

Dell, C. I., 1959, A study of the mineral composition of sand in southern Ontario: Canadian Jour. Soil Sci., v. 39, p. 185-196.

Denny, C. S., and Lyford, W. H., 1963, Surficial geology and soils of the Elmira-Williamsport region, New York and Pennsylvania: U.S. Geol. Survey

Prof. Paper 379, 60 p.

Denny, C. S., and Postel, A. W., 1964, Rapid method of estimating lithology of glacial drift of the Adirondack Mountains, N. Y.: U.S. Geol. Survey Prof. Paper 501, p. 143-145.

Dreimanis, A., 1957a, Stratigraphy of the Wisconsin glacial stage along the northwestern shore of Lake Erie: Science, v. 126, p. 166-168.

_____ 1957b, Depth of leaching in glacial deposits: Science, v. 126, p. 403-404.

_____ 1958, Wisconsin stratigraphy at Port Talbot on the north shore of Lake Erie, Ontario: Ohio Jour. Sci., v. 58, p. 65-84.

_____ 1959a, Rapid macroscopic fabric studies in drill-cores and hand specimens of till and tillite: Jour. Sed. Petrology, v. 29, p. 459-463.

_____ 1959b, Measurements of depth of carbonate leaching in service of Pleistocene stratigraphy: Geol. Fören. Stockholm Förh., v. 81, p. 478-484.

_____ 1959c, Proposed local stratigraphy of the Wisconsin glacial stage in the area south of London, southwestern Ontario, in Friends of Pleistocene (E. Sec.) 22d reunion, Field Trip Guide: p. 24-30.

_____ 1960, Pre-classical Wisconsin in the eastern portion of the Great Lakes region, North America: Internat. Geol. Cong., 21st, Copenhagen 1960, Rept. Session 4, p. 108-119.

_____ 1961, Tills of southern Ontario, in Legget, R. F., ed., Soils in Canada: Royal Soc. Canada Spec. Pub. 3, p. 80-96.

_____ 1962, Quantitative gasometric determination of calcite and dolomite by using Chittick apparatus: Jour. Sed. Petrology, v. 32, p. 520-529.

_____ 1966, Lake Arkona-Whittlesey and post-Warren radiocarbon dates from "Ridgetown Island" in southwestern Ontario: Ohio Jour. Sci., v. 66, p. 582-586.

_____ 1967, Cary-Port Huron Interstade in eastern North America and its correlatives: Geol. Soc. America Spec. Paper 115, p. 259.

_____ 1968, Extinction of mastodons in eastern North America: testing a new climate-environmental hypothesis: Ohio Jour. Sci., v. 68, p. 257-272.

_____ 1969a, Late-Pleistocene lakes in the Ontario and Erie basins: Great Lakes Research, 12th Conf. Proc., p. 170-180.

_____ 1969b, Selection of genetically significant parameters for investigation of tills: Poznań, Zeszyty Naukowe U.A.M. Geografia 8, p. 15-29.

_____ 1970, Last ice age deposits in the Port Stanley map-area, Ontario (401/11): Canada Geol. Survey Paper 70-1, pt. A, p. 167-169.

_____ 1971, The last ice age in the eastern Great Lakes region, North America, in Ters, M., ed., Études sur le Quaternaire dans le Monde: VIII INQUA, v. 1, p. 69-75.

Dreimanis, A., and Karrow, P. F., 1965, Southern Ontario, in INQUA 7th Congress, Guidebook for field conference G: Nebraska Acad. Sci., p. 90-110.

Dreimanis, A., and Reavely, G. H., 1953, Differentiation of the lower and the upper till along the north shore of Lake Erie: Jour. Sed. Petrology, v. 23, p. 238-259.

Dreimanis, A., and Vagners, U. J., 1969, Lithologic relation of till to bedrock, in Wright, H. E., Jr., ed., Quaternary geology and climate: Natl. Acad. Sci. Pub. 1701, p. 93-98.

Dreimanis, A., Reavely, G. H., Cook, R. J. B., Know, K. S., and Moretti, F. J., 1957, Heavy mineral studies in tills of Ontario and adjacent areas: Jour. Sed. Petrology, v. 27, p. 148-161.

Dreimanis, A., Terasmae, J., and McKenzie, G. D., 1966, The Port Talbot Interstade of the Wisconsin glaciation: Canadian Jour. Earth Sci., v. 3, p. 305-325.

Droste, J. B., 1956, Clay minerals of calcareous tills in northeastern Ohio: Jour. Geology, v. 64, p. 187-190.

Droste, J. B., and Doehler, R. W., 1957, Clay mineral composition of calcareous till in northwestern Pennsylvania: Illinois Acad. Sci. Trans., v. 56, p. 194-198.

Drumm, F., 1963, Mammoths and mastodons: Ice age elephants of New York: New York State Mus. and Sci. Service Educ. Leaflet 13, 31 p.

Dryer, C. R., 1894, The drift of the Wabash-Erie region—a summary of results: Indiana Dept. Geol. and Nat. Resources Ann. Rept. 18, p. 38-90.

Evans, E. B., 1970, Pleistocene beach ridges of northwestern Pennsylvania (M.A. thesis): Bowling Green, Ohio, Bowling Green State Univ.

Fairchild, H. L., 1909, Glacial waters in central New York: New York State Mus. and Sci. Service Bull. 127, 66 p.

____ 1932, New York moraines: Geol. Soc. America Bull., v. 45, p. 627-662.

Farrand, W. R., Zahner, R., and Benninghoff, W. S., 1969, Cary-Port Huron Interstade: Evidence from a buried byrophyte bed, Cheboygan County, Michigan: Geol. Soc. America Spec. Paper 123, p. 249-262.

Flemal, Ronald C., Hinkley, Kenneth C., and Hesler, James L., 1973, DeKalb mounds: A possible Pleistocene (Woodfordian) pingo field in north-central Illinois: Geol. Soc. America Mem. 136, p. 229-250.

Flint, R. F., 1963, Status of the Pleistocene Wisconsin Stage in central North America: Science, v. 135, p. 402-404.

____ 1970, Glacial and Quaternary geology: New York, John Wiley & Sons, Inc., 892 p.

Flint, R. F., and Rubin, M., 1955, Radiocarbon dates of pre-Mankato events in eastern and central North America: Science, v. 121, p. 649-658.

Flint, R. F., Colton, R. B., Goldthwait, R. P., and Willman, H. B., 1959, Glacial map of the United States east of the Rocky Mountains: Geol. Soc. America.

Forman, S. A., and Brydon, J. E., 1961, Clay mineralogy of Canadian soils, in Legget, R. F., ed., Soils of Canada: Royal Soc. Canada Spec. Pub. 3, p. 140-146.

Forsyth, J. L., 1963, Ice age census: Ohio Conservation Bull. 27, no. 9, p. 13, 16-19.

____ 1965, Age of the buried soil in the Sidney, Ohio, area: Am. Jour. Sci., v. 263, p. 521-597.

____ 1967, Glacial geology of the East Liberty quadrangle, Logan and Union Counties, Ohio: Ohio Div. Geol. Survey Rept. Inv. 66.

Forsyth, J. L., and LaRocque, J. A. A., 1956, Age of the buried soil at Sidney, Ohio: Geol. Soc. America Bull., v. 67, p. 1696.

Frenzel, B., 1967, Die Klimaschwankungen des Eiszeitalters: Braunschweig, F. Vieweg & Sohn, 291 p.

Frye, J. C., Glass, H. D., and Willman, H. B., 1962, Stratigraphy and mineralogy of the Wisconsinan loesses of Illinois: Illinois Geol. Survey Circ. 334, 55 p.

Gilbert, G. K., 1871, On certain glacial and postglacial phenomena of the Maumee Valley: Am. Jour. Sci., v. 1, p. 339-345.

____ 1873, Surface geology of the Maumee Valley: Ohio Div. Geol. Survey Rept.

Inv., v. 1, p. 536–556.

Goldthwait, J. W., 1910, An instrumental survey of the shore lines of the extinct lakes Algonquin and Nipissing in southwestern Ontario: Canada Geol. Survey Mem. 10, 57 p.

Goldthwait, R. P., 1958, Wisconsin age forests in western Ohio: I-Age and glacial events: Ohio Jour. Sci., v. 58, p. 209–219.

_____ 1959a, Leached, clay-enriched zones in post-Sangamonian drift in southwestern Ohio and southeastern Indiana: A reply: Geol. Soc. America Bull., v. 70, p. 927–928.

_____ 1959b, Scenes in Ohio during the last ice age: Ohio Jour. Sci., v. 59, p. 193–216.

_____ 1968, Two loesses in central southwest Ohio, in Bergstrom, R. E., ed., The Quaternary of Illinois: Univ. Illinois College Agr., Spec. Pub. 14, p. 41–47.

_____ 1970, Mountain glaciers of the Presidential Range in New Hampshire: Arctic and Alpine Research, v. 2, p. 85–102.

Goldthwait, R. P., and Forsyth, J. L., 1962, Midwestern Friends of the Pleistocene Field Guide, Columbus, Ohio, 40 p.

_____ 1965, Ohio, in INQUA 7th Congress, Guidebook for field conference G: Nebraska Acad. Sci., p. 64–82.

Goldthwait, R. P., and Rosengreen, T., 1969, Till stratigraphy from Columbus southwest to Highland county, Ohio: Geol. Soc. America (North-Central Sec.), 3d Ann. Meeting, Field Trip No. 2, p. 2-1 to 2-17.

Goldthwait, R. P., White, G. W., and Forsyth, J. L., 1961, Glacial map of Ohio: U.S. Geol. Survey Misc. Geol. Inv. Map I-316.

Goldthwait, R. P., Dreimanis, A., Forsyth, J. L., Karrow, P. F., White, G. W., 1965, Pleistocene deposits of the Erie Lobe, in Wright, H. E., Jr., and Frey, D. G., eds., The Quaternary of the United States: Princeton, N.J., Princeton Univ. Press, p. 85–97.

Gooding, A. M., 1961, Illinoian and Wisconsin history in southeastern Indiana, in Guidebook for field trips, Cincinnati meeting: Geol. Soc. America, p. 99–130.

_____ 1963, Illinoian and Wisconsin glaciations in the Whitewater Basin, southeastern Indiana, and adjacent areas: Jour. Geology, v. 71, p. 665–682.

_____ 1971, Character of late Wisconsin tills in eastern Indiana: Indiana Geol. Survey.

Gooding, A. M., and Ogden, J. M., III, 1965, A radiocarbon-dated pollen sequence from the Wells mastodon site near Rochester, Indiana: Ohio Jour. Sci., v. 65, p. 1–11.

Gooding, A. M., Thorp, J., and Gamble, E., 1959, Leached, clay-enriched zones in post-Sangamonian drift in southwestern Ohio and southeastern Indiana: Geol. Soc. America Bull., v. 70, p. 921–925.

Gross, D. L., 1967, Mineralogical gradations within Titusville Till and associated tills of northwestern Pennsylvania (M.S. thesis): Urbana, Ill., Univ. Illinois, 77 p.

Gross, D. L., and Moran, S. R., 1971, Grain size and mineralogical gradations within tills of the Alleghany Plateau, in Goldthwait, R. P., ed., Till, a symposium: Columbus, Ohio State Univ. Press, p. 251–274.

Gunn, C. B., 1967, Provenance of diamonds in the drift of the Great Lakes region, North America (M.Sc. thesis): London, Ont., Western Ontario Univ., 132 p.

Harrison, W., 1957, A clay-till fabric: its character and origin: Jour. Geology, v. 65, p. 275–308.

____ 1960, Original bedrock composition of Wisconsin till in central Indiana: Jour. Sed. Petrology, v. 30, p. 432–446.

Hinde, G. J., 1877, Glacial and interglacial strata of Scarboro Heights and other localities near Toronto, Ontario: Canadian Jour. (new ser.), v. 15, p. 388–413.

Hobson, C. D., and Terasmae, J., 1969, Pleistocene geology of the buried St. Davids gorge, Niagara Falls, Ontario: Geophysical and palynologic studies: Canada Geol. Survey Paper 68-67, 16 p.

Hobson, C. D., Herdendorf, C. E., and Lewis, C. F. M., 1969, High resolution reflection seismic survey in western Lake Erie: Great Lakes Research, 12th Conference Proc., p. 210–224.

Holmes, C. D., 1941, Till fabric: Geol. Soc. America Bull., v. 52, p. 1301–1354.

____ 1952, Drift dispersion in west-central New York: Geol. Soc. America Bull., v. 63, p. 993–1010.

____ 1960, Evolution of till-stone shapes, central New York: Geol. Soc. America Bull., v. 71, p. 1645–1660.

Hough, J. L., 1958, Geology of the Great Lakes: Urbana, Univ. Illinois Press, 313 p.

____ 1963, The prehistoric Great Lakes of North America: Am. Scientist, v. 51, p. 84–109.

Janssens, A., 1963, A contribution to the Pleistocene geology of Champaign County, Ohio (M.Sc. thesis): Columbus, Ohio State Univ., 96 p.

Johnston, W. A., 1916, The Trent Valley outlet of Lake Algonquin and the deformation of the Algonquin water-plane in Lake Simcoe district, Ontario: Canada Geol. Survey Mus. Bull. 23, 27 p.

____ 1926, The age of the upper great gorge of Niagara River: Royal Soc. Canada Trans., 3d ser., v. 22, sec. 4, p. 13–29.

Kapp, R., 1964, A radiocarbon dated pollen profile from Sunbeam Prairie bog, Darke County, Ohio: Am. Jour. Sci., v. 262, p. 259–266.

Kapp, R., and Gooding, A., 1964, Pleistocene vegetational studies in the White-water Basin, southeastern Indiana: Jour. Geology, v. 72, p. 307–326.

Karrow, P. F., 1963, Pleistocene geology of the Hamilton-Galt area: Ontario Dept. Mines Geol. Rept. 16, 68 p.

____ 1967, Pleistocene geology of the Scarborough area: Ontario Dept. Mines Geol. Rept. 46, 108 p.

____ 1969, Stratigraphic studies in the Toronto Pleistocene: Geol. Assoc. Canada Proc., v. 20, p. 4–16.

Karrow, P. F., and Terasmae, J., 1970, Pollen bearing sediments of the St. Davids buried valley till at the Whirlpool, Niagara River gorge, Ontario: Canadian Jour. Earth Sci., v. 7, p. 539–542.

Karrow, P. F., Clark, J. R., and Terasmae, J., 1961, The age of Lake Iroquois and Lake Ontario: Jour. Geology, v. 69, p. 659–667.

Kempton, J. P., and Goldthwait, R. P., 1959, Glacial outwash terraces of the Hocking and Scioto River valleys, Ohio: Ohio Jour. Sci., v. 59, p. 135–151.

Lajtai, E. Z., 1969, Stratigraphy of the University Subway, Toronto, Canada: Geol. Assoc. Canada Proc., v. 20, p. 17–23.

LaRocque, J. A. A., 1966, Pleistocene mollusca of Ohio: Ohio Div. Geol. Survey Bull. 62, pt. 1, p. 1–111.

____ 1967, Pleistocene mollusca of Ohio: Ohio Div. Geol. Survey Bull. 62, pt. 2, p. 113–356.

LaRocque, J. A. A., 1968, Pleistocene mollusca of Ohio: Ohio Div. Geol. Survey Bull. 62, pt. 3, p. 357-553.

—— 1970, Pleistocene mollusca of Ohio: Ohio Div. Geol. Survey Bull. 62, pt. 4, p. 555-800.

LaRocque, J. A. A., and Forsyth, J. L., 1957, Pleistocene molluscan faunules of the Sidney cut, Shelby County, Ohio: Ohio Jour. Sci., v. 57, p. 81-89.

Leonard, A. B., 1953, Molluscan faunules in Wisconsinan loess at Cleveland, Ohio: Am. Jour. Sci., v. 251, p. 369-376.

Leverett, F., 1895, On the correlation of New York moraines with raised beaches of Lake Erie: Am. Jour. Sci., v. 50, p. 1-20.

—— 1902, Glacial formations and drainage features of the Erie and Ontario basins: U.S. Geol. Survey Mon. 41, 529 p.

—— 1931, Quaternary system, in Cushing, H. P., Leverett, F., and Van Horn, F. R., Geology and mineral resources of the Cleveland district, Ohio: U.S. Geol. Survey Bull. 818, p. 57-81.

Leverett, F., and Taylor, F. B., 1915, Pleistocene of Indiana and Michigan and the history of the Great Lakes: U.S. Geol. Survey Mon. 53, 816 p.

Lewis, C. F. M., 1969, Late Quaternary history of lake levels in the Huron and Erie basins: Great Lakes Research, 12th Conference Proc., p. 250-270.

Lewis, C. F. M., Anderson, T. W., and Berti, A. A., 1966, Geological and palynological studies of early Lake Erie deposits: Great Lakes Research, 9th Conference Proc., Univ. Michigan, p. 176-191.

MacClintock, P., 1954, Leaching of Wisconsin glacial gravels in eastern North America: Geol. Soc. America Bull., v. 65, p. 309-383.

MacClintock, P., and Apfel, E. T., 1944, Correlation of the drifts of the Salamanca re-entrant, New York: Geol. Soc. America Bull., v. 55, p. 1143-1164.

MacClintock, P., and Dreimanis, A., 1964, Reorientation of till fabric by overriding glacier in the St. Lawrence Valley: Am. Jour. Sci., v. 262, p. 133-142.

MacClintock, P., and Stewart, D. P., 1963, Glacial geology of the St. Lawrence Lowlands: New York State Mus. and Sci. Service Bull. 394, 152 p.

McDonald, B. C., and Shilts, W. W., 1971, Quaternary stratigraphy and events in southeastern Quebec: Geol. Soc. America Bull., v. 82, p. 683-698.

Malott, C. A., 1922, Physiography of Indiana, in Logan, W. N., Handbook of Indiana geology: Indiana Dept. Conservation Pub. 21, pt. 2, p. 59-256.

May, R. W., and Dreimanis, A., 1973, Differentiation of glacial tills in southern Ontario, Canada, based on their Cu, Zn, Cr, Ni geochemistry: Geol. Soc. America Mem. 136, p. 221-228.

Meritt, R. S., and Muller, E. H., 1959, Depth of leaching in relation to carbonate content of till in central New York State: Am. Jour. Sci., v. 257, p. 465-480.

Miller, N. G., and Benninghoff, W., 1969, Plant fossils from a Cary-Port Huron interstadial deposit and their paleoecological interpretation: Geol. America Spec. Paper 123, p. 225-248.

Mirynech, E., 1967, Pleistocene and surficial geology of the Kingston-Coburg-Tweed area, Ontario, in Geology of parts of Eastern Ontario and Western Quebec: Geol. Assoc. Canada Guidebook, Kingston, Ont., p. 183-198.

Mörner, N.-A., 1971, The Plum Point Interstadial: age, climate and subdivision: Canadian Jour. Earth Sci., v. 8, p. 1423-1431.

Mörner, N.-A., and Dreimanis, A., 1973, The Erie Interstade: Geol. Soc. America Mem. 136, p. 107-134.

Moss, J. H., and Ritter, D. F., 1962, New evidence regarding the Binghampton substage in the region between the Finger Lakes and Catskills, New York:

Am. Jour. Sci., v. 260, p. 81–106.

Muller, E. H., 1964, Quaternary section at Otto, New York: Am. Jour. Sci., v. 262, p. 461–478.

—— 1965, Quaternary geology of New York, in Wright, H. E., Jr., and Frey, D. G., eds., The Quaternary of the United States: Princeton, N.J., Princeton Univ. Press, p. 99–112.

Ogden, J. G., III, 1965, Pleistocene pollen records from eastern North America: Bot. Rev., v. 31, p. 481–504.

—— 1966, Forest history of Ohio: I. Radiocarbon dates and pollen stratigraphy of Silver Lake, Logan County, Ohio: Ohio Jour. Sci., v. 66, p. 387–400.

—— 1967, Radiocarbon and pollen evidence for a sudden change in climate in the Great Lakes region approximately 10,000 years ago, in Cushing, E. J., and Wright, H. E., Jr., eds., Quaternary paleoecology: New Haven, Conn., Yale Univ. Press, p. 117–127.

—— 1969, Correlation of contemporary and late Pleistocene pollen records in the reconstruction of postglacial environments in northeastern North America: Mitt. Intern. Verin Limnol., v. 17, p. 64–77.

Ostry, R. C., and Deane, R. E., 1963, Microfabric analyses of till: Geol. Soc. America Bull., v. 74, p. 165–168.

Prest, V. K., 1970, Quaternary geology of Canada, in Douglas, R. J. W., ed., Geology and economic minerals of Canada: Canada Geol. Survey Econ. Geol. Rept. 1, 5th ed., p. 675–764.

Rosengreen, T., 1970, The glacial geology of Highland County, Ohio (Ph.D. thesis): Columbus, Ohio State Univ., 163 p.

Scudder, S. H., 1895, The Coleoptera hitherto found fossil in Canada: Canada Geol. Survey Contrib. Canadian Paleontology, v. 2, pt. 1, p. 27–56.

Shepps, V. C., 1953, Correlation of the tills of northeastern Ohio by size analysis: Jour. Sed. Petrology, v. 23, p. 34–48.

—— 1958, "Size factors," a means of analysis of data from textural studies of till: Jour. Sed. Petrology, v. 28, p. 482–485.

Shepps, V. C., White, G. W., Droste, J. B., and Sitler, R. F., 1959, Glacial geology of northwestern Pennsylvania: Pennsylvania Geol. Survey Bull. G32, 59 p.

Sherzer, W. H., 1902, Ice work in southeast Michigan: Jour. Geology, v. 10, p. 194–216.

Sitler, R. F., 1963, Petrography of till from Ohio and Pennsylvania: Jour. Sed. Petrology, v. 33, p. 365–379.

—— 1968, Glacial till in oriented thin section: Internat. Geol. Cong., 23d, Prague 1968, v. 8, p. 283–295.

Sitler, R. F., and Chapman, C. A., 1955, Microfabrics in till from Ohio and Pennsylvania: Jour. Sed. Petrology, v. 25, p. 262–269.

Skeels, M. A., 1962, The mastodons and mammoths of Michigan: Michigan Acad. Sci., Arts and Letters Paper 47, p. 59–65.

Skinner, S. I. M., and Halstead, R. L., 1958, Note on rapid method for determination of carbonates in soils: Canadian Jour. Soil Sci., v. 38, p. 187–188.

Smith, J. M., 1949, A study of the Wisconsin glaciation of southeastern Indiana and southwestern Ohio (M.Sc. thesis): Oxford, Ohio, Miami Univ., 74 p.

Spencer, J. W. S., 1907, The falls of Niagara: Canada Geol. Survey Pub. 970, 490 p.

Steiger, J. R., and Holowaychuck, N., 1971, Particle-size and carbonate analysis of glacial till and lacustrine deposits in western Ohio, in Goldthwait, R. P., ed., Till, a symposium: Columbus, Ohio State Univ. Press, p. 275–289.

Stuiver, M., 1969, Yale natural radiocarbon measurements IX: Radiocarbon, v. 11, p. 545-658.

Taylor, F. B., 1913, The moraine systems of southwestern Ontario: Royal Canadian Inst. Trans. 10, p. 1-23.

Teller, J. T., 1967, The glacial geology of Clinton County, Ohio: Ohio Geol. Survey Rept. Inv. 67 (map).

_____ 1970, Early Pleistocene glaciation and drainage in southwestern Ohio, southeastern Indiana, and northern Kentucky (Ph.D. thesis): Cincinnati, Ohio, Univ. Cincinnati.

Terasmae, J., 1958, Non-glacial deposits in the St. Lawrence Lowlands, Quebec: Canada Geol. Survey Bull. 46, p. 13-34.

_____ 1960, A palynological study of the Pleistocene interglacial beds at Toronto, Ontario: Canada Geol. Survey Bull. 56, p. 23-41.

_____ 1965, Surficial geology of the Cornwall and St. Lawrence Seaway Project areas, Ontario: Canada Geol. Survey Bull. 121, 54 p.

Terasmae, J., and Mirynech, E., 1964, Postglacial chronology and the origin of deep lake basins in Prince Edward County, Ontario: Great Lakes Research Pub. 11, p. 161-169.

Thomas, E. S., Goldthwait, R. P., Sears, P. B., Clisby, K. H., LaRocque, A., and Wood, A. E., 1952, The Orleton farms mastodon: Ohio Jour. Sci. v. 52, p. 1-28.

Thomas, J. B., 1970, Illinoian-Wisconsin Pleistocene deposits at Eaton, Ohio: Geol. Soc. America Bull., v. 81, p. 3433-3436.

Thornbury, W. D., 1937, Glacial geology of southern and south-central Indiana: Indiana Dept. Conservation Div. Geol. Pub.

_____ 1940, Weathered zones and glacial chronology in southern Indiana: Jour. Geology, v. 48, p. 449-475.

Totten, S. M., 1960, Quartz/feldspar ratios of tills in northeastern Ohio (M.S. thesis): Urbana, Univ. Illinois.

_____ 1965, Multiple tills near Shenandoah, Richland County, Ohio: Ohio Jour. Sci., v. 65, p. 353-357.

Vagners, U. J., 1966, Lithologic relationship of till to carbonate bedrock in southern Ontario (M.Sc. thesis): London, Ont., Univ. Western Ontario, 154 p.

_____ 1970, Mineral distribution in tills in central and southern Ontario (Ph.D. thesis): London, Ont., Univ. Western Ontario, 277 p.

van der Hammen, T., Maarleveld, G. C., Vogel, J. C., and Zagwijn, W. H., 1967, Stratigraphy, climatic succession and radiocarbon dating of the last glacial in the Netherlands: Geologie en Mijnbouw, v. 46, p. 79-95.

Van Wyckhouse, R. J., 1966, A study of test borings from the Pleistocene of the southeastern Michigan glacial lake plain (M.Sc. thesis): Detroit, Mich., Wayne Univ., 84 p.

Von Engeln, O. D., 1961, The Finger Lake region: its origin and nature: Ithaca, N.Y., Cornell Univ. Press, 156 p.

Wall, R. E., 1968, A sub-bottom reflection survey in the central basin of Lake Erie: Geol. Soc. America Bull., v. 79, p. 91-106.

Wayne, W. J., 1959, Stratigraphic distribution of Pleistocene land snails in Indiana: Sterkiana, v. 1, p. 9-12.

_____ 1963, Pleistocene formations of Indiana: Indiana Geol. Survey Bull. 25, 85 p.

_____ 1965, The Crawfordsville and Knightstown Moraines in Indiana: Indiana

Geol. Survey Rept. Prog. 28, 15 p.

—— 1966, Ice and land. A review of the Tertiary and Pleistocene history of Indiana, in Natural features of Indiana: Indiana Acad. Sci., Indianapolis, Ind., State Library, p. 21-39.

—— 1967, Periglacial features and climatic gradient in Illinois, Indiana, and western Ohio, east-central United States, in Cushing, E. J., and Wright, H. E., Jr., eds., Quaternary paleoecology: New Haven, Conn., Yale Univ. Press, p. 393-414.

—— 1968, The Erie Lobe margin in east-central Indiana during the Wisconsin glaciation: Indiana Acad. Sci. Proc., v. 77, p. 277-291.

Wayne, W. J., and Zumberge, J. H., 1965, Pleistocene geology of Indiana and Michigan, in Wright, H. E., Jr., and Frey, D. G., eds., The Quaternary of the United States: Princeton, N.J., Princeton Univ. Press, p. 63-83.

Westgate, J. A., and Dreimanis, A., 1967, The Pleistocene sequence at Zorra, southwestern Ontario: Canadian Jour. Earth Sci., v. 4, p. 1127-1143.

White, G. W., 1953, Sangamon soil and early Wisconsin loesses at Cleveland, Ohio: Am. Jour. Sci., v. 251, p. 362-368.

—— 1960, Classification of Wisconsin glacial deposits in northeastern Ohio: U.S. Geol. Survey Bull. 1121-A, 12 p.

—— 1967, Glacial geology of Wayne County, Ohio: Ohio Geol. Survey Rept. Inv. 62, 39 p.

—— 1968, Age and correlation of Pleistocene deposits at Garfield Heights (Cleveland), Ohio: Geol. Soc. America Bull., v. 79, p. 749-752.

—— 1969, Pleistocene deposits of the northwestern Allegheny Plateau, U.S.A.: Geol. Soc. London Quart. Jour., v. 124, p. 131-151.

—— 1971, Thickness of Wisconsin tills in Grand River and Kilbuck lobes, in Goldthwait, R. P., ed., Till, a symposium: Columbus, Ohio State Univ. Press, p. 149-163.

—— 1973, History of investigation and classification of Wisconsinan drift in north-central United States: Geol. Soc. America Mem. 136, p. 3-34.

White, G. W., and Totten, S. M., 1965, Wisconsin age of the Titusville Till (formerly called "Inner Illinoian"), northwestern Pennsylvania: Science, v. 148, p. 234-235.

White, G. W., Totten, S. M., and Gross, D. L., 1969, Pleistocene stratigraphy of northwestern Pennsylvania: Pennsylvania Geol. Survey, 4th ser., Gen. Geol. Rept. G 55, 88 p.

Wilding, L. P., Drees, L. R., Smeck, N. E., and Hall, G. F., 1971, Mineral and elemental composition of Wisconsin-age till deposits in west-central Ohio, in Goldthwait, R. P., ed., Till, a symposium: Columbus, Ohio State Univ. Press, p. 290-317.

Willman, H. B., and Frye, J. C., 1970, Pleistocene stratigraphy of Illinois: Illinois Geol. Survey Bull. 94, 204 p.

Wilson, A. W. G., 1905, A forty-mile section of Pleistocene deposits north of Lake Ontario: Canadian Inst. Trans., v. 8, pt. 1, p. 11-21.

Winchell, N. H., 1872, The surface geology of northwestern Ohio: Am. Assoc. Adv. Sci. Pub., p. 152-186.

Wright, F. M., 1970, Pleistocene stratigraphy of the Farmersville and the northern part of Middletown quadrangles, southwestern Ohio (M.A. thesis): Oxford, Ohio, Miami Univ., 189 p.

Wright, G. F., 1889, The ice age in North America: New York, D. Appleton & Co., 622 p.

Zumberge, J. H., 1960, Correlation of Wisconsin drifts in Illinois, Indiana, Michigan and Ohio: Geol. Soc. America Bull., v. 71, p. 1177–1188.

Zumberge, J. H., and Benninghoff, W. S., 1969, A. Mid-Wisconsin peat in Michigan, U.S.A: Pollen et Spores, v. 11, p. 585–601.

Manuscript Receibed by The Society December 27, 1971

GEOLOGICAL SOCIETY OF AMERICA
MEMOIR 136
© 1973

The Erie Interstade

Nils-Axel Mörner

University of Stockholm, Stockholm, Sweden

A. Dreimanis

University of Western Ontario, London, Ontario, Canada

ABSTRACT

The type section for the Erie Interstade shows a well-developed beach between two layers of offshore sands. The section is underlain by a major late Wisconsin till, the Catfish Creek Till, and is overlain by the Port Stanley Till. The position of the buried beach 3 to 4 m above present Lake Erie level indicates that the interstadial lake, here named Lake Leverett, was lower than previously estimated and that it drained eastward, probably via the Mohawk Lowland, during an ice recession into the Ontario Basin.

The Erie Interstade correlates well with a world-wide amelioration of the climate at about 15,500 yrs B.P. that separates two glacial maxima of the late Wisconsin-Weichselian, the older 20,000 to 17,000 yrs B.P. and the younger 14,800 to 14,400 yrs B.P.

INTRODUCTION

The name Erie Interstadial was assigned by Dreimanis (1958) to a major late Wisconsin retreat of the Erie Lobe from central Ohio into the eastern portion of Lake Erie, followed by a readvance back to Ohio. This retreat is indicated by the high amount of lacustrine clays and silts incorporated from a proglacial lake bottom into the overlying tills. These were named the Port Stanley Till in southwestern Ontario

(DeVries and Dreimanis, 1960) and the Hiram Till south of Lake Erie (White, 1960). The underlying silty sandy till differs in texture and lithology from the overlying till (Dreimanis and Reavely, 1953; Dreimanis and Karrow, 1965). It was named Catfish Creek Till in Ontario (DeVries and Dreimanis, 1960) and Kent Till and Navarre Till south of Lake Erie (White, 1960, 1961).

The presence of lacustrine silts underneath the upper till had been reported from northern Ohio by Claypole (1887) and Leverett (1902). The lowest level of the interstadial lake in the Erie Basin was first thought to be 5 m (16 ft) below the Lake Warren level (Dreimanis, 1968, 1970) after thé ice receded northeast of Lake Erie, which drained eastward (Dreimanis, 1970, Fig. 3, map VI).

In 1970 a well-developed beach between the Catfish Creek and Port Stanley Tills was found just above the present Lake Erie shore. This indicated a drop of the interstadial lake level to the elevation of 178 m (584 ft) above sea level, or about 44.5 m (146 ft) below the Warren Beach. This site is here chosen as the type locality for the Erie Interstade.

TYPE SECTION

The type section is in a cliff on the north shore of Lake Erie, about 40 km (25 mi) south-southwest of London, Ontario. More precisely it is 3.2 km (2.0 mi) northeast of Plum Point and 1.4 km (0.9 mi) southwest of Talbot Creek (Figs. 1 and 2).

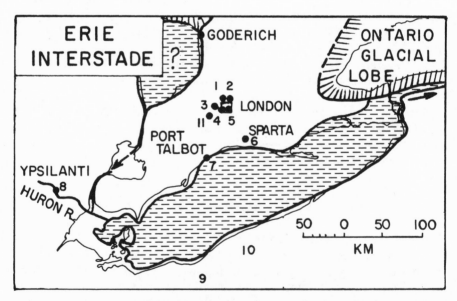

Figure 1. Location of the sites discussed. Patterned area: probable lowest phase of the lowest Erie interstadial lake (Lake Leverett) with eastward outlet. Localities 1 to 5: buried beaches of higher lake phases (Table 3); 6: buried beach at Sparta; 7: type section of the Lake Leverett beach; 8: buried valleys at Ypsilanti, Michigan; 9 and 10: areas with lacustrine deposits covered by the upper till; 11: Dingman Creek section.

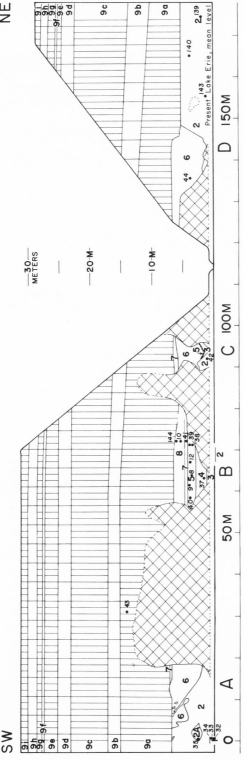

Figure 2. Type section of Erie Interstade. The sand and gravel deposits of the Erie Interstade (3 to 6), underlain by the Catfish Creek Drift (1 to 2) and overlain by the Port Stanley Drift (7 to 9), are visible in four main exposures (A–D) between slump-covered areas (criss-crossed). Exposures B and C are enlarged in Figure 3. The dots with small numbers refer to sample sites listed in Table 2.

In the area of the type section the cliff rises almost vertically to a height of 30 m above the present lake level, which is 174.5 m (572 ft) above sea level. Sandy and gravelly lake and beach deposits are exposed between two tills for a distance of 150 m (Fig. 2). These strata are undisturbed in the central part of the section, but are contorted at both ends.

A 200-m-long portion of the cliff was investigated (Fig. 2). The stratigraphy below the upper till, the Port Stanley Till, was measured in detail about every 3 m along the cliff. The entire section, including the details of the Port Stanley drift, was measured at the 55 and 180 m marks.

The late Wisconsin stratigraphy of the section is shown in Figures 2 and 3 and in Table 1. Southwestward from this section, the lowermost of the late Wisconsin tills, the Catfish Creek Till, rises gradually above lake level. The till is underlain by mid-Wisconsin and early Wisconsin deposits, the type sections for which are at Bradtville and Plum Point, 1.8 km (1.1 mi) and 2.9 km (1.8 mi), respectively, southwest of the section described here (Dreimanis and others, 1966; Dreimanis, 1970).

The section consists of a complete sequence of the late Wisconsin deposits in the central Lake Erie area, and the major units are described below.

Catfish Creek Drift (Layers 1 and 2)

A dirty gravel interbedded with sandy flow-till was found by trenching below the Catfish Creek Till at the zero point of the section (Fig. 2);

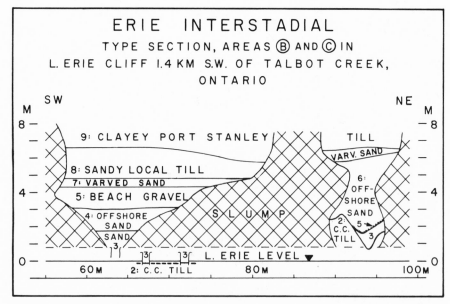

Figure 3. Enlarged portion of the main exposures of the Erie Interstade type section (areas B and C). See Table 1 for descriptions of layers 2 to 9, and Figures 4 and 5 for photographs.

TABLE 1. LATE WISCONSIN DEPOSITS EXPOSED IN THE LAKE ERIE CLIFF 1.4 KM SOUTHWEST
OF TALBOT CREEK*

Second Late Wisconsin Glaciation (Cary Stade)
Port Stanley Drift
9. Silty and clayey Port Stanley Till (9a, 9c, 9e, 9g, 9i) interbedded with glacio-lacustrine silts (9b, 9d, 9f, 9h).
8. Local sandy Port Stanley Till.
7. Proglacial, lacustrine, varved, fine sand.

Erie Interstade
Lake Leverett Deposits
6. Offshore sand and fine gravel.
5. Beach gravel containing balls of till, compressed peat and gyttja, and some formless lumps of the Catfish Creek Till.
4. Offshore sand and gravel.
3. Sand with thin Catfish Creek Till laminae and inclusions: offshore or glacio-lacustrine sediment.

First Late Wisconsin Glaciation (Tazewell Stade)
Catfish Creek Drift
2. Sandy and silty Catfish Creek Till, containing large balls of greenish-gray till.
1. Silty and sandy gravel, probably outwash, interbedded with flow till.

*See Figures 2 and 3.

together they constitute layer 1. The carbonate content of the silty matrix of this gravel and of the flow-till is similar to that of the Catfish Creek Till (Table 2). Pebble lithology, except for the presence of more greenish-gray, fine-grained Devonian sandstone and less dolostone, is also similar.

The Catfish Creek Till (layer 2) separates layer 1 from the sandy and gravelly sediments of the Erie Interstade, except at both ends of the exposure, where it directly underlies the Port Stanley Till. From the northeastern to the southwestern side of the section (Fig. 2) the surface of the Catfish Creek Till rises from 5 m to 7 m above Lake Erie, and the surface can be traced southwestward for more than 4 km and northeastward for about 1 km. This rise is interrupted by the 140-m-wide depression that is filled by the interstadial deposits. In the depression, the till surface is uneven, ranging from near the present lake level to 2.7 m above it.

As in other exposures along the Lake Erie north shore (Dreimanis, 1961; Dreimanis and Karrow, 1965; Dreimanis and others, 1966), the Catfish Creek Till in this section is silty and pebbly. Nearly half of the pebbles are limestone; the other most common rocks are Precambrian igneous and metamorphic rocks and dolostone (Table 2).

About 175 m southwest of the zero point of the interstadial section, the Catfish Creek Till is divided into two beds by a boulder pavement. The pebble lithology of the lower bed (Table 2) is similar to that of the till (layer 2) that underlies the interstadial depression. Both have a relatively higher sandstone content and a higher limestone/dolostone ratio than the upper bed of the Catfish Creek Till. Therefore, the upper bed of the Catfish Creek Till appears to have been removed by erosion from the area of the depression prior to the deposition of the interstadial layers 3 and 4.

TABLE 2. ANALYSES OF LITHOLOGY, CARBONATES, AND TEXTURE

Layer	Material	Sample no.	Lithology of pebbles (8-16 mm) Percent							Carbonates (-63 μ)		Texture* (-2 mm)		
			Ls.	Dol.	Chrt.	Ss.	Sh.	Ig.-Met.	$\frac{Ls.}{Dol.}$	Percent	$\frac{Calc.}{Dol.}$	Sand	Silt	Clay
I. Port Stanley Till														
9.	typical	43	22	4	5	5	62	1	5.5	31	2.2	4	59	37
8.	local	10	40	31	12	6	4	5	1.3	40	0.9	59	29	12
	local	144	46	24	8	9	2	11	1.8	39	0.8	63	29	8
I. Lake Leverett Deposits														
6.	upper offshore deposit													
	sand	41										93	6	1
	gravel	40										99†	1	0
5.	beach gravel	38	43	23	11	5	1	17	1.8	52	1.6	100†	0	0
	beach sand	39										100†	0	0
4.	lower offshore deposit													
	sand	37								55	1.5	100	0	0
I. Catfish Creek Drift														
2.	greenish-gray till	{35	42	11	4	9	5	15	3.8	28	1.5	24	51	25
		42	49	11	3	7	22	13	4.5	31	1.7	22	55	23
	gray till	{139	47	14	6	11	8	14	3.5	36	1.5	25	49	26
		140								33	1.5	22	49	29
		143								28	1.5	27	48	25
1.	flow till	32	48	8	5	14	9	14	6.0	37	1.1	63	35	2
	gravel	33	48	14	4	17	4	12	3.4	36	1.6			
	gravel	34	48	13	3	16	2	18	3.7					

II. Till and clay pebbles in Lake Leverett beach gravel (layer 5)

Catfish Creek Till	36								36	1.7	39	39	22
Southwold Till	9								43	0.5	25	59	16
	40								41	0.6	35	51	14
Port Talbot interstadial lacustrine clay	8								33	0.9	1	56	43
Bradtville Till	12								22	3.2	(Clayey silt till)		

III. Reference samples from the area southwest of the interstadial site

Catfish Creek Till 150 to 170 m southwest

upper bed	45	49	19	7	4	4	17	2.6	37	1.2	36	46	18
	142	46	17	3	9	4	21	2.7	34	1.1	34	46	20
lower bed	11	56	12	8	11	3	10	4.7	34	1.4	29	47	24
	141								36	1.1	55	35	10
Southwold Till at its type locality at Bradtville cottages 1.8 km southwest													
	69-103	44	22	5	10	0	19	2.5	39	0.8	27	59	14
	69-105	39	22	10	8	1	20	1.8	39	0.4	40	47	13
	69-106	43	22	7	4	1	22	2.0	38	0.4	53	42	5
	69-107	43	24	4	2	1	26	1.4	38	0.9	36	50	14
	69-108	43	25	8	10	1	13	1.7	39	0.6	43	52	5
Dunwich Till at its type locality 1.8 km southwest (Dreimanis and others, 1965)													
buff till		24	46	2	2	2	23	0.6	41	0.3	55	34	11

*Sand-silt boundary: 63 μ, silt-clay boundary: 2 μ.
†Sand, granules, and pebbles.
Note: I and II are from the Erie Interstade type section; III is from the adjoining area.

The calcite/dolomite ratio of the till matrix in layer 2 is slightly higher than previously reported for the Catfish Creek Till (Dreimanis, 1961; Dreimanis and others, 1966). However, it is still in the range found for this till by W. Stankowski (written commun.) who sampled two 12-m-thick sections at 25-cm intervals about 1 km southwest of the site presently under discussion. This ratio (1.4 to 1.7, Table 2) is quite different from that of the texturally similar but older Southwold Till and Dunwich Till (ratios 0.3 to 0.8, according to Dreimanis and others, 1966; Dreimanis, 1970).

At the zero point of the section, the Catfish Creek Till contains a large inclusion, about a meter in diameter, of greenish-gray till (sample 70-35, Table 2), and the till at the 90-m mark (Fig. 3) is also greenish-gray when moist. This greenish tinge may be due to incorporation of greenish-gray Devonian shales and sandstones, or of the Port Talbot I Interstade green clay, or both.

Erie Interstade Deposits (Layers 3 to 6)

The depression in the top of the Catfish Creek Till is filled with sandy and gravelly deposits of the Erie Interstade (layers 3 to 6, Figs. 2 to 5). Because of slumping, the interstadial sediments crop out only in four major areas, here called A, B, C, and D (Fig. 2).

Layer 3 occurs in areas B and C. It is fine to medium sand, mixed with till laminae and inclusions. In area B it looks like an ice-marginal deposit with a varvelike alternation of sand and thin till layers. However, in area C it appears more like a mixture of offshore deposits consisting of layers and lumps of till in sand and gravel. If layer 3 is an offshore deposit, it is easier to explain the presence of the erosional depression in the Catfish Creek Till. We suggest that the interstadial lake level dropped to an elevation 3.0 to 3.5 m above present Lake Erie and that a back-eroding shore and down-eroding river were responsible for the depression in the Catfish Creek Till. The mouth of the river, a possible predecessor of Talbot Creek, later moved northeastward, and a spitlike beach was formed on the southwest side (layers 4 and 5).

Layer 4 is a sand and gravel deposit with fine cross-bedding and some well-rounded lumps of Catfish Creek Till (Fig. 4). It must have been deposited in shallow water just off the shore.

Layer 5 is a beach deposit (Figs. 3 and 4). Its elevation is 3.0 to 4.3 m above mean Lake Erie level or 178 m (584 ft) above the present sea level. The beach is formed of more or less horizontally bedded gravel and sand with fine cross-laminae. It contains abundant lumps of Catfish Creek Till that are either well rounded or formless. Rounded lumps were also found of Southwold Till; Port Talbot peat, gyttja, and clay; and Bradtville Till, just as along the present-day beach of Lake Erie. Some of them were flat, others cylindrical. The granulometric and carbonate analyses of some till and clay pebbles are listed in Table

Figure 4. Left: Erie Interstade type section, area B, layers 4 to 9 (the man is digging in layer 3). Measuring rod 160 cm. Right: Details of layers 4, 5, and 7. Measuring rod 85 cm. Layer 4, the offshore gravelly sand, is cross-bedded indicating subaquatic deposition. Layer 5, the beach, is horizontally bedded with cross-laminated individual beds. It contains numerous pebbles of till and some lumps of peat and gyttja (for example, the dark spot 10 cm below the large flat light-colored "cobble"). The swallow nests are excavated in layer 7.

2. One of the peat balls was radiocarbon dated at 37,840 + 3,255 to 2,310 yrs B.P. (St-3438), which would place the peat in the Port Talbot Interstade.

In area C, layer 5 is represented merely by a thin layer of gravel or pebbles, partly resting directly on the till surface, partly cutting across a small local depression in the Catfish Creek surface. This thin layer was deposited probably just off the shore and just below the mean lake level.

The pebble lithology of the beach gravel (layer 5) is very similar to that of the upper bed of the Catfish Creek Till, suggesting that most of these pebbles were derived by littoral erosion of the till.

Layer 6 represents the rising lake level. It is an offshore deposit of sand and gravel, most completely exposed in area C. It is not present in area B. In area C it grades upward from gravel to sand and contains pebbles of till and lumps of peat and gyttja in its lower part. In area D, the offshore sediments are sand without littoral structures. In area A, this sand is partly contorted.

Layer 7 is fine to medium sand with four thin layers of silt (Fig.

3), and is apparently a coarse-textured varved sediment that was deposited in a proglacial lake. In area C, layer 6 grades into layer 7. The interstadial lake level must have been rising rapidly after the eastward discharge was blocked by glacial advance. The small number of varves may suggest that the top of the sediment was eroded later by the advancing Erie Lobe.

In summary, layers 3 and 4 represent a falling lake level followed by formation of a beach (layer 5), and then by a rise of the lake level (layers 6 and 7). This interstadial sequence is depicted in Figure 6.

Port Stanley Tills (Layers 8 and 9)

When the advancing ice reached the area of this section, it eroded a nearly horizontal surface across the older deposits, and interstadial sediments were left only in the depression (Fig. 2). On both sides of the depression the Port Stanley Till directly overlies the Catfish Creek Till and the boundary is sharp and distinct. In area A, overthrust slices of sand (layer 7) and Catfish Creek Till and two wedges of Port Stanley Till are intruded into layer 7. These wedges were probably injected in extension fractures when the glacier overrode the sand (Dreimanis, 1969b). They trend north 10° to 20° east, suggesting that the Port Stanley glacier moved from east-southeast toward west-northwest.

Layer 8 is unusual for the Port Stanley drift. It is more sandy than

Figure 5. Erie Interstade type section. Area B, east end, layers 7 to 9. Note the westward rising sand-filled shear planes in the sandy Port Stanley Till (layer 8) and the columnar jointing in the silty Port Stanley Till (layer 9). Measuring rod 1.5 m (5 ft).

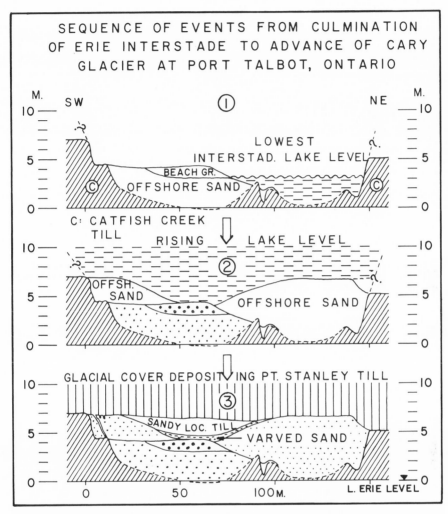

Figure 6. Reconstruction of the sequence of deposition at the Erie Interstade type section (Fig. 2). The first stage (above) is at the lowest Erie interstadial lake level (Lake Leverett). A river mouth is displaced northeastward, the old river bed is filled with sediments (layers 3 to 4), and a spitlike beach (layer 5) is developed in the southwest part. Lake level was 3.0 to 3.5 m above present Lake Erie level. The second stage (middle) gives the time when the lake level rose and offshore sediments (layer 6) were deposited. The third stage (below) depicts the episode when the second late Wisconsin ice covered the section. A thin bed of varved sand (layer 7) was deposited while the ice was advancing, and a local sandy till (layer 8) was deposited when ice first covered the section.

other tills in this area. The till is transected by numerous westward rising shear planes containing the sand of layer 7 (Fig. 5). It appears that the interstadial sand became incorporated in the base of the glacier while it was moving northwestward along the sand-filled depression and was redeposited with other basal drift material after short transport. The carbonate content and the pebble lithology (Table 2) show that the local sandy till also incorporated gravel, Catfish Creek Till, and

Southwold Till. The incorporation of various local materials in the base of the Port Stanley and Catfish Creek Tills was observed in several exposures along the north shore of Lake Erie.

Layer 9a is typical Port Stanley Till—a clayey silty till, rich in local Devonian shale pebbles and having high limestone/dolostone and calcite/dolomite ratios (Table 2).

The layers of stratified drift in the upper and middle part of the Port Stanley Till (9b, 9d, 9f, 9h, in Fig. 2) are common along the entire cliff between Port Glasgow and Port Bruce. They represent younger oscillations of the ice margin and will be described separately (Mörner and Dreimanis, in prep.).

LAKE LEVEL, ICE RECESSION, AND AGE

The elevation of the interstadial beach proper (layer 5) indicates that the lake level dropped to 178 m (584 ft) above the present sea level at this isobase latitude in the Erie Basin. The level is 44.5 m (146 ft) below the Warren shore at the same latitude. The vertical change from offshore to beach to offshore deposits suggests that the beach represents the lowest interstadial lake level. The lake must have drained eastward, most probably via the Mohawk Lowland, as no outlets that low are known across Michigan or Indiana. Furthermore, the Erie-Ontario Lobe must have receded into the Ontario Basin. In the Huron Basin, a contemporaneous lake existed at least as far north as Goderich, judging from the superposition of a clayey and silty till equivalent to the Port Stanley Till on the more sandy Catfish Creek Till (Fig. 1).

The sediments of the Erie Interstade have a wide distribution in southwestern Ontario and northern Ohio. However, the boundary for the recession, as yet, cannot be closely fixed. A suggested position is given in Figure 1.

Since the interstadial deposits were first described by Dreimanis (1958), beach and deltaic deposits have been found (Table 3) in the London area at elevations similar to those of Lake Maumee II and III, and at elevations 4.6 m (15 ft) below the Lake Warren level in the Sparta area (Dreimanis, 1967, 1969a). However, those levels correspond to early or late lake stages during the Erie Interstade and not to its lowest phase.

Kunkle (1963) has described three buried channels in the Ann Arbor-Ypsilanti area in Michigan (Fig. 1): one below the pre-Cary Till, one between pre-Cary and Cary Tills, and one between Cary Till and younger gravel. The channel between the pre-Cary and Cary Tills most probably is equivalent to the Erie Interstade. It seems to be graded to a lake level lower than present Lake Erie.

If the Erie Interstade lake level at the type section is extrapolated as a shoreline parallel to the Lake Warren shoreline, it falls at a level 9 m (30 ft) below the present Lake Erie level in the area of horizontal

TABLE 3. ERIE INTERSTADE BEACHES BURIED UNDERNEATH TILL IN LONDON AREA, ONTARIO

Number on Figure 1	Location	Elevation of the till/gravel or sand contact above sea level*	
		(ft)	(m)
1.	Arva, Highway 4 cut 0.1 km north of Medway Creek	885	270
2.	Arva, gravel pit 0.2 km north of Medway High School	885	270
3.	Gravel pit between CNR and CPR railway tracks 4.3 km west-southwest of Hyde Park Corner	885	261
4.	London, Gordon Ave., 0.2 km south of Commissioner's Road	875	267
5.	London, Manor Park Gravel Pit at the south end of Highwood Ave.	830	253

*Measured from the Canadian Army Survey Establishment Topographic maps 40 P/3c and 40I/14f, scale 1:25,000, contour interval 10 ft.

shorelines at the southwestern end of Lake Erie. It is proposed here to assign the name Lake Leverett to this interstadial lake.

In the Dingman Creek river banks southwest of London (Fig. 1, location 11), a layer of sand and gravel possibly deposited in the Erie Interstade was found at an elevation of about 238 m (781 ft) above sea level (Mörner and others, in prep.). It seems to be a stream deposit, and it is covered by varved clay (50 to 60 varves were counted). The varves are thin and suggest glacial ponding with the ice still far away. The change from sand and gravel to varved clay indicates a sudden rise of lake level, probably as high as Lake Maumee. The varved clay grades upward into till that was deposited in water.

In the type section of the Erie Interstade, four sandy varves (layer 7) were deposited in a lake that most probably was not higher than later Lake Warren. The rise of the lake level shown in the Dingman Creek locality seems to have occurred just after the glacier overrode the area of the type section. If this assumption is correct, it means that it took about 60 yrs for the ice to advance over a distance of 34 km (from Port Talbot to London, Ontario). This means that the ice advanced 560 m per yr. Such a fast advance is to be expected, especially as a 500-m-per-yr advance and 170-m-per-yr recession have been demonstrated for the much younger oscillation that built the Tillsonburg-Sparta Moraine (Mörner and Dreimanis, in prep.). As the rate of ice advance of about 600 m per yr measures the lateral movement of the glacier, the advance along the direction of the main ice flow may have been twice as fast.

After the Erie Interstade, the Erie and Huron Lobes advanced to the Powell and Union City Moraines in Ohio and Indiana. The distance from Niagara to the Powell Moraine, which was formed by the Erie Lobe, is about 480 km. With a 600-m-per-yr rate of advance, it would have taken 800 yrs. As most of the distance is in the direction of the main ice flow, this probably is a maximum value for the time required. More details on the tills and moraines in Ohio and Indiana are given by Dreimanis and Goldthwait (this volume, p. 71–105).

The onset of the Erie Interstade was probably 17,000 to 16,500 yrs B.P., when the ice must have started to recede. The maximum date for the preceding advance to the Reesville Moraine is 17,340 ± 390 yrs B.P. (OWU-256; Goldthwait and Rosengreen, 1969). The retreat during the Erie Interstade ended at about 15,600 yrs B.P., and the ice then began to readvance. It probably reached its maximum at about 14,800 yrs B.P. (Mörner, 1970b).

CORRELATIONS WITH OTHER AREAS

The Erie Interstade is an interval of climatic amelioration separating two major late Wisconsin cold peaks—the older 22,000 to 17,000 yrs B.P. (Mt. Olive, Cuba, and Reesville Moraines in Ohio), and the younger about 14,800 yrs B.P. (Powell and Union City Moraines in Ohio and Indiana; Dreimanis and Goldthwait, this volume, p. 71–105). This threefold sequence is recorded in many other places (Mörner, 1970c, Fig. 1), and we believe that the Erie Interstade corresponds to a world-wide climatic amelioration (Table 4).

North America

The ice marginal changes from coast to coast indicate several glacial maxima during the late Wisconsin, or its equivalents, the Woodfordian, the Pinedale, and the Fraser glaciations. Two of the major maxima are separated by synchronous recession of the glaciers contemporaneous with the Erie Interstade. The outer margin of the late Wisconsin drift over North America (according to the summary map of Prest, 1969) seems to have been deposited mainly by the earliest advance in the northeast, east, and midwest and by the second maximum of glaciation in the west and northwest.

In New England, the outer border of the first and second maxima of late Wisconsin glaciation must have been very close. Schafer and Hartshorn (1965) placed both the Ronkonkoma and the Harbor Hill Moraines in the late Wisconsin, but Kaye (1964) considered the Ronkonkoma Moraine to be early Wisconsin. The second glacial maximum is fairly well dated in New England by two radiocarbon dates. One is a date of 15,300 ± 800 yrs B.P. (W-1187, Martha's Vineyard) on tundra plants in clay below sand, which nearby is overlain by till. Another is a date of 14,240 ± 240 yrs B.P. (Y-950/51, Schafer and Hartshorn, 1965) on basal organic sediments in Rogers Lake, southern Connecticut,

TABLE 4. REGIONAL CORRELATIONS AND RADIOCARBON DATES (B.P.) OF THE WISCONSIN-WEICHSELIAN GLACIAL MAXIMA AND THE ERIE INTERSTADE

Area	First glacial maximum	Erie Interstade	Second glacial maximum
North America			
St. John, New Brunswick New England	maximum extent advance about 20,000	gyttja 16,500 ± 320 unknown	alluvial clay > 14,300 ± 270 maximum extent? < 15,300 ± 800 > 14,900 ± 450
New York State–Pennsylvania	maximum extent	recession	Valley Heads Moraine > 14,900 ± 450
South of Lake Erie	Cuba-Reesville Moraine 20,000 to 17,500	Erie Interstade about 15,500	Powell-Union City Moraine > 14,300 ± 450
Lake Michigan Lobe	maximum extent	recession	Valparaiso or Marseilles Morainic System
Washington State	Evans Creek Stade	recession	Vashon Stade; maximum extent just after 15,000 ± 300
Europe			
Kattegatt area	C line, maximum extent 20,000 to 18,000	C/D recession about 15,500	D line about 14,800
North Germany	Brandenburg-Frankfurt Moraine 20,000	Kühlung Interstadial	Pomeranian Moraine about 15,000
East Baltic	Bologoye/Edrovo, about 20,000	Ula Interstadial (?) about 16,000	Vepsovo Stadial about 15,000
Great Britain	Late Midlandian maximum about 20,000	marine transgression	Drumlin Readvance and marine regression about 15,000
Alps	Killwanger phase about 20,000	unknown	Schlieren or Zurich phase > 14,470 ± 385
France and Belgium	three cold peaks 23,000 to 17,500	Lascaux Interstadial 17,500 to 15,500	cold peak about 15,000
Southern Hemisphere			
New Zealand (South Island)	Kumara-2B glacial maximum about 21,000 to 18,000	interstadial about 16,500	Kumara-3A about 15,500
New Zealand (North Island)	Hinuera-1 Formation about 23,000 to 20,000	interstadial	Hinuera-2 Formation about 16,000 to 12,000
South America	cold peak	interstadial	cold peak
Other records			
Greenland 0[18] curve	cold peak	distinct interstadial	distinct cold peak
Deep-sea core R10-10	cold peak 20,300 ± 900	transgression −60 m	cold peak 15,820 ± 600
Eustatic changes	−85 to 90 m; maximum regression		regression to −70 m
Synthesis	Stade 20,000 to 17,500	Interstade 16,500 to 15,500	Stade about 15,000

just above the till. Prest (1969) correlated the Ronkonkoma Moraine with the first and the Harbor Hill Moraine with the second late Wisconsin glacial maximum.

In the Hudson Valley, a readvance to the Wallkill Moraine occurred about 15,000 yrs ago (Connally and Sirkin, 1970a, 1970b).

In northwestern New York State three drift sheets are recognized: the Olean, the Almond or Binghamton, and the Valley Heads drifts (Connally, 1964). The oldest, the Olean drift, was deposited by a glacier that advanced southwesterly, whereas the Almond and the Valley Heads drifts belong to the Ontario Lobe. The Almond, or Binghamton, drift probably represents the first late Wisconsin maximum, and the Valley Heads advance, which overrode the Almond Moraine in the central Finger Lakes region, is younger. The age of the Valley Heads Moraine or its eastward extension, the Lake Escarpment Moraine, is indicated by a radiocarbon date of 14,900 ± 450 yrs B.P. (I-4216; Calkin, 1970) of "woody detritus" 2 cm above outwash, in a locality just south of the moraine. After the first glacial maximum at the Almond-Binghamton Moraine, the ice must have receded into the Ontario Basin, allowing the Erie Interstadial lake to drain eastward via the Mohawk Lowland. The position of glacial margin for this retreat (Fig. 1) is based upon the farthest upglacial localities of Port Stanley Till over Catfish Creek Till in the Ontario Lobe area (Karrow, 1963, 1968) and the Seaforth Moraine Till over the till which is equivalent to the Catfish Creek Till in the Huron Lobe area (Dreimanis, unpub. data). The Ontario Lobe then advanced southward again and built the Valley Heads Moraine. In northwestern Pennsylvania, the Kent Till, which was deposited prior to the Erie Interstade, is separated by a paleosol from the overlying Lavery Till (White and others, 1969) that was deposited by the Erie Lobe after the Erie Interstade.

From the area just southwest of Saint John in New Brunswick (eastern Canada), Gadd (1970, and oral commun.) has reported a C-14 dated lake sequence that may totally change the previous interpretation (for example, Prest, 1969) of deglaciation in this region. In a kettle lake in the area above the marine limit, Gadd found the following sequence from the base: drift, gyttja, 1 m clay, and gyttja. The lower gyttja was dated 16,500 ± 320 yrs B.P. (GSC-1272). The clay seems to be an alluvial clay similar to the much younger North European "Dryas clays" that formed during a cold climate. Consequently, the clay probably corresponds to the second late Wisconsin glacial maximum, the margin of which, however, must have been still farther inland. The gyttja below the clay shows a warmer interstadial climate equivalent to the Erie Interstade.

The ice marginal positions in Ohio and Indiana are discussed above (Goldthwait and others, 1965; Mörner, 1970b; Dreimanis and Goldthwait, this volume, p. 71-105). The outermost limit of the late Wisconsin drift in that area is clearly formed by several glacial maxima belonging to the oldest group (22,000 to 17,000 yrs B.P.), whereas the margin of the second glacial maximum (the Powell and Union City Moraines) is well inside. The second glacial maximum must be older than 14,300 ± 450 yrs B.P. (W-198, Goldthwait, 1958), because this date refers to the Wabash Moraine, which is at least several hundred years younger than the Powell-Union City Moraines (Mörner, 1970b) and younger than 14,780 ± 192 yrs B.P.—a date (OWU-83, Ogden and Hay, 1965) obtained on spruce wood 3 m deep in till at Liberty, Ohio, inside the Powell Moraine.

In the Lake Michigan Lobe area, either the Valparaiso or the Marseilles Moraine System, both of which mark significant readvances (Willman and Frye, 1970), may be the equivalent of the glacial advance after the Erie Interstade. Absence of radiocarbon dates makes the choice difficult.

According to Wright (1970, p. 157), about 14,500 yrs ago "the Des Moines Lobe and its Grantsburg sublobe in Iowa and Minnesota reached an all-time maximum, in part covering an area that was bared by an 80-mile retreat of the Superior Lobe, which, incidentally, made only a very minor readvance at this time." In Iowa, the latest part of the Cary advance is dated by two radiocarbon age determinations on spruce wood in the southern portion of the Des Moines Lobe—14,700 ± 400 yrs B.P. (W-155) and 14,470 ± 400 yrs B.P. (W-512), the latter on a tree rooted in place in loess (Ruhe, 1969).

In South Dakota, North Dakota, and Montana, five late Wisconsin advances of the continental ice are recorded (Lemke and others, 1965). However, none of these advances is as yet well dated, and it is uncertain which correspond to the two glacial maxima discussed here.

In the alpine glaciation of the Rocky Mountains, Richmond (1965) distinguishes three stades of the Pinedale glaciation, which is equivalent to the late Wisconsin of the Laurentide Ice Sheet. The younger of the two interstades is dated at 12,000 yrs B.P., but the older interstade has not been dated in the glaciated areas. Morrison (1970) has recently concluded that a short-lived but deep recession of nonglacial Lake Bonneville was followed by a rise of the lake to the Bonneville shoreline and overflow at Red Rock Pass about 14,000 yrs B.P.

In Washington State, the Fraser glaciation, which is equivalent to late Wisconsin, began with an alpine glaciation, the Evans Creek Stade, which reached its maximum about 20,000 yrs B.P. After the alpine ice had receded, the Cordilleran ice advanced during the Vashon Stade and reached its maximum extension about, or shortly after, 15,000 yrs B.P. (Easterbrook, 1968). A C-14 date from Seattle shows that the ice reached that area about 15,000 ± 300 yrs B.P. (Mullineaux and others, 1965). Pollen studies in the Mount Olympia area (Heusser, 1964, 1969) and studies of deep-sea sediments off Oregon (Griggs and others, 1970) clearly show a threefold division of the late Wisconsin—a stade corresponding to the Evans Creek Stade, an interstade corresponding to the recession between the two glacial maxima, and a stade corresponding to the Vashon Stade. Heusser (1969) dated the interstade at about 18,000 to 16,700 yrs B.P., but these dates seem to be about a thousand years too old judging from the glacial chronology of this area (Easterbrook, 1968) and the global climatic changes (Mörner, 1970c, Fig. 1).

In northwestern North America, the maximum late Wisconsin advances occurred about 14,000 yrs B.P. (Denton, 1970), and the record of the preceding interval which might correspond to the Erie Interstade, is either buried or destroyed by erosion.

Northern Europe (The Scandinavian Ice Sheet)

Mörner (1969a) studied the recession of ice in the Kattegatt Sea region and correlated the Main Stationary line (the C line) in Denmark with the Brandenburg-Frankfurt complex in Germany and the East Jylland line (the D line) with the Pomeranian line. From the rate of recession, shore-level displacement, and some indirect radiocarbon dates, he calculated the age of the East Jylland stadial to be about 14,800 yrs B.P. For the preceding interstadial in northern Germany the name Kühlung Interstadial was proposed; for Denmark the Pre-D or C/D Interstadial was used temporarily (Mörner, 1970b). The first glacial maximum, the C line, was reached at about 20,000 yrs B.P. (Cepek, 1965), and the retreat began at an unknown date, probably about 18,000 yrs B.P. (Mörner, 1969a). Galon (1969, Table 1) found a similar division for the south Baltic region—a first double cold maximum, an interstadial, and a second cold maximum.

For the eastern Baltic region, Chebotareva (1969, Fig. 3) constructed a glaciation curve very similar to the one established by Mörner (1969a, Fig. 42) for southern Scandinavia. The first glacial maximum seems here to be represented by two separate stages, the Bologye and Edrovo. The second late Valdaj glacial maximum, the Vepsovo Stage, corresponds to the Pomeranian line in Germany. It was preceded by the Ula Interstadial (Chebotareva, 1969; Serebryanny, 1969; Vaitekunas, 1969). Several radiocarbon dates have been determined from deposits thought to represent the Ula Interstadial (Serebryanny, 1969, Table 1). However, the dates range from 11,500 to 18,300 yrs B.P. According to Serebryanny and others (1969), the explanation is that the deposits "were accumulated in an ancient lacustrine basin at some distance from the ice margin" and the deposits "had no relation to the Pomeranian ice-marginal formation." In general, the entire Scandinavian ice cap seems to register two major late Weichselian glacial maxima separated by ice recession during an interstadial (Mörner, 1970c, Fig. 1).

If the Scandinavian ice of the first glacial maximum fused with the Scottish ice in the North Sea, which seems highly probable, the margin of the second glacial maximum must have swung around the Norwegian coast and grounded on the outer edge of the Norwegian Channel.

The interstadial between the two late Weichselian glacial maxima was

neither warm enough nor long enough to change the vegetation significantly. It is not registered, for example, in the pollen sequences of
the Netherlands (van der Hammen and others, 1967).

Great Britain

Even the Irish and Scottish ice caps seem to register two late Weichselian glacial maxima. The first of them, the late Midlandian maximum,
formed the outermost limit of Weichselian drift in England and southern
Ireland. The second glacial maximum is apparently represented by the
Drumlin Readvance in Ireland, which shows a major reactivation of
the ice and formed the outermost limit of Weichselian glaciation in northwestern Ireland (Synge and Stephens, 1966). The age has been estimated
at about 15,000 yrs B.P. The Drumlin Readvance in Ireland most probably
corresponds to the Carlingford, Scottish, Lammermuir, and Aberdeen
Readvances in England and eastern Scotland. The southern and southeastern limit of the first glacial maximum lies clearly outside the limit of
the second glacial maximum.

Southern Europe and the Alps

The same threefold division found in other areas seems also to be
recognizable in the Alps. Woldstedt's (1958) correlation between the
moraines in northern Germany and the moraines in the Alps, indicates
that the first late Würm glacial maximum corresponds to the Killwangen
phase and the second to the Schlieren phase in the Zürich area. According
to Hantke (1959), the Zürich phase, which is next youngest after the
Schlieren and is dated 14,470 ± 385 yrs B.P., is correlative with the
Pomeranian phase in northern Germany.

Leroi-Gourham (1965) has shown that pollen analyses from deposits
found in some French caves and in a Spanish bog give evidence of
two late Würm cold peaks separated by a distinct interstade, the Lascaux
Interstadial. Radiocarbon dates of the Lascaux Interstadial itself (exclud-
ing dates from the Laugerie Interstadial) range from 17,000 to 16,000
yrs B.P. The Lascaux Interstadial is followed by a stadial reaching its
maximum at about 15,000 yrs B.P. The first late Würm cold period
is represented by a complex of cold and warm stages—a cold maximum
just before 23,000 yrs B.P.; an interstadial, the Tursac Oscillation, at
about 23,000 yrs B.P.; a cold maximum at about 20,000 to 22,000 yrs
B.P., probably corresponding to the Brandenburg maximum of the Scandinavian Ice Sheet and the Cuba Moraine of the Laurentide Ice Sheet
south of Lake Erie; an interstadial, the Laugerie, dated at 19,000 to
17,500 yrs B.P., probably corresponding to the soil beneath the deposits
of the Reesville Moraine south of Lake Erie; and finally a cold peak
at about 18,000 to 17,500 yrs B.P., probably corresponding to the Frankfurt
maximum of the Scandinavian Ice Sheet and the Reesville Moraine of

the Laurentide Ice Sheet south of Lake Erie. Bastin (1970) recognized the Lascaux Interstadial also in Belgium—in the pollen diagrams from Maisieres.

New Zealand

The glacial chronology of South Island (Suggate, 1965; Suggate and Moar, 1970) clearly shows two glacial maxima of late Wisconsin-Weichselian age—the late Kumara-2 advance and Kumara-3 advance. The late Kumara-2 advance is younger than 22,300 ± 350 yrs B.P. and older than 16,600 ± 390 yrs B.P. The Kumara-3 advance is younger than 16,600 ± 390 yrs B.P. and older than 14,100 ± 220 yrs B.P., and according to Suggate (1965) divided into one "early" and one "later" advance separated by a minor retreat at about 14,800 ± 230 yrs B.P. The Kumara-2 and Kumara-3 advances are separated by a distinct retreat of the glaciers.

On North Island, Schofield (1965) has found the same threefold division in nonglacial deposits of the Hinuera Formation in the south Auckland area. The Hinuera-1 deposition occurred during the cold climate of the Takapu Stadial at 23,000 to 20,000 yrs B.P. Between Hinuera 1 and 2 there is an undated interstadial with warm climate. The Hinuera-2 deposition occurred during a second stadial of cold climate between 16,000 and 12,000 yrs B.P.

South America

Van der Hammen (1961, Fig. 2) shows a curve of average annual temperature for Sabana de Bogotá, with three cold peaks at depths of 4.2, 5.2, and 6.6 m. The cold peaks at −6.6 m and −5.2 m most probably correspond to the first and the second glacial maxima here discussed. They are separated by a distinct interstadial. The cold peak at −4.2 m seems to correspond to the North European and North American glacial advances at about 13,000 yrs B.P. (Dreimanis, 1966; Mörner, 1970b), and is preceded by the Susacá Interstadial (van der Hammen and Vogel, 1966).

Other Areas

We believe that the same threefold division discussed here is present in many other areas. However, the information from some areas is yet too scanty, undated, or not precise enough to allow close correlations.

O^{18} CURVES OF PRESENT ICE CAPS

Dansgaard and others (1969) have presented a detailed O^{18} curve from the Greenland Ice Sheet that shows two distinct late Wisconsin cold peaks separated by a distinct warm period. Even though the time scale is not firm, the amplitude of the temperature oscillations is significant and we can distinguish the two glacial maxima and the intervening intersta-dial (Mörner, 1971a, Fig. 3).

Epstein and others (1970) have presented an O^{18} curve from the Antarc-tic Ice Sheet. The main cold and warm periods of the Wisconsin Stage are recognizable, in particular the cold peak 20,000 yrs B.P. However, the analyses are too few to tell anything about the detailed changes during the interval discussed here.

DEEP-SEA SEDIMENTS

The deep-sea cores are usually not detailed enough to show changes as small and rapid as the ones here discussed. However, two cold peaks separated by a warm period seem to be present in some of Emiliani's paleotemperature curves, for example cores 235A (Emiliani, 1966), 280 (Emiliani, 1958; compare with Olausson, 1960), and P6304-9 (Emiliani, 1966). In most of Ericson's and Wollin's climatic curves based on Fora-minifera these changes are not recognizable. However, core R10-10 (Eric-son and Wollin, 1956) seems to show two cold peaks C-14 dated at 20,300 ± 900 yrs B.P. and 15,820 ± 600 yrs B.P. As already mentioned, the deep-sea cores off the coast of Oregon (Griggs and others, 1970) clearly indicate the threefold division here discussed. Some of the deep-sea cores from the Arctic also show the threefold division, especially cores Arlis 11-2 and Arlis 11-1 (Herman, 1969, Fig. 5).

OCEAN LEVEL

From a climatic-glaciologic point of view, we would expect to find two low sea levels corresponding to the two late Wisconsin-Weichselian glacial maxima, the first one being the lower sea level and the younger a little higher. However, the reconstruction of the eustatic changes is a complicated story that involves questions about relative sea-level changes resulting from local crustal warping, ocean volume changes, and hydro-isostatic deformation of shelves and ocean floors.

Mörner (1969a, 1969b, 1970d, 1971b) has demonstrated that an uplifted area is probably the best to give a detailed picture of the eustatic rise after the last glacial maximum, if the shoreline covers a large area over which the eustatic and isostatic factors can be calculated and checked. Flint (1973) calculated the volume of the last glaciation at its maximum extent as equivalent to an ocean level change of 120 m. Mörner (1970a) stated that the shelves and ocean floors most probably were hydro-isostati-cally deformed by the load of water added after an ice age, but probably not by the removal of water before an ice age, as suggested by Flint. The ocean volume change must have been larger than the final eustatic change. We would expect, therefore, that the eustatic rise was only about two-thirds of the ocean volume increase after an ice age (Mörner, 1970a, 1971b).

The eustatic curve presented by Mörner (1969a, 1970d, 1971b) is calculated to give the true eustatic changes after possible hydro-isostatic subsidence of the shelves and ocean floors and after other crustal changes affected the ocean level. According to him, ocean level rose 25 to 30 m from a level at −85 to −90 m during the first glacial maximum of the late Wisconsin to an interstadial level at −60 m (about 15,500 yrs B.P.) and then dropped 10 m to a −70 m level during the second glacial maximum. These changes agree with the changes in climate.

Curray's (1965) sea-level curve for the Gulf of Mexico lies very close to Mörner's curve back to the time of the second glacial maximum (Mörner, 1970a, Fig. 7). During the first glacial maximum Curray's curve places sea level at −124 m, or 55 m lower than during the second glacial maximum. Such a big difference is quite improbable from a glaciologic-climatic point of view, and Curray's curve is therefore suspected to be erroneous prior to the second glacial maximum. In this connection it is interesting to note that, off West Africa, McMaster and others (1970) did not find shore features below −90 m after detailed mapping of the shelf morphology. A coral *(Porites benardi)* from a depth of −103 to −111 m in that area was radiocarbon dated as 18,750 ± 350 yrs B.P. (I-3678).

Milliman and Emery (1968) present a sea-level curve for the North American Atlantic shelf that shows the maximum late Wisconsin sea level lowering to −130 m at about 15,000 yrs B.P., which is the second glacial maximum, not the first as would be expected. At 18,000 to 20,000 yrs B.P., the first glacial maximum, their curve shows sea level at −90 to −55 m. This is definitely not in concordance with the extent of the continental ice sheets, and we therefore doubt that the position suggested for the first glacial maximum is correct. Garrison and McMaster (1966) found a terrace along the Atlantic shelf (from Newfoundland to South Carolina) at −145 m. This terrace probably corresponds to the first glacial maximum, which means a difference between the first and second glacial maxima of only 15 m (Mörner, 1969a, 1970d, 1971b). A 15-m difference in sea level agrees well with the extent of the continental ice caps.

Milliman and Emery (1968) showed that their curve agreed fairly well with time-depth indications from several other shelves of the world, but it was clearly below the Curray curve. This is a strong indication of a hydro-isostatic deformation not only of most shelves but also of the ocean floors in general (Mörner, 1970a).

In Table 5, some of the "eustatic" levels for the late Wisconsin are compared. Only the Mörner curve is detailed enough to show minor oscillations, and it clearly indicates the threefold division of the glacial and climatic changes registered in different areas of the world

TABLE 5. EUSTATIC SEA LEVELS ACCORDING TO DIFFERENT AUTHORS

Reference	Second glacial maximum 14,800 yrs B.P.	Interstade 15,500 yrs B.P.	First glacial maximum 17,000 to 20,000 yrs B.P.
Mörner (1969a)	−70 m	−60 m	−85 to 90 m
Curray (1965)	−70 m		−124 m
Shepard (1963)	−80 m		−132 m
Milliman and Emery (1968)	−130 m		−90 to 55 m
Garrison and McMaster (1966)			−145 m

CONCLUSIONS

The two major continental ice sheets, the Laurentide and the Scandinavian, clearly show a threefold division during the late Weichselian or late Wisconsin glaciation—a first glacial maximum at 20,000 to 17,000 yrs B.P., an interstadial at 16,500 to 15,500 yrs B.P., and a second glacial maximum at 14,800 yrs B.P. Similar sequences are found in the glaciations of Great Britain, New Zealand, and probably also of the Alps. Climatic records from France-Spain, South America, and the Greenland Ice Sheet indicate the same threefold division. In some deep-sea cores the same sequence can be recognized. Eustatic changes also seem to record the threefold division.

We therefore conclude that the threefold division was caused by global climatic changes.

ACKNOWLEDGMENTS

This study has been supported by the National Research Council of Canada, Grant A4215; some of the field data were obtained as part of the Geological Survey of Canada project 490052, and are published here with the permission of the Director of the Geological Survey of Canada. We would like to express our thanks to J. S. Hancock and A. O. Grins for their participation at the field and laboratory investigations and to all those colleagues who have discussed with us the correlations of the Erie Interstade.

REFERENCES CITED

Bastin, B., 1970, La chronostratigraphie du Würm en Belgique, à la lumière de la palynologie des loess et limons: Soc. Géol. Belgique, Ann., v. 93, no. 3, p. 545–580.

Calkin, P. E., 1970, Strand lines and chronology of the glacial Great Lakes in northwestern New York: Ohio Jour. Sci., v. 70, p. 78–96.

Cepek, A. G., 1965, Geologische Ergebnisse der ersten radio-carbondatierungen von Interstadialen in Lausitzer Urstromtal: Geologie, v. 14, p. 625–657.

Chebotareva, N. S., 1969, Recession of the last glaciation in northeastern European

USSR, *in* Wright, H. E., Jr., ed., Quaternary geology and climate: Proc. INQUA VII, v. 16, Washington, D.C., Natl. Acad. Sci. Pub. 1701, p. 79-83.

Claypole, E. W., 1887, The lake age in Ohio: Edinburgh Geol. Soc. Trans. (quoted from Leverett, 1902, p. 604).

Connally, G. G., 1964, Garnet ratios and provenance in the glacial drift of western New York: Science, v. 144, p. 1452-1453.

Connally, G. G., and Sirkin, L. A., 1970a, The Wisconsinan history of the Huron-Champlain lobe [abs.]: Geol. Soc. America, Abs. with Programs (Ann. Mtg.), v. 2, no. 7, p. 524-525.

―――― 1970b, Late glacial history of the upper Wallkill Valley, New York: Geol. Soc. America Bull., v. 81, p. 3297-3306.

Curray, J. R., 1965, Late Quaternary history, continental shelves of the United States, *in* Wright, H. E., Jr., and Frey, D. G., eds., The Quaternary of the United States: Princeton, N. J., Princeton Univ. Press, p. 723-735.

Dansgaard, W., Johnsen, S. J., Møller, J., and Langway, C. C., 1969, One thousand centuries of climatic record from Camp Century on the Greenland Ice Sheet: Science, v. 166, p. 378-381.

Denton, G. H., 1970, Late Wisconsin glaciation in northwestern North America: Ice recession and origin of Paleo-Indian Clovis complex [abs.]: Bozeman, Mont., AMQUA Abs. (First Mtg.), p. 34-35.

DeVries, H., and Dreimanis, A., 1960, Finite radiocarbon dates of the Port Talbot interstadial deposits in southern Ontario: Science, v. 131, p. 1738-1739.

Dreimanis, A., 1958, Wisconsin stratigraphy at Port Talbot on the north shore of Lake Erie, Ontario: Ohio Jour. Sci., v. 58, p. 65-84.

―――― 1961, Tills of southern Ontario, *in* Legget, R. F., ed., Soils in Canada: Royal Soc. Canada Spec. Pub. 3, p. 80-96.

―――― 1966, Lake Arkona-Whittlesey and post-Warren radiocarbon dates from "Ridgetown Island" in southwestern Ontario: Ohio Jour. Sci., v. 66, p. 582-586.

―――― 1967, Pre-Maumee lake stages of Wisconsin Ice Age in Lake Erie Basin [abs.]: Toronto, Ont., Abstracts, 10th Conf. Great Lakes Research, p. 33.

―――― 1968, Extinction of mastodons in eastern North America: Testing a new climate-environmental hypothesis: Ohio Jour. Sci., v. 68, p. 257-272.

―――― 1969a, Late Pleistocene lakes in the Ontario and the Erie Basins: Ann Arbor, Mich., Internat. Assoc. Michigan Univ. Great Lakes Research, Proc. 12th Conf., p. 170-180.

―――― 1969b, Till wedges as indicators of direction of glacial movement [abs.]: Geol. Soc. America, Abs. with Programs for 1969, Pt. 7 (Ann. Mtg.), p. 52-53.

―――― 1970, Last ice-age deposits in the Port Stanley map-area, Ontario (40 I/11): Canada Geol. Survey Paper 70-1 (A), p. 167-169.

Dreimanis, A., and Goldthwait, R. P., 1973, Wisconsin glaciation in the Huron, Erie, and Ontario Lobes: Geol. Soc. America Mem. 136, p. 71-105.

Dreimanis, A., and Karrow, P. F., 1965, Southern Ontario, *in* Schultz, C. B., and Smith, H.T.V., eds., Guidebook for Field Conference G, Great Lakes-Ohio River Valley (INQUA VII Congress, 1965): Nebraska Acad. Sci., p. 90-110.

Dreimanis, A., and Reavely, G. H., 1953, Differentiation of the lower and the upper till along the north shore of Lake Erie: Jour. Sed. Petrology, v. 23, p. 238-259.

Dreimanis, A., Terasmae, J., and McKenzie, G. D., 1966, The Port Talbot Interstade of the Wisconsin glaciation: Canadian Jour. Earth Sci., v. 3, no. 3, p. 305-325.

Easterbrook, D. J., 1968, Pleistocene stratigraphy of Island County: Washington Dept. Water Resources Water Supply Bull., v. 25, 34 p.

Emiliani, C., 1958, Paleotemperature analysis of core 280 and Pleistocene correlations: Jour. Geology, v. 66, no. 3, p. 264-275.

_____ 1966, Paleotemperature analysis of Caribbean cores P6304-8 and P6304-9 and a generalized temperature curve for the past 425,000 years: Jour. Geology, v. 74, no. 1, p. 109-126.

Epstein, S., Sharp, R. P., and Gow, A. J., 1970, Antarctic Ice Sheet: Stable isotope analyses of Byrd Station cores and interhemispheric climatic implication: Science, v. 168, p. 1570-1572.

Ericson, D. B., and Wollin, G., 1956, Micropaleontologic and isotopic determination of Pleistocene climates: Micropaleontology, v. 2, no. 3, p. 257-270.

Flint, R. F. (in press), Position of sea level in a Glacial Age, in INQUA VIII Congress in Paris 1969: Quaternaria, v. XV.

Gadd, N. R., 1970, Quaternary geology, southwest New Brunswick (21 G): Canada Geol. Survey Paper 70-1 (A), p. 170-172.

Galon, R., 1969, The course of deglaciation in the Peribalticum, in Last Scandinavian Glaciation in Poland: Inst. Geogr., Polish Acad. Sci., Geogr. Studies 74, p. 201-212.

Garrison, L. E., and McMaster, R. L., 1966, Sediments and geomorphology of the Continental Shelf off southern New England: Marine Geology, v. 4, p. 273-289.

Goldthwait, R. P., 1958, Wisconsin age forests in western Ohio: Ohio Jour. Sci., v. 58, no. 4, p. 209-230.

Goldthwait, R. P., and Rosengreen, T., 1969, Till stratigraphy from Columbus southwest to Highland County, Ohio: Geol. Soc. America, Abs. with Programs for 1969, North-Central Sec., Pt. 2, p. 2-1 to 2-17.

Goldthwait, R. P., Dreimanis, A., Forsyth, J. L., Karrow, P. F., and White, G. W., 1965, Pleistocene deposits of the Erie Lobe, in Wright, H. E., Jr., and Frey, D. G., eds., The Quaternary of the United States: Princeton, N.J., Princeton Univ. Press, p. 85-97.

Griggs, G. B., Kulm, L. D., Duncan, J. R., and Fowler, G. A., 1970, Holocene faunal stratigraphy and paleoclimatic implication of deep-sea sediments in Cascadia Basin: Palaeogeography, Palaeoclimatology, Palaeoecology, v. 7, p. 5-12.

Hantke, R., 1959, Zür Phasenfolge der Hochwürmeiszeit des Linth- und des Reuss-Systems, verglichen mit derjenigen des Inn- und des Salzach-Systems sowie mit der nordeuropänische Vereisung: Mitteilungen Geol. Instit. Eidg. Techn. Hochschule u. Univ. Zürich, Ser. B., Nr. 14, p. 390-342.

Herman, Y., 1969, Arctic Ocean Quaternary microfauna and its relation to paleoclimatology: Palaeogeography, Palaeoclimatology, Palaeoecology, v. 6, p. 251-276.

Heusser, C. J., 1964, Palynology of four bog sections from western Olympic Peninsula, Washington: Ecology, v. 45, p. 23-40.

_____ 1969, Pleistocene environments and chronology of the western Olympic Peninsula, Washington [abs.]: Geol. Soc. America, Abs. with Programs for 1969, Pt. 7 (Ann. Mtg.), p. 99.

Karrow, P. F., 1963, Pleistocene geology of the Hamilton-Galt area: Ontario Dept. Mines Geol. Rept. 16, 68 p.

Karrow, P. F., 1968, Pleistocene geology of the Guelph area: Ontario Dept. Mines Geol. Rept. 61, 38 p.

Kaye, C. A., 1964, Outline of Pleistocene geology of Martha's Vineyard, Massachusetts: U.S. Geol. Survey Prof. Paper 500-C, p. C134-C139.

Kunkle, G. R., 1963, Lake Ypsilanti: A probable late Pleistocene low-lake stage in the Erie Basin: Jour. Geology, v. 71, p. 72-75.

Lemke, R. W., Laird, W. M., Tipton, M. J., and Lindvall, R. M., 1965, Quaternary geology of Northern Great Plains, in Wright, H. E., Jr., and Frey, D. G., eds., The Quaternary of the United States: Princeton, N. J., Princeton Univ. Press, p. 15-27.

Leroi-Gourham, A., 1965, The Würm climate during the Upper Paleolithic from 36,000 to 8,000 B.C. [abs.]: Boulder, Colo., Abstracts, INQUA VII Congress, p. 290 and multigraphed illustrations.

Leverett, F., 1902, Glacial formations and drainage features of the Erie and Ohio Basins: U.S. Geol. Survey Mon. 41, 802 p.

McMaster, R. L., LaChance, T. P., and Ashrat, A., 1970, Continental shelf geomorphic features off Portugese Guinea, Guinea, and Sierra Leone (West Africa): Marine Geology, v. 9, no. 3, p. 203-213.

Milliman, J. D., and Emery, K. O., 1968, Sea levels during the past 35,000 years: Science, v. 162, p. 1121-1123.

Mörner, N.-A., 1969a, The late Quaternary history of the Kattegatt Sea and the Swedish West Coast; deglaciation, shore-level displacement, chronology, isostasy, and eustasy: Sveriges Geol. Undersökning Årsb., C-640, 487 p.

―――― 1969b, Eustatic and climatic changes during the last 15,000 years: Geologie en Mijnbouw, v. 48, no. 4, p. 389-399.

―――― 1970a, Isostasy and eustasy. Late Quaternary isostatic changes of southern Scandinavia and general isostatic changes of the world: Western Ontario Univ. Dept. Geology Contr. 171 (also see Abstracts, International Symposium on Recent Crustal Movements and Associated Seismicity: Royal Soc. New Zealand Trans., p. 25-27).

―――― 1970b, Late Wisconsin ice marginal changes in the Erie Lobe area and comparisons with the late Weichselian sequence of southern Scandinavia [abs.]: Geol. Soc. America, Abs. with Programs (Ann. Mtg.), v. 2, no. 7, p. 751-752.

―――― 1970c, Comparison between late Weichselian and late Wisconsin ice marginal changes: Eiszeitalter u. Gegenwart, v. 21, p. 173-176.

―――― 1970d, Late Quaternary isostatic, eustatic and climatic changes: Quaternaria, p. 12-13.

―――― 1971a, The position of the ocean level during the interstadial at about 30,000 B.P.: A discussion from a climatic-glaciologic point of view: Canadian Jour. Earth Sci., v. 8, p. 132-143.

―――― 1971b, Eustatic changes during the last 20,000 years and the method of separating the isostatic and eustatic factors in an uplifted area: Palaeogeography, Palaeoclimatology, Palaeoecology, v. 9, p. 153-181.

Morrison, R. B., 1970, Conflicting pluvial-lake evidence on climatic changes between 14 and 9 millenia ago, with particular reference to Lakes Lahonton, Bonneville, and Searles [abs.]: Bozeman, Mont., AMQUA Abs. (First Mtg.), p. 97-98.

Mullineaux, D. R., Waldron, H. H., and Rubin, M., 1965, Stratigraphy and chronology of late interglacial and early Vashon glacial time in the Seattle area, Washington: U.S. Geol. Survey Bull. 1194-0, p. 01-010.

Ogden, J. G., III, and Hay, R. J., 1965, Ohio Wesleyan University natural radiocarbon measurements II: Radiocarbon, v. 7, p. 166-173.

Olausson, E., 1960, Description of sediment cores from North Atlantic: Swedish Deep-Sea Exp. Rept., 1947-1948, v. 7, no. 6, p. 226-286.

Prest, V. K., 1969, Retreat of Wisconsin and Recent ice in North America: Canada Geol. Survey Map 1257A.

Richmond, G. M., 1965, Glaciation of the Rocky Mountains, in Wright, H. E., Jr., and Frey, D. G., eds., The Quaternary of the United States: Princeton, N. J., Princeton Univ. Press, p. 217-230.

Ruhe, R. V., 1969, Quaternary landscapes in Iowa: Ames, Iowa, Iowa State Univ. Press, 255 p.

Schafer, J. P., and Hartshorn, J. H., 1965, The Quaternary of New England, in Wright, H. E., Jr., and Frey, D. G., eds., The Quaternary of the United States: Princeton, N. J., Princeton Univ. Press, p. 113-128.

Schofield, J. C., 1965, The Hinuera Formation and associated Quaternary events: New Zealand Jour. Geology and Geophysics, v. 8, p. 772-791.

Serebryannyy, L. R., 1969, L'apport de la radio-chronométrie à l'étude de l'histoire tardi-quaternaire des régions de glaciation ancienne de la plaine russe: Rev. Géographie Phys. et Géologie Dynam., v. 11, no. 3, p. 293-320.

Serebryannyy, L. R., Raukas, V. N., and Punning, J.-M.K., 1969, On the upper Pleistocene history of glaciation in the northwestern part of Russian Plain: Acad. Sci. Inst. Geogr. USSR, Data Glaciol. Studies, Chron. Discuss., v. 15, p. 167-181.

Shepard, F. P., 1963, Thirty-five thousand years of sea level, in Clements, T., ed., Essays in marine geology in honor of K. O. Emery: Los Angeles, Calif., Univ. Southern California Press, p. 1-10.

Suggate, R. P., 1965, Late Pleistocene geology of the northern part of the South Island, New Zealand: New Zealand Geol. Survey Bull. 77, 99 p.

Suggate, R. P., and Moar, N. T., 1970, Revision of the chronology of the Late Otira Glacial: New Zealand Jour. Geology and Geophysics, v. 13, p. 742-746.

Synge, F. M., and Stephens, N., 1966, Late- and post-glacial shorelines and ice limits in Argyll and north-east Ulster: Inst. British Geographers Trans., v. 39, p. 101-145.

Vaitekunas, P., 1969, On a stratigraphic subdivision of the Neo-pleistocene in the glacial area of Soviet Baltic Provinces [abs.]: Abstracts, INQUA VIII Congress, Paris, p. 255.

van der Hammen, T., 1961, The Quaternary climate changes of North and South America: New York Acad. Sci. Annals, v. 95, no. 1, p. 676-683.

van der Hammen, T., and Vogel, J. C., 1966, The Susacá-Interstadial and the subdivision of the Late-Glacial: Geol. en Mijnbouw, v. 45, p. 33-35.

van der Hammen, T., Maarleveld, G. C., Vogel, J. C., and Zagwijn, W. H., 1967, Stratigraphy, climatic succession and radiocarbon dating of the last glacial in the Netherlands: Geol. en Mijnbouw, v. 46, p. 79-95.

White, G. W., 1960, Classification of Wisconsin glacial deposits in northeastern Ohio: U.S. Geol. Survey Bull. 1121-A, 12 p.

_____ 1961, Classification of glacial deposits in the Kilbuck Lobe, northeast-central Ohio: U.S. Geol. Survey Prof. Paper 424-C, p. 71-73.

White, G. W., Totten, S. M., and Gross, D. L., 1969, Pleistocene stratigraphy of northwestern Pennsylvania: Pennsylvania Geol. Survey, 4th Ser., Gen. Geol. Rept. 655, 88 p.

Willman, H. B., and Frye, J. C., 1970, Pleistocene stratigraphy of Illinois: Illinois Geol. Survey Bull. 94, 204 p.

Woldstedt, P., 1958, Das Eiszeitalter II: Stuttgart, F. Enke.

Wright, H. E., Jr., 1970, Retreat of the Laurentide ice sheet from 14,000 to 9,000 years ago [abs.]: Bozeman, Mont., AMQUA Abstracts (First Mtg.), p. 157-159.

MANUSCRIPT RECEIVED BY THE SOCIETY DECEMBER 27, 1971

GEOLOGICAL SOCIETY OF AMERICA
MEMOIR 136
© 1973

Wisconsinan Climatic History Interpreted from Lake Michigan Lobe Deposits and Soils

JOHN C. FRYE

H. B. WILLMAN

Illinois State Geological Survey, Urbana, Illinois 61801

ABSTRACT

The climatic history of the Illinois region during Wisconsinan time is interpreted from the character and extent of the glacial deposits and buried soils of the Lake Michigan Lobe. The Illinoian-Sangamonian history of the lobe is briefly discussed. The history is complex; it includes three major glacial episodes during the Illinoian, and a succession of glacial advances and retreats during the Wisconsinan. The major interglacial stages were times of near climatic equilibrium, and the soil profiles indicate temperatures were higher than those of the present. The glacial stages, on the other hand, were characterized by sharply fluctuating climates, with episodes of glaciation alternating with minor intervals of stability, soil formation, and temperatures approaching those of the major interglacial stages.

INTRODUCTION

Many reconstructions of the climatic history of Pleistocene time have been made from a variety of criteria and with widely diverse conclusions. The criteria used have included marine and nonmarine invertebrates, land mammals, pollen, tree rings, isotope ratios in shells and in glaciers, sea-level fluctuations, rates of marine sedimentation, alluvial and eolian

135

cycles, desiccation cycles, fluctuations of interior lake levels, soil morphology, periglacial phenomena, and glacial advances and retreats. It is our purpose here (1) to examine the glacial history of the Lake Michigan Lobe as interpreted from its deposits, buried soils, and moraines, and (2) to construct from the data a climatic history, primarily for Wisconsinan time, but beginning with the Illinoian. We do not contend that such a history is representative of world-wide or continent-wide conditions. Rather, the evidence from this region is presented so that comparisons can be made with climatic interpretations derived from other regions.

The glacial history of this area is not simple, nor does it fit the classic interpretation of four episodes of glacial advance and retreat. Rather, the Illinoian and Wisconsinan history of the Lake Michigan Lobe includes many advances and retreats of the ice front and cool episodes of glaciation separated by warmer intervals of varying durations.

Data that can be used to deduce the climate of the nonglacial intervals are primarily derived from the soil profiles that developed on the relatively stable surfaces during periods that were not marked by major deposition or, except along the margins of large valleys, by major erosion.

It seems evident that glaciers advanced during periods of relatively low temperatures and high rates of precipitation and retreated during episodes of lower rates of precipitation or higher temperatures, or both. The relative importance of these two factors cannot be determined from the sediments. Nevertheless, it is reasonable to assume that glacial retreat coincided with times of higher temperatures, and in this discussion glacial advance and retreat are interpreted as indicative of temperature changes.

LAKE MICHIGAN LOBE

The Lake Michigan Lobe has been defined from the glacial deposits of both Illinoian and Wisconsinan ages, but it is not recognized with certainty in earlier episodes of glaciation. The lobe was the mass of glacial ice that had an axial flow southward through what is now the basin of Lake Michigan. In northern and east-central Illinois it was diverted westward by interference of other glacial lobes advancing toward the southwest from the region of what is now Lake Erie and Lake Huron (Fig. 1). During Illinoian time the lobe reached and crossed the present position of the Mississippi River in central western Illinois, but it was the Erie Lobe, advancing from the northeast, that reached the maximum southern glacial limit in southern Illinois. Deposits of glaciers from the two source regions have been distinguished by their characteristic mineral composition, as well as by their morailnal patterns (Willman and Frye, 1970).

Although the purpose of this paper is to discuss the climatic cycles recorded by the Lake Michigan Lobe during Wisconsinan time, the Illinoian history of the lobe should be reviewed. The deposits of the Illinoian

Stage are classed in three substages, the Liman, Monican, and Jubileean, each of which is based upon a closely associated group of tills and attendant deposits (Fig. 2). A strongly developed soil, the Pike Soil, forms a plane of demarcation between the first two substages, and a less well developed soil has been observed between the upper two substages. The extent of glaciers in the Lake Michigan Lobe was greatest during the Liman Substage, and successively less during each of the succeeding substages. The retreat of glaciers between each pair of pulsatory glacial episodes was extensive, as indicated by deposits of outwash and silt and by buried soils and the molluscan faunas contained in the intercalated silts (Leonard and others, 1971). The ice front probably retreated beyond the present southern shore of Lake Michigan during these intervals.

WISCONSINAN STAGE .

The Wisconsinan Stage is subdivided into the Altonian, Farmdalian, Woodfordian, Twocreekan, and Valderan Substages (Fig. 2). The first three encompass most of the Wisconsinan time and the deposits of the Lake Michigan Lobe in Illinois with which we are primarily concerned. Of the many pulsations of the lobe during Wisconsinan time, the earliest and the latest were the least extensive. The maximum extent of the lobe was attained during the early Woodfordian at 19,000 to 20,000 radiocarbon yrs B.P. (Fig. 2).

The Lake Michigan Lobe during Wisconsinan time, as during Illinoian, was diverted westward (south of the present lake basin) by glaciers advancing toward the southwest from the areas of Lake Erie and Lake Huron (Fig. 1). The configuration of the lobe during Altonian time is not known in detail because, except in a relatively small area in north-central Illinois, the glacial deposits of Altonian age were overridden by Woodfordian glaciers.

Altonian Substage

The stratigraphy of the glacial deposits of the Altonian Substage in Illinois is better known from subsurface data obtained by core borings (Kempton, 1963; Frye and others, 1969) than from observation of surface outcrops. However, the stratigraphic sequence developed from subsurface data for the glacial deposits correlates closely with the stratigraphic sequence developed for the loess deposits beyond the Wisconsinan glacial limits (Willman and Frye, 1970).

The earliest episode of Altonian glaciation is represented by tills that extend only a short distance beyond the limit of present Lake Michigan and are not exposed at the surface in Illinois (Fig. 1). Their counterpart in the loess sequence along the Illinois Valley is represented by the

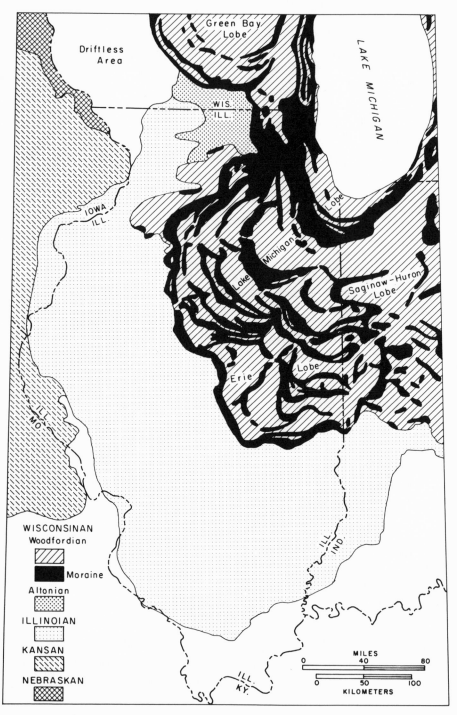

Figure 1. Generalized glacial map showing the relation of the Lake Michigan Glacial Lobe to the older drifts.

deeply weathered Markham Member of the Roxana Silt, which consists of thin, widespread, loesslike silt and colluvial silt. The Markham Member commonly is included within the profile of the Chapin Soil, the top of which coincides with the top of the member (Fig. 2). Mineralogic study of this unit in the Illinois Valley shows the presence of loess derived from outwash from the Lake Michigan Lobe. The Chapin Soil is a strongly developed soil profile that signifies an interval of warm to hot climate separating the earliest Altonian glacial pulse from the major glacial advances of later Altonian time. It is the most significant interruption of loess deposition in the Altonian.

The relatively long interval of soil development recorded in the Chapin Soil was followed by an interval of loess deposition, the McDonough Loess Member of the Roxana Silt. Although this loess, extensively deposited along the Illinois and middle Mississippi River valleys, clearly indicates an episode of glaciation, tills that can be certainly correlated with it have not as yet been identified in northern Illinois. This short episode of loess deposition was followed by formation of the Pleasant Grove Soil (Fig. 2). As this soil is much less strongly developed than the Chapin Soil, the McDonough Loess Member is more closely related to the overlying Meadow Loess Member than to the beds below.

The maximum advance of the Altonian Lake Michigan Lobe is recorded in the Argyle Till Member of the Winnebago Formation, which is the surficial till in a significant area of north-central Illinois that was not overridden by subsequent Woodfordian glaciers. Following the retreat of the Argyle glacier, silts rich in organic material accumulated widely and have been penetrated by many core borings. These deposits, classed as the Plano Silt Member of the Winnebago Formation, have been radiocarbon dated from 32,600 to 41,000 yrs B.P. (Willman and Frye, 1970). Following deposition of the Plano Silt a less extensive glacial advance (the last of the Altonian Substage) deposited the Capron Till Member of the Winnebago Formation. The Argyle, Plano, and Capron Members are equivalent to the Meadow Loess Member of the Roxana Silt. Shells from the middle of the Meadow Loess Member have been dated at 35,000 to 37,000 radiocarbon yrs B.P. (Willman and Frye, 1970).

Farmdalian Substage

The Farmdalian Substage is represented by the Farmdale Soil (Fig. 2) and by peat and silt rich in organic material that accumulated above the Capron Till Member of the Winnebago Formation and the Meadow Loess Member of the Roxana Silt. These deposits, classed as the Robein Silt, and contemporary lacustrine deposits, record the absence of glaciers from northeastern and central Illinois. During deposition of the Robein Silt, glaciers of the Lake Michigan Lobe probably had retreated from the southern part of the present lake basin.

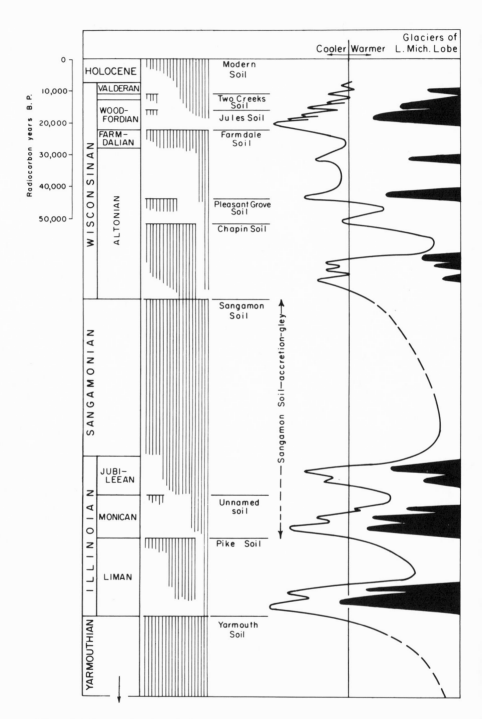

Figure 2. Diagram showing relative temperature variations in relation to glacial advances and buried soils since Yarmouthian time. The vertical line through the temperature curve represents present temperature in the Driftless Area of northwestern Illinois.

Woodfordian Substage

The complex history of Woodfordian glacial advance and retreat of the Lake Michigan Lobe is shown graphically by the time-space diagram in Figure 3. This diagram is plotted to scale along a straight line from the southernmost extent of the Lake Michigan Lobe to the present shore of Lake Michigan at Highland Park, in southern Lake County, Illinois (Willman and Frye, 1970, Pl. 1). As shown by the diagram, loess deposition preceded the advancing glacier and produced the deposits now classed as Morton Loess. The rock-stratigraphic units within the tills (Willman and Frye, 1970), and the correlations with the stratigraphy, mineral zonation, and pertinent radiocarbon dates of the Peoria Loess beyond the limits of Woodfordian glaciation (Frye and others, 1968), also are shown.

The till members of the Wedron Formation have been differentiated on the basis of distinctive differences in lithology and in mineral composition, and the boundaries between these rock units coincide with major episodes of glacial retreat and readvance. The most extensive pulses of the ice front are based on detailed stratigraphic data that show the presence of till sheets separated by water-laid deposits many miles inside the glacial border (see, for example, Willman and Payne, 1942) and on the sharp contrasts in composition of the individual till units that can be correlated with the tills of the major moraines. Also noteworthy is the correlation of the Jules Soil, which occurs at the contact between loess mineral Zones II and III, with the most extensive Woodfordian glacial withdrawal and with the stratigraphic position of the most striking contrast in till composition.

The most significant glacial retreat of the Lake Michigan Lobe during the Woodfordian occurred between deposition of the Tiskilwa and Malden Till Members of the Wedron Formation (Fig. 3). This interval of glacial withdrawal is equivalent to the Jules Soil, as is shown by tracing the mineral zones in the loess (Frye and others, 1968), and is referred to informally as the Jules Soil interval. Also, although radiocarbon dates at this stratigraphic position have not been obtained from the till sequence, dates from the loess related to the mineral zones place the age of the Jules Soil between $17,950 \pm 550$ (W-1055) and $17,100 \pm 300$ (W-730) radiocarbon yrs B.P. The interval between the Yorkville and Wadsworth Till Members of the Wedron Formation was called the St. Charles intraglacial by Leighton (1960) and probably correlates with the Erie Interstadial of Dreimanis (1958) and of Mörner and Dreimanis (this volume, p. 107-134). The diagram (Fig. 3) suggests that several other readvances may be of equal significance.

With the exception of the Jules Soil, the glacial pulsations are not stratigraphically recognized in the loess sequence. The most rapid loess deposition occurred during the first half of the Woodfordian. Even then the loess accumulated at a rate of only a few millimeters per year

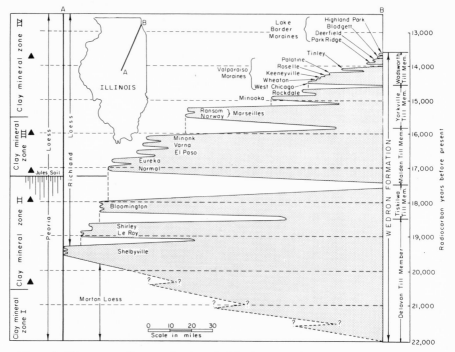

Figure 3. Time-space diagram showing glacial advance and retreat during Woodfordian time along line AB (northeast of Lincoln, Logan County, to Highland Park, Lake County). Stipple pattern indicates presence of glacial ice. Names of moraines and rock-stratigraphic units that occur along the line are shown. Beyond the limit of glaciation, the position of the Jules Soil and the clay mineral zones in the loess are indicated. Pertinent radiocarbon dates are shown by solid triangles.

in the thickest sequences, and its physical characteristics were modified by its continuing accumulation as the A-horizon of the surface soil.

It is not possible to trace the stratigraphic position of the maximum glacial advance into the loess deposited beyond the tills. Therefore, vertical cutoff is used to differentiate the Woodfordian loess sequence beyond the glacial limit as Peoria, the loess below the till as Morton, and the loess above it as Richland (Fig. 3).

A more detailed subdivision of the Woodfordian deposits is based on the assemblage of deposits (till, outwash, ice-contact deposits, and lacustrine sediments) related to each of the pulsations of the ice front. As these units are based largely on their relation to the moraines, they are classified as morphostratigraphic units, are given the same names as the moraines, and are called "drifts." At least 32 moraines were deposited in Illinois during the Woodfordian, and 23 of them are along the line of the cross section (Fig. 3).

Following the deposition of the Valparaiso Morainic System, the Lake Michigan Lobe conformed closely to the outline of the present basin of the lake. Subsequent glacial pulses built the Tinley and Lake Border Moraines (Fig. 3). These were followed by extensive retreat and by the readvance that deposited the Port Huron Moraine in Michigan, after which the glacial margin retreated to the head of the lake, or farther. It was during this retreat that the Two Creeks Forest grew in Wisconsin. It was later overridden by the last advance of the Lake Michigan Lobe, the Valders, which made a prominent lobe in the lake basin and partly overlapped the Port Huron Moraine.

The retreat of the Valders glacial front marked the disappearance of the Lake Michigan Lobe at about 9,500 radiocarbon yrs B.P. (Black, 1970).

CRITERIA FOR CLIMATIC INTERPRETATIONS

The criteria used in this study for the reconstruction of the climatic record during late Pleistocene time include the character of buried soil profiles within the sediments, fossil mollusks, the succession of glacial advances and retreats, the character of the deposits resulting from glacial action, and periglacial features.

Soil Profiles

The profiles of zonal, or normal, soils develop in the mineral matter below subaerial surfaces as a direct result of the climate and the actions of plants and animals. Their characteristics are strongly influenced by such factors as the microtopography and the texture, mineral composition, structure, and permeability of the deposits. In many cases it is possible to reconstruct, at least in a general way, the nature of these physical factors when the soil was at the surface and thus to compare and contrast soil profiles that developed in similar situations. The plant and animal population on and in the soil is strongly influenced by the climate and the physical factors. Therefore, if we are able to hold constant the physical factors of parent material and microtopography, the characteristics of a buried soil profile may be used as a climatic indicator for the period of its development. The use of buried soils as climatic indicators has been reviewed by Thorp (1965), Ruhe (1965), Morrison (1967), and Frye and Leonard (1967).

As zonal soil profiles may develop over long periods of time, during which the climate has fluctuated, they may to some degree integrate the total climatic effects of the period of their development. However, as high temperatures and high precipitation rates strongly accelerate the rate of soil development, it is probable that the buried profiles, as we see them, do not indicate average climate, but more nearly indicate much shorter episodes of extreme climatic conditions (Morrison, 1967).

Climatic interpretations are much more uncertain from intrazonal soil profiles, where one genetic factor has been so overpowering that it tends to mask all other factors. This is particularly true of the accretion-gley profiles that commonly occur on the undrained or very poorly drained areas of the till-plain surfaces. The accretion-gley soils clearly indicate areas that were intermittently wet and into which fine sediment was moved from adjacent low-angle slopes. This condition may be controlled by microtopography and low permeability of the surficial deposits rather than by excessive rainfall, and it gives little indication of temperature.

The movement of calcium carbonate and clay in a soil profile are clues to climate and to the length of the period of soil formation. In Illinois, the major buried soils are leached of carbonates to as much as 20 ft in depth, and significant secondary accumulations of carbonates, or "caliche," below the leached B-horizon are quite rare, thus indicating at least moderately high precipitation rates. Accumulations of fine clay in the B-horizons, on the other hand, are the general rule. Soil color in the B-horizons of zonal soils that were moderately to well drained have been used as an indication of temperature. Strongly red B-horizons, by analogy with modern soils, indicate hot climates that existed over significant periods of time.

Fossil Mollusks

Although uneven in their distribution, shells of fossil mollusks occur in sediments interstratified with and immediately adjacent to glacial deposits. By analogy with the habitats of similar living species, fossil mollusks can be used as a clue to the climate at the time and place where they were entrapped in sediments (Leonard and Frye, 1960; Frye and Leonard, 1967; Leonard and others, 1971). Species that are restricted to water habitats are of lesser value, but terrestrial species that have known and restricted ecologic requirements may be quite useful in deciphering paleoclimates. Ecologic requirements of snails allow an interpretation of the existing cover of vegetation as well as the ranges of temperature.

Fossil Pollen

Wisconsinan climatic history has been interpreted from studies of pollen in lake deposits on the Illinoian till plain about 50 mi south of the Wisconsinan glacial limit in central Illinois (Grüger, 1970). Grüger suggested that the lowest zone, a *Pinus-Picea* phase, was late Illinoian. A rich deciduous forest above this was interpreted as Sangamonian. The overlying prairie-dominated zone was interpreted as Altonian. A younger conifer phase with *Pinus* and later *Picea* was correlated with the Farmdalian and Woodfordian, the *Picea* correlating with the farthest extent of the Woodfordian glacier in Illinois. This was followed by the presence of deciduous trees and an abundance of nonarboreal plants thought to represent late Wisconsinan and Holocene. The absence of a loessial increment in the 14.6 ft (4.5 m) of organic mud (gyttja) beneath an increment dated at 38,100 ± 1,000 radiocarbon yrs B.P. (ISGS-11) in one of the lakes added support to the late Illinoian age of the basal *Pinus-Picea* pollen zone (Jacobs, 1970). An alternate interpretation, more consistent with the climatic history interpreted from the loess deposits of the Illinois Valley, is that the cold interval correlated with the Illinoian is the early Altonian glacial episode, and that the warm interval correlated with the Sangamonian is the warm climatic episode that developed the Chapin Soil.

Glacial Advance and Retreat

Episodes of glacial advance, moraine building, and retreat record the effects of climatic changes. During the major interglacial intervals, continental glaciers in the interior of North America disappeared entirely. Although there is general agreement that the central interior part of the continent was free of glacial ice during the Yarmouthian and Sangamonian times, there is less agreement concerning the complete disappearance of glaciers during the shorter intervals interspersed through the Illinoian and Wisconsinan. The fact that glaciers repeatedly melted back many tens or hundreds of miles and then readvanced is shown by the stratigraphy of the deposits they left, by the changed composition from one till layer to the next, and at some places by the presence of fossils in the silt beds separating the several deposits of glacial till.

The significance of end moraines has been questioned because some ridges formerly classed as end moraines have been found to be buried ridges that were overridden by glaciers which left only veneers of till. However, indications are that many end moraines have been built at the margin of a slightly fluctuating glacier, and in some cases the ice front remained in approximately the same position long enough to deposit more than 200 ft of till. This relation has been demonstrated by the contrasting composition of the tills, by data from borings, and by stratigraphic tracing, as well as by observation of the landform that resulted.

The sequence of rapid glacial advance, moraine building, and rapid frontal retreat is related to fluctuations of climate. The use of such a sequence to interpret climatic history, however, is not simple or direct. It is not possible to determine the relative importance of temperature and precipitation, or to determine whether the retreat was mainly caused by climatic factors in the marginal region or resulted from changes in the nourishment of the interior of the glacial mass. In spite of these uncertainties, it seems reasonable to postulate lower temperatures during times of glacier advance and higher temperatures during times of glacier retreat.

Although movements in adjacent Wisconsinan lobes in Illinois were not entirely synchronous (Willman and Frye, 1970), this may result from differences in distance to accumulation centers, in amount of snowfall, and in terrain crossed by the glaciers. Some of the morainic areas in extreme northeastern Illinois that are characterized by thin drift and stagnation features may have been produced by surging, as described by Wright (this volume, p. 251–276). However, in most of Illinois, the absence of stagnation features and the rhythmic nature of the pulsations within the lobes favor a basic climatic control. Re-entrants and protuberances in the ice front, such as tongues in the valleys and in gaps through moraines, are related to the local topography.

The pulses of the ice front related to the individual Woodfordian moraines have been estimated to average only about 190 yrs, and some probably were much shorter (Willman and Frye, 1970). Although we relate these more or less rhythmic movements of the ice front to climatic cycles, it is doubtful that these cycles significantly affected the climate in the areas bordering the glaciers. No cyclical variations in the fauna of the Peoria Loess, which was deposited during these pulsations, have been noted.

In contrast to its common occurrence in the basal till of the earliest Woodfordian glacier in Illinois, wood is rarely found in the till deposited by the readvances during the Woodfordian. The cycles were too short and the ice front moved too rapidly for forest or even thickets of woody vegetation to become established on the till plain in the 10 to 20 mi overridden in many readvances. In some cases the zone free of forests may have extended more than 100 mi from the ice front. Grasses and other plants may have formed a ground cover on the surfaces overridden during the readvances, but there is no evidence to support this. The discharge of outwash into the valleys appears to have been continuous through the pulsing movements of the ice front because the texture of the Peoria Loess is remarkably uniform and no cyclical variations have been noted.

Loess Deposits

Deposits of loess have been used by some workers as climatic indicators. Although arctic or desert conditions are not indicated by the loess deposits, they do serve as evidence of wind direction and vegetal cover. The dominance of westerly winds is indicated by thicker loess on the east sides of the valleys than on the west. However, perhaps a third as much loess occurs on the west sides of the valleys, which indicates wind shifts comparable to those accompanying the passing of high- and low-pressure weather systems such as we have at present. Because the areas of maximum thickness occur southeast of the broader expanses of the valleys, the loess has been thought to have been deposited mainly during the fall and winter when northwesterly winds dominated and less flooding of the outwash surfaces in the valleys permitted more drying and more wind erosion of the silt deposits. However, the fact that loess accumulation was slow, averaging less than 1 mm per yr in the upland areas, removes the need for a seasonal episode of great dust storms to account for the loess. Except in the very thick loess on the bluffs, which in places contains lenses of dune sand, the vertical uniformity in grain size and the uniform decrease in grain size away from the bluffs suggest a steady accumulation from many intervals of wind deposition rather than annual cycles. Furthermore, the accumulation of the loess was not rapid enough to destroy the vegetation and the abundant snail fauna.

Periglacial Features

Frost features such as ice-wedge fillings and cryoturbations occur locally in northern Illinois (Frye and Willman, 1958) and in Wisconsin (Black, 1965). Such features indicate low average temperatures. Although they can be related to major episodes of glaciation, it unfortunately is difficult to relate them to specific pulses of the glaciers.

CLIMATIC HISTORY OF THE LAKE MICHIGAN LOBE

A hypothetical temperature curve for the Illinoian and the Wisconsinan is presented in Figure 2. The vertical temperature line represents present climatic conditions and, as the curve departs from this line toward warmer or cooler, only relative temperatures are indicated.

As has been pointed out, the estimation of the range of air temperatures beyond the margin of a glacier is at best precarious. The extension of major continental glaciers into regions of moderate climate is indicated by the presence of shells of snails of a temperate habitat in the base of tills and by the absence of any evidence of permafrost in the south half of Illinois. In Figure 2, the temperature history is considered to be that of the Driftless Area of northwestern Illinois. We assume that lowered temperatures preceded and accompanied the advance of glaciers of the Lake Michigan Lobe and that temperatures increased during the time of glacial retreat. It is also assumed that temperatures probably reached lower levels during maximum southern advance of the ice than when glacial advances stopped farther north. However, changes in rate of precipitation and length of glacial interval are also variables that must have had an influence on the extent of the glaciers. The temperature curve is based on climatic conditions beyond the glacial limit; obviously, latitude and geographic configuration of the glacier margins would cause significant local differences.

Although it is impractical to discuss all the data used in plotting the curve (Fig. 2), some of the more significant items should be mentioned. The climates of Yarmouthian and Sangamonian times are judged to have been roughly comparable because of the similarities of soil profile morphology, but the more deeply developed profile of the Yarmouth Soil implies it developed during a greater span of time. Following the Yarmouthian, the first known episode of Illinoian glacial advance, the Liman, reached farthest south. For this reason, a minimum cold temperature is plotted. Following the Liman glaciation, there was a moderately long interval of deglaciation, during which the Pike Soil developed. The Pike Soil, although known from only a few exposures, is a well-developed profile, with a red-brown solum resembling the much more deeply developed Yarmouth and Sangamon Soils. The character of the Pike Soil indicates that the climate of the late Liman time approached that of the longer interglacial intervals.

The development of the Pike Soil was followed by Monican glaciation, which was somewhat less extensive than the Liman glaciation, although equally complex. Only a minor soil, lacking a deeply developed solum, developed after Monican glaciation and prior to the glacial advances of Jubileean age. The scarcity of exposures of this soil and its relatively weak development render it difficult to interpret, but it is reasonable to conclude that it indicates a temperature increase much less than that shown by the Pike Soil.

The minor warming of late Monican time was followed by the glacial advances of Jubileean time. These glaciers stopped short of the south limit of the preceding two and, as a result, Jubileean glaciation is judged to represent a somewhat less rigorous climate.

The physical characteristics of the Illinoian glacial deposits, particularly those of the first two substages, contrast sharply with the characteristics of Woodfordian glacial deposits. The Illinoian glaciers deposited relatively thin sheets of till, built poorly defined and discontinuous moraines, and produced a smaller quantity of water-laid deposits. These features, in combination with preserved crevasse fillings, kames, and eskers, the incomplete filling of the major bedrock valleys of the ancient Mississippi River system, and the extreme flatness of the till plain surface, indicate that large areas of glacial ice stagnated after the glaciers reached their maximum extent. In contrast, the Woodfordian glaciers in Illinois left evidence of stagnation in only limited areas. These relations suggest that the cold intervals of the Illinoian terminated more abruptly than was true during the Woodfordian.

The Sangamon Soil represents a prolonged interglacial interval. The strongly developed red-brown solum of the in situ profile and the thick accretion gleys that accumulated in poorly drained areas indicate long-continued surface stability and, during at least part of the time, temperatures significantly higher than those of the present. Climatic fluctuations that may have taken place during this relatively long time interval have not so far been detected in the physical record.

For the early part of the Altonian Substage of the Wisconsinan, several closely spaced glacial advances are known only from the subsurface of northeastern Illinois (Frye and others, 1969). Although these glaciers reached only a short distance beyond the limits of the present Lake

Michigan Basin, the climatic effects were recorded throughout the region by the initiation of erosional dissection of the Sangamon surface and by a thin but extensive layer of deposits consisting of a mixture of sheetwash, colluvium, and loess (the Markham Member, Roxana Silt). The Chapin Soil developed in this unit resembles the much thicker underlying Sangamon Soil, and its development required a climate with a temperature approaching that of the Sangamonian but enduring for a much shorter interval of time.

No glacial till has been positively identified in Illinois that correlates with the minor temperature fluctuation (shown in Fig. 2) that followed the development of the Chapin Soil and terminated with the Pleasant Grove Soil. This fluctuation is based entirely upon the presence of the McDonough Loess Member of the Roxana Silt and the Pleasant Grove Soil of the Illinois and middle Mississippi River valleys. However, the cold episode immediately following the development of the Pleasant Grove Soil is based on the Argyle Till Member of the Winnebago Formation, which was deposited by the most extensive Altonian glacial advance in Illinois.

The time immediately following deposition of the Argyle was at least as cool as the present. This is deduced from the organic material in the Plano Silt Member of the Winnebago Formation that occupies this stratigraphic position, and from the absence of any reddish-brown or brown in situ soil profiles. Furthermore, the cold cycle that followed in late Altonian time is considered minor because of the character and limited extent of the Capron Till Member of the Winnebago Formation on which it is based.

During succeeding Farmdalian time, glaciers retreated at least within the Lake Michigan Basin, but there is no direct evidence that there was deglaciation of the continent. Climatic evidence is derived from the widespread silt and organic deposits, paleontologic data, and the lack of a red-brown solum in the in situ soil profiles. These data are interpreted to indicate that, after the warming trend that dissipated the last Altonian glacier, Farmdalian time was marked by a cooling trend that culminated in the advance of the first Woodfordian glacier.

The climate at the beginning of Woodfordian time was cold and continued to become colder until it rivaled the temperature of the earliest Illinoian. This is indicated by the extent of the glacier, its relatively rapid advance, and by the presence of periglacial features that developed at the same time in northern Illinois (Frye and Willman, 1958). A particularly frigid climate is considered to have existed north of the Green River Lobe in the northwestern part of Illinois (Frye and others, 1969), and it caused a strong erosional topography that contrasts with the relatively undissected Illinoian till plain south of the lobe.

The behavior of the Woodfordian glaciers was in strong contrast to that of the Illinoian. The retreating Woodfordian glaciers left a record of pulsating readvance and retreat. Stratigraphic evidence (Fig. 3) indicates that during some of these retreats the glacier front withdrew more than 100 mi (Fisher, 1925; Willman and Payne, 1942) and subsequently quickly readvanced as much as 80 mi. The climatic reversals that produced these pulses must have been sharp and severe.

As shown in Figure 3, after the glacier front reached its maximum limit at the Shelbyville Moraine, its total travel distance before finally retreating beyond the southwest shore of Lake Michigan was approximately 800 mi during 6,000 radiocarbon yrs of elapsed time. Because the times when the glacier front was essentially stable (as represented by the moraines) probably consumed about half this time, the rate of ice-front movement during the Woodfordian pulses averaged somewhat more than 1,500 ft per yr (Willman and Frye, 1970). This is a much higher rate of ice-front movement than has been estimated for other regions (Horberg, 1955; Goldthwait and others, 1965). The conclusion that Woodfordian climate had stronger and more rapid fluctuations than climates of earlier intervals of time is supported by the amount of outwash and loess deposits that greatly exceeded the deposits of earlier glacial episodes, the common occurrence of coarse cobbly gravel outwash, and the evidence of intensive eolian scour.

The severity of the climate at times during the middle to late Woodfordian is also indicated by local occurrences of permafrost features (Sharp, 1942; Horberg, 1949; Wayne, 1967; Flemal and others, this volume, p. 229–250).

After the most intense cold of the Woodfordian, from 19,000 to 20,000 yrs B.P., there was a pulsating trend toward climatic amelioration that resulted in repeated retreats, readvances, and moraine building. Each of the major readvances was followed by the deposition of large morainic systems. This relation suggests that the sequence of cycles of approximately 200 yrs duration that produced the individual moraines was imposed on a few longer climatic cycles.

The episode of most climatic significance during the Woodfordian followed the building of the Bloomington system of moraines and was the time of development of the Jules Soil. Although this soil does not contain a textured B-horizon, its A-horizon and local carbonate accumulation at the base of the solum suggest relatively warm conditions.

Glacial retreat during the short-lived Twocreekan interval must be interpreted as a warmer time because forest invaded the former till plain before the readvance that deposited the Valders till. Valderan time may have been relatively cold as indicated by the development of a prominent lobe of the glacier in the Lake Michigan Basin.

As new data become available, climatic curves will be modified. In fact, the one in this paper represents a significant modification of one that we presented a few years ago (Frye and Willman, 1961). However, it seems useful to offer our current interpretations in the hope that attempts to reconcile differences with other regions will lead to better understanding of the climatic history of the Lake Michigan Lobe.

REFERENCES CITED

Black, R. F., 1965, Ice-wedge casts of Wisconsin: Wisconsin Acad. Sci., Arts and Letters Trans., v. 54, p. 187-222.

_____ 1970, Chronology and climate of Wisconsin and Upper Michigan—14,000 to 9,000 radiocarbon years ago [abs.]: Am. Quaternary Assoc., Abstracts 1970, p. 12.

Dreimanis, Alexis, 1958, Wisconsin stratigraphy at Port Talbot on the north shore of Lake Erie, Ontario: Ohio Jour. Sci., v. 58, p. 65-84.

Fisher, D. J., 1925, Geology and mineral resources of the Joliet quadrangle: Illinois Geol. Survey Bull. 51, 160 p.

Flemal, Ronald C., Hinkley, Kenneth C., and Hesler, James L., 1973, DeKalb mounds: A possible Pleistocene (Woodfordian) pingo field in north-central Illinois: Geol. Soc. America Mem. 136, p. 229-250.

Frye, J. C., and Leonard, A. B., 1967, Buried soils, fossil mollusks, and late Cenozoic paleoenvironments, in Essays in paleontology and stratigraphy: Kansas Univ. Dept. Geol. Spec. Pub. 2, p. 429-444.

Frye, J. C., and Willman, H. B., 1958, Permafrost features near the Wisconsin glacial margin in Illinois: Am. Jour. Sci., v. 256, p. 518-524.

_____ 1961, Continental glaciation in relation to McFarlan's sea-level curves for Louisiana: Geol. Soc. America Bull., v. 72, p. 991-992.

Frye, J. C., Glass, H. D., and Willman, H. B., 1968, Mineral zonation of Woodfordian loesses of Illinois: Illinois Geol. Survey Circ. 427, 44 p.

Frye, J. C., Glass, H. D., Kempton, J. P., and Willman, H. B., 1969, Glacial tills of northwestern Illinois: Illinois Geol. Survey Circ. 437, 47 p.

Goldthwait, R. P., Dreimanis, Alexis, Forsyth, J. L., Karrow, P. F., and White, G. W., 1965, Pleistocene deposits of the Erie Lobe, in Wright, H. E., Jr., and Frey, D. G., eds., The Quaternary of the United States: Princeton, N.J., Princeton Univ. Press, p. 85-97.

Grüger, Eberhard, 1970, The development of the vegetation of southern Illinois since late Illinoian time (Prelim. Rept.): Rev. Géographie Phys. et Géologie Dynam., v. 12, fasc. 2, p. 143-148.

Horberg, C. L., 1949, A possible fossil ice wedge in Bureau County, Illinois: Jour. Geology, v. 57, p. 132-136.

_____ 1955, Radiocarbon dates and Pleistocene chronological problems in the Mississippi Valley Region: Jour. Geology, v. 63, p. 278-286.

Jacobs, A. M., 1970, Persistence of lake basins of southern Illinois, U.S.A., from Late Illinoian time to the present: Rev Géographie Phys. et Géologie Dynam., v. 12, fasc. 2, p. 137-142.

Kempton, J. P., 1963, Subsurface stratigraphy of the Pleistocene deposits of central northern Illinois: Illinois Geol. Survey Circ. 356, 43 p.

Leighton, M. M., 1960, The classification of the Wisconsin glacial stage of the north-central United States: Jour. Geology, v. 68, p. 529-552.

Leonard, A. B., and Frye, J. C., 1960, Wisconsinan molluscan faunas of the Illinois Valley Region: Illinois Geol. Survey Circ. 304, 32 p.

Leonard, A. B., Frye, J. C., and Johnson, W. H., 1971, Illinoian and Kansan molluscan faunas in Illinois: Illinois Geol. Survey Circ. 461, 24 p.

Mörner, N.-A., and Dreimanis, A., 1973, The Erie Interstade: Geol. Soc. America Mem. 136, p. 107-134.

Morrison, R. B., 1967, Principles of Quaternary soil stratigraphy, in Quaternary soils: VII Congr. INQUA Proc., v. 9, Univ. Nevada, Reno, p. 1-69.

Ruhe, R. V., 1965, Quaternary paleopedology, *in* Wright, H. E., Jr., and Frey, D. G., eds., The Quaternary of the United States: Princeton, N.J., Princeton Univ. Press, p. 755-764.

Sharp, R. P., 1942, Periglacial involutions in northeastern Illinois: Jour. Geology, v. 50, p. 113-133.

Thorp, J., 1965, The nature of the pedologic record in the Quaternary: Soil Sci., v. 99, p. 1-8.

Wayne, W. J., 1967, Periglacial features and climatic gradient in Illinois, Indiana, and western Ohio, east-central United States, *in* Quaternary paleoecology: VII Congr. INQUA Proc., v. 7, p. 393-414.

Willman, H. B., and Frye, J. C., 1970, Pleistocene stratigraphy of Illinois: Illinois Geol. Survey Bull. 94, 204 p.

Willman, H. B., and Payne, J. N., 1942, Geology and mineral resources of the Ottawa, Marseilles and Streator quadrangles: Illinois Geol. Survey Bull. 66, 388 p.

Wright, H. E., Jr., 1973, Tunnel valleys, glacial surges, and subglacial hydrology of the Superior Lobe, Minnesota: Geol. Soc. America Mem. 136, p. 251-276.

MANUSCRIPT RECEIVED BY THE SOCIETY DECEMBER 27, 1971

GEOLOGICAL SOCIETY OF AMERICA
MEMOIR 136
© 1973

Superior and Des Moines Lobes

H. E. WRIGHT, JR.

Department of Geology and Geophysics, University of Minnesota,
Minneapolis, Minnesota 55455

CHARLES L. MATSCH

Department of Geology, University of Minnesota,
Duluth, Minnesota 55812

EDWARD J. CUSHING

Department of Botany, University of Minnesota,
Minneapolis, Minnesota 55455

ABSTRACT

The Superior Lobe of the Wisconsin glaciation was initially localized by the deep lowlands of the Lake Superior Basin, cut in relatively nonresistant late Precambrian red sandstone. It advanced southwest out of this lowland and crossed a low divide leading to the Minneapolis Lowland, which is underlain by Precambrian and Cambrian sandstones. The conspicuous drumlins of central Minnesota, and the rugged St. Croix Moraine that borders the drumlins on the west and loops across the Minneapolis Lowland, delimit the major stillstand of the Superior Lobe. Discovery of red drift with diagnostic rock types from the Lake Superior Basin (agate, amygdaloidal basalt, red and purple felsite, red sandstone) in southwestern Minnesota indicates that the Superior Lobe once extended farther southwest down the Minneapolis Lowland to the Minnesota River valley and beyond during a pre-Wisconsin or early Wisconsin glaciation.

As the Superior Lobe wasted from the St. Croix Moraine, a series

153

of sharp subparallel tunnel valleys were cut into the drumlin plain and even into the underlying bedrock by subglacial streams driven to high velocity by the hydrostatic pressure resulting from the load of many hundreds of meters of active ice. Subsequent thinning and stagnation of the Superior Lobe opened the tunnel valleys to atmospheric pressure and converted the subglacial streams from major erosional streams to small depositional streams, which formed discontinuous eskers along many of the tunnel valleys.

After distant retreat the Superior Lobe readvanced three times out of its basin, twice after proglacial lakes had produced a supply of red clay to be overridden. These latter two readvances may represent surges of the ice lobe resulting from the buildup of basal meltwater behind the frozen toe of the ice lobe.

The Des Moines Lobe, originating in the Red River Valley of Manitoba and western Minnesota, moved southeastward down the Minnesota River valley and thence northeastward up the Minneapolis Lowland, overriding a segment of the St. Croix Moraine and extending across the state to Wisconsin in the form of the Grantsburg Sublobe. Its drift is characteristically gray to yellowish-brown and highly calcareous. The ice incorporated masses of Superior Lobe drift as it overrode the St. Croix Moraine, stringing them out to produce a complex of foliated red and gray drift.

The Grantsburg Sublobe blocked the Mississippi River and other drainage from central Minnesota to form glacial Lake Grantsburg about 16,000 yrs ago. By that time the Superior Lobe had withdrawn completely from central Minnesota, for it supplied meltwater (and red clay) only on the east, down the St. Croix River and its upper tributaries.

Meanwhile, the main Des Moines Lobe, which thus far supplied ice only to the Grantsburg Sublobe, thickened sufficiently to spill southward out of the Minnesota Valley across a low divide into Iowa. This produced the lobe that reached Des Moines 14,000 yrs ago. Beheading of the Grantsburg Sublobe in this manner caused the stagnation of the latter, which then wasted to form the Anoka Sandplain in its stead.

The St. Louis Sublobe protruded from the Des Moines Lobe in northwestern Minnesota at a later date (about 12,000 yrs ago). Its meltwater flowed down the St. Louis River toward Lake Superior, but it was diverted southward into the St. Croix drainage by the still-existing Superior Lobe. Final wastage of the entire Des Moines Lobe produced glacial Lake Agassiz in northwestern Minnesota and adjacent North Dakota and Manitoba.

INTRODUCTION

Digitation of the periphery of the Laurentide Ice Sheet was largely controlled by the existence of preglacial lowlands, which in turn reflect the relative resistance of the bedrock to erosional processes (Figs. 1 and 2). All the major ice lobes of the Great Lakes region show this

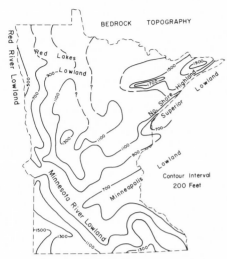

Figure 1. Generalized bedrock topography of Minnesota showing location of major lowlands, which are controlled in their position primarily by the bedrock geology (Fig. 2).

dependence, and the Des Moines Lobe carries the pattern farther west.

The Superior Lobe was localized by poorly resistant late Precambrian red sandstone and shale, which incidentally imparted a distinctive red color to most of the drift. The Des Moines Lobe protruded from the ice sheet in the Winnipeg Lowland of southern Manitoba, which is localized by early Paleozoic carbonate rocks dipping westward off the Canadian Shield. It filled a lowland cut in west-dipping Cretaceous shales along the Minnesota-Dakota border and followed southeastward along the Minnesota River valley, which contained localized Cretaceous sediments next to Precambrian crystalline rocks; it then crossed a low divide southward into Iowa to the Des Moines River valley, whose preglacial trend was probably determined by the structure of Paleozoic rocks. Its drift is gray where fresh, but the oxidized form is almost yellow.

Glacier ice centered in the Winnipeg Lowland was active during Nebraskan and Kansan glaciations, for all the drifts of these ages in Iowa have the diagnostic fragments of Cretaceous shale and Paleozoic carbonate. The more complicated relations worked out for the Wisconsin glaciation, however, provide more understanding of the lobation of the ice masses and of the interactions of the Des Moines and Superior Lobes particularly, as well as of other ice lobes that sometimes covered the terrain between these two.

The principal advances of these two major ice lobes have been recognized for almost a century, since the early days of Winchell and Upham. The advances were thought by these pioneer glacial geologists to have been at least partly synchronous, and the interlayering of red and yellow drifts in the Minneapolis area was attributed to juxtaposition of the two active ice lobes. The long northeastward protuberance of the Des Moines Lobe, termed by them the Chisago Lobe but now called the Grantsburg Sublobe, was recognized as a late feature that developed during a waning phase of ice activity, and the vast sand area (Anoka Sandplain) that occupies much of the same ground as did the Grantsburg Sublobe was correctly seen as a glaciofluvial feature related to ice wastage on a broad front.

The glacial history of these two major lobes was first worked out in general detail by Leverett (1932) and Sardeson (1916) and then in part refined and corrected by Cooper (1935) in the area of the Anoka

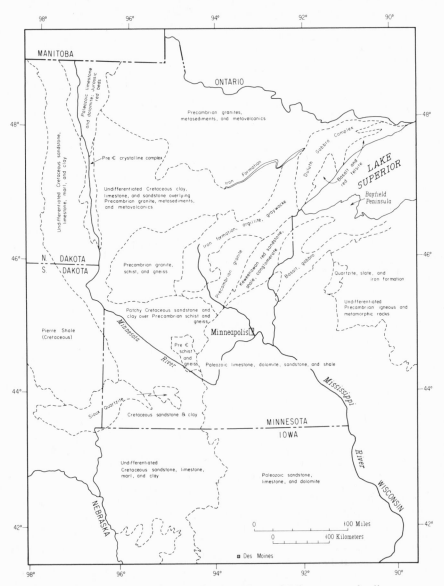

Figure 2. Generalized map of the bedrock geology of Minnesota and adjacent areas. Distinctive rock types cause the drifts of the Superior and Des Moines Lobes to differ markedly in color, texture, and stone lithology. Compiled from the geologic maps of Minnesota (1971), Iowa (1969), North Dakota (1969), eastern South Dakota (1971), and Wisconsin (1970), issued by the respective geological surveys.

Sandplain. Subsequent studies, carried out largely with the support of the Minnesota Geological Survey, have fitted these two lobes into a broader framework (Wright and Ruhe, 1965; Wright and Watts, 1969). This paper indicates, in addition, current lines of study that contribute more details to the history and elucidate stratigraphic and geomorphic features not previously described.

SUPERIOR LOBE

General

The form of the Superior Lobe during its several advances was dictated primarily by the configuration of the bedrock topography (Fig 1). The lobe was rooted in the western part of the Lake Superior Basin, which is marked by a trough along the Minnesota shore that includes almost 300 m of water and another 300 m of glacial till and lake sediment. The western Lake Superior Basin is localized by a syncline of late Precambrian (Keweenawan) sandstone, bounded by resistant volcanic rocks on the north side and by sandstone and volcanics on the Bayfield and Keweenaw Peninsulas on the south side (Fig. 2). This basin is the deepest lowland of the Great Lakes region, and it provided a reservoir of ice that spread far southwestward to southern Minnesota. In a larger sense the entire Lake Superior Basin, including the generally shallower eastern part, provided ice for other lobes as well—the Chippewa, Green Bay, and Lake Michigan Lobes—and smaller lobes delineated by Black (1969) in northern Michigan.

The main trough of the western Lake Superior Basin terminates southwestward near Sandstone, Minnesota, where bedrock exposed at the surface forms a divide leading gradually southward to the Minneapolis Lowland. The Minneapolis Lowland is arranged en echelon to the south; it also trends southwest, being localized by the soft Keweenawan and Cambrian sandstones. It is bounded on the northwest largely by the Precambrian crystalline terrane of central Minnesota, and on the southeast by Ordovician carbonate rocks. Both the Lake Superior Basin and the Minneapolis Lowland were occupied by the Superior Lobe at times.

St. Croix Phase

General. During its maximum advance in the Wisconsin glaciation the Superior Lobe completely filled the western Lake Superior Basin (Fig. 3). It was confluent on the northwest with the Rainy Lobe, which covered the upland of northeastern Minnesota. Together, the two adjacent lobes operated as one as they spread southwestward side by side. The Superior Lobe topped the low divide near Sandstone and then filled much of the Minneapolis Lowland. Having a greater reservoir of ice behind it than the Rainy Lobe, it fanned somewhat to the west at the expense of the Rainy Lobe. But the two lobes came to a common terminus in central Minnesota at the St. Croix Moraine. This ice advance is termed the St. Croix phase—the second of five Wisconsin ice advances recognized in Minnesota (Fig. 3).

The drift of the Superior Lobe in its St. Croix phase may be identified by its red color and its sandy texture. Both features result from the incorporation of Precambrian red sandstone, which underlies the western

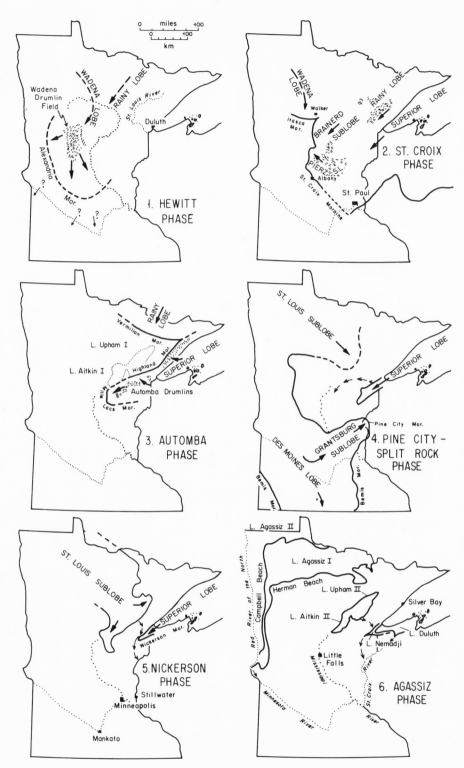

Figure 3. Sketches showing the areas covered during successive phases of ice movement in Minnesota during the Wisconsin glaciation.

Lake Superior Basin and much of the terrain to the southwest. The drift of the Rainy Lobe in the St. Croix phase, on the other hand, is gray or brown, depending on whether the dominant rock type is gabbro (in the northeast) or various metamorphic and granitic rocks that oxidize readily to brown (in central Minnesota).

St. Croix Moraine. The St. Croix Moraine forms a long curved belt of rugged, forested, lake-dotted terrain, delineating the combined Superior and Rainy Lobes in the St. Croix phase (Fig. 4). It starts in north-central Minnesota near Walker, where it makes an interlobate junction with the Itasca Moraine of the Wadena Lobe, and it extends southward for 160 km in a fairly straight course, with an average breadth of 10 km, to near Albany in Stearns County. In this segment it is composed of the brown sandy drift of the Rainy Lobe. Its outwash to the west overlaps the Wadena Drumlin Field, which is a manifestation of the Hewitt phase of Wisconsin glaciation on the part of the Wadena Lobe (Wright, 1962).

Southward from Albany the St. Croix Moraine curves slightly to the southeast, and for 150 km it is largely buried by younger drift of the Grantsburg Sublobe of the Des Moines Lobe. The buried drift in this segment is the red sandy drift of the Superior Lobe.

Beyond the overlap of the Grantsburg Sublobe drift, the St. Croix Moraine curves eastward through Minneapolis and St. Paul. There it is fringed on the south by outwash plains that grade to the 900 ft (275 m) terrace of the Mississippi River below St. Paul. The moraine then curves northeastward toward the St. Croix River, which it crosses north of Stillwater. Beyond there it trends straight northeastward into Wisconsin.

In the St. Croix phase the combined Superior and Rainy Lobes were about 250 km broad, and the perimeter from the Itasca Moraine near Walker to the Chippewa Lobe in Wisconsin was about 500 km.

Drumlin Fields. Another major geomorphic feature associated with the St. Croix phase of the Superior and Rainy Lobes is the drumlin fields, formed when the ice stood at the St. Croix Moraine (Fig. 3). Two main drumlin fields can be delineated, both made dominantly of brown sandy till and thus within the scope of the Rainy Lobe rather than the Superior Lobe per se. The Brainerd Drumlin Field trends southeastward near Brainerd. The Pierz Drumlin Field trends to the west and fans to the northwest and southwest, crossing the course of the Mississippi River west of Little Falls (Schneider, 1961). South of the Pierz Drumlin Field the drift of the Superior Lobe is largely buried by that of the Grantsburg Sublobe, and any drumlins that exist in that area are obscured.

Tunnel Valleys. Perhaps the most striking landforms connected with the St. Croix phase of the Superior Lobe are the tunnel valleys. These, we believe, were eroded in the substratum by streams flowing to the southwest in tunnels at the base of the Superior Lobe under great hydrostatic pressure (Wright, this volume, p. 251-276). The pattern of tunnel

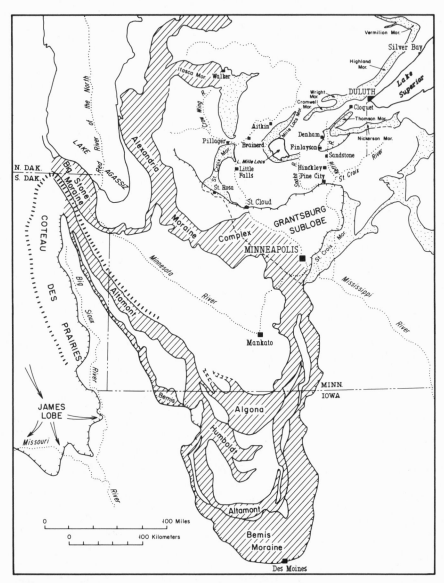

Figure 4. Map of the main Des Moines and Superior Lobes in Minnesota and Iowa and their principal moraines. On the west the Coteau des Prairies separated the Des Moines Lobe from the James Lobe, and on the east the Des Moines Lobe ice overlapped the Alexandria Moraine, the Wadena Drumlin Field (of the Wadena Lobe), and the St. Croix Moraine (of the Superior Lobe).

valleys extends in a fan shape from the divide area west of Sandstone for almost 150 km to the Mississippi River and the St. Croix Moraine. Many of the tunnel valleys are characterized by eskers, formed when the ice thinned to stagnation and the hydrostatic pressure was reduced. In the southern portion the pattern of tunnel valleys and eskers is partly obscured by a cover of younger drift, but the trenches show up as strings of lakes and swamps. Apparently the valleys became plugged with dead ice before they were buried by the drift of younger ice masses. The dead ice persisted locally for thousands of years, perhaps buried by debris melted out of the ice itself or by drift of the overriding ice, while the terrain became stabilized with vegetation (Florin and Wright, 1969). The rate of melting of such ice blocks apparently depended on chance removal of the protective mantle, exposing the ice to the sun. The ponds and lakes thus formed preserve in their basal sediments the plant detritus of the superglacial forest floor.

Dating. The St. Croix phase of the Superior Lobe was formerly correlated with the middle Wisconsin (Cary) moraines of the Lake Michigan Lobe, which formed about 14,500 yrs ago. But radiocarbon dates on basal organic sediments from lakes on drifts younger than the St. Croix phase reach 16,000 yrs, and a date from lake sediments in an interdrumlin depression of the St. Croix phase itself is 20,500 ± 400 yrs B.P. (I-5443). It seems unlikely that this interdrumlin lake also represents an ice-block depression, because in the process of drumlin formation the rock debris is probably not carried far up from the ice floor and is thus not in a position to protect any underlying ice from surficial melting. Consequently, this date probably represents close to the maximum age of the St. Croix phase, which is thus correlative with the southernmost Wisconsin moraines of the Lake Michigan Lobe and other lobes to the east.

Automba Phase

After its wastage from the St. Croix Moraine the Superior Lobe was restricted in its geographic scope but significant in its local effects. When the Superior Lobe stood at the St. Croix Moraine it was joined on its northwest side to the Rainy Lobe, which at that time moved southwestward along the upland north of Lake Superior and extended to central Minnesota, forming drumlins along the way. The pair of lobes retreated, but the Rainy Lobe portion, being thinner, withdrew much farther—probably into Canada—whereas the Superior Lobe portion retreated barely into the Lake Superior Basin.

The ice then readvanced in the Automba phase (Fig. 3). The Rainy Lobe in this readvance reached the Vermilion Moraine, which makes an interlobate junction with the Highland Moraine of the Superior Lobe near Isabella in northeasternmost Minnesota. The Highland Moraine caps the entire length of upland north of Lake Superior, and it is backed on the inside by fluted terrain in both bedrock and drift, molded as the ice rose up the North Shore slope. As the Superior Lobe reached the southwestern end of the basin, near Duluth, it spread westward more extensively, forming the Automba Drumlin Field. It produced the Wright and Cromwell Moraines on its northwestern flank, along with proglacial lakes Upham I and Aitkin I to its north. It climbed up the 60 m Denham Escarpment that on the west bounds the saddle of the

Sandstone Divide, and it reached a terminus at the Mille Lacs Moraine, which closes the very large and round Lake Mille Lacs on the north, west, and south. The northern portion of the lobe in the Automba phase is thus well defined by the drumlins and by the Highland Moraine and its connections southwestward to the Mille Lacs Moraine.

The south side of the lobe is not delineated by a prominent moraine, however. The lack of a clear-cut limit for the ice is related to another question about the form of the lobe. The basic problem is to explain why the ice climbed westward over the Denham Escarpment and flowed to the Mille Lacs area, rather than continuing a southwesterly course across the low rock divide near Sandstone, which leads into the Minneapolis Lowland. Possibly the more southerly course was blocked by masses of stagnant ice left by the Superior Lobe in its retreat in the St. Croix phase that preceded the Automba phase—it is known from subsequent events that at least some ice blocks lasted for several thousand years longer. If the ice blocks were laden with rock debris, their plasticity may have been reduced enough to resist remobilization and incorporation into the readvancing ice, which thus could have been diverted to the west toward the Mille Lacs area.

Alternatively, one could postulate that the Automba advance represents a minor event interrupting the wastage of the combined Superior and Rainy Lobes of the St. Croix phase. When the Superior Lobe stood at the St. Croix Moraine, it was joined on its northwest side to the Rainy Lobe, which at that time moved southwestward along the upland north of Lake Superior and extended also to the St. Croix Moraine—the segment in central Minnesota from about Albany north to Walker. With climatic change, the ice supply diminished and the ice thinned. The Rainy Lobe side of the combined ice mass, riding across the upland and being initially thinner, became inactive, and eventually withdrew to the Vermilion Moraine. The Superior Lobe side became somewhat thinner in its outer portion in the Minneapolis Lowland, and tunnel valleys and then eskers were formed as the ice thinned and stagnated. But ice was continually supplied in large quantities from the Lake Superior Basin, and it spread laterally into the area just vacated by the Rainy Lobe, thus forming the line of the Mille Lacs, Wright, Cromwell, and Highland Moraines.

In this alternative interpretation of the sequence of events, the Automba phase of the Superior and Rainy Lobes does not, therefore, represent a readvance of the ice after distant retreat, but rather a time of stability in glacial regime after an interval of wastage. The radiocarbon age of these events is not known.

Split Rock and Nickerson Phases

The Superior Lobe drift of the Split Rock and Nickerson phases (Fig. 3) differs from that of the St. Croix and Automba phases in consisting primarily of red clayey till—a result of overriding of proglacial lake beds. Apparently the Superior Lobe withdrew at the end of the Automba phase far enough into the Lake Superior Basin so that a proglacial lake formed at its front, and the proglacial lake must have been deep enough for clays to be deposited. If the natural dam of such a proglacial lake was the sill near Sandstone, at an elevation of 330 m, then a sizable lake of 30 m depth—certainly enough for deposition of clay—would have been possible with ice withdrawal about 30 km northeast of the Sandstone dam. With this arrangement, the amount of clay deposited in the lake, and thus the volume of material reworked into clayey till, was largely proportional to the length of time during which the ice fed its debris into the lake. On the other hand, the deep boring through the Lake Superior sediments off Silver Bay shows clayey tills alternating with lake clays in the lower part of the section (Farrand, 1969), and it is possible that the ice withdrew as much as 150 km before readvancing.

The clayey till of the Split Rock phase forms a generally thin veneer of drift on previously existing landforms, scarcely masking them. For example, it is well displayed as a discontinuous cap on the eskers and tunnel valleys of the Finlayson area. The ice formed a narrow tongue extending to the very head of the Lake Superior Basin, but not across the Sandstone Divide. The tongue was only about 20 km wide and extended about 100 km beyond Duluth, being delimited at an elevation of about 380 m (1,250 ft) by the higher ground of the Automba Drumlin Field and by the bedrock escarpment near Denham, and on the southeast side by poorly defined moraines or drift plains of the Automba or St. Croix phases. It sent a small protuberance westward up the Split Rock Valley, which forms a re-entrant just north of the Denham Escarpment, and it even formed a few low drumlins there, which contrast in form, size, and orientation to the Automba drumlins, against which they are juxtaposed and on which they are probably superimposed. This little lobe produced a small lake at its front, with an outlet westward to the Snake River.

North of the Split Rock Valley, the northwest flank of the lobe was marked by a series of five conspicuous eskers. They fed a broad coalescing outwash fan that drained partly westward to the headwaters of the Snake River and partly northward to the St. Louis River, which at that time must have flowed westward, ultimately to the Mississippi River, rather than eastward to Lake Superior. These great eskers, which are as much as 35 m high, may represent the extrusion of subglacial water that had accumulated beneath the thick ice in the Lake Superior Basin—water that during the St. Croix phase had produced the tunnel valleys farther south. Other subglacial water still discharged at the point of the lobe, for it produced the great outwash valley train that follows a tunnel valley to the Hinckley outwash plain and delta into glacial Lake Grantsburg. Points of discharge of subglacial streams reflect the hydrostatic pressure gradient and thus the slope of the ice surface; the steeper surface slope on the northwest flank near Cloquet compensated for the high lift required (to 390 m elevation compared to 330 m at the point of the lobe west of Sandstone).

The Nickerson phase of the Superior Lobe, likewise marked by red clayey till, displays a distinct moraine near its V-shaped terminus—the Nickerson Moraine on the southeastern side and the Thomson Moraine on the northwestern side (Fig. 3). The ice reached an elevation limit of 360 m (1,200 ft) at the margins of the narrow ice tongue. A well-developed outwash plain leads from the terminus to the Kettle River. This ice tongue blocked the St. Louis River, which by this time had developed its eastward course as a result of advance of the St. Louis Sublobe from the west. As the latter retreated, the St. Louis River drained a large proglacial lake (Lake Upham II), whose outlet water was clear of sediment and capable of eroding a drainage channel around the edge of the Thomson Moraine—the first of several drainage channels that formed by erosion into the Thomson Moraine (Wright and others, 1970).

Whereas the age of the Automba phase can only be conjectured, radiocarbon dates for the Split Rock phase are more directly available. A kettle lake in the end moraine of the Split Rock phase near the Sandstone divide (White Lily Lake) has basal lake sediments dating to 15,250 ± 220 yrs B.P. (I-5051). An outwash fan of the Split Rock phase, in front of the Cloquet Esker and Thomson Moraine, has a kettle lake (Kotiranta Lake) dated to 16,150 ± 550 yrs B.P. (W-1973). This date was previously rejected as being too old (Wright and Watts, 1969), but the relation of the Split Rock phase to features of the Des Moines Lobe, discussed below, permit this date to be accommodated. The Split Rock phase thus can be dated as at least 16,000 yrs old rather than only 14,000 yrs old. This is certainly a major revision from its still earlier correlation with the Valders till (circa 11,500 B.P.), which was made on the basis of similarity of the red clayey tills.

Dating of the Nickerson phase must be approached less directly. The key dates are from the upper part of the sediments of glacial Lake Aitkin II—11,710 ± 325 yrs B.P. (W-502) and 11,560 ± 400 yrs B.P. (W-1141)—which formed during retreat of the St. Louis Sublobe of the Des Moines Lobe in its Alborn phase. This lake drained into glacial Lake Upham II, which had as an outlet the St. Louis River. The latter flowed east toward the Lake Superior Basin, but it was diverted southward into a series of channels by the Superior Lobe at the Thomson Moraine of the Nickerson phase (Wright and others, 1970). The latter must there-fore have climaxed before the dates mentioned—a figure of 12,000 yrs B.P. is reasonable. Basal lake-sediment dates on the Nickerson Moraine itself are 10,400 to 10,800 yrs B.P. (Wright and Watts, 1969), but the basal sediments consist of plant detritus characteristic of ice-block lakes, so the dates are minimal. A peat that fills the last of the diversion channels mentioned above was studied in the hope that its basal date would indicate the time of ice withdrawal, for such a drainage channel should not be underlain by dead ice. The dates 10,630 ± 500 yrs B.P. (W-1677) and 10,420 ± 300 yrs B.P. (W-1714) (Wasylikowa and Wright, 1970), if they accurately record the abandonment of this last channel, imply that the entire Thomson Moraine lasted until this time, backed by the still active Superior Lobe. The Nickerson-Thomson Moraine of the Nickerson phase thus can be dated as about 12,000 to 10,500 yrs B.P.

As the Superior Lobe withdrew farther into the basin, a series of small proglacial lakes formed at its southern margin. Lake Nemadji, at the point of the lobe, utilized the diversion outlets of the St. Louis River just described, leading to the St. Croix River (Fig. 3). With further retreat, these lakes coalesced to form Lake Duluth, which drained to the St. Croix River via the Brule River in northwestern Wisconsin, resulting in a shoreline at an elevation of about 1,000 ft (300 m) near the west point of the basin, or about 50 ft below that of Lake Nemadji. Carbon dates from near the Porcupine Mountains in northwestern Michigan imply that Lake Duluth lasted from 10,230 to 9,600 yrs ago (Hack, 1965). Subsequent lake levels, with drainage to the east into the Lake Michigan Basin, are described by Farrand (1969).

North of Minnesota the Superior Lobe also left a record through minor advances into the Lake Nipigon Basin and other lowlands in Ontario (Zoltai, 1965). One of these, which formed the Marks Moraine, produced a proglacial lake in which red clays were deposited. The lake drained westward, carrying its red clay for 100 km to glacial Lake Agassiz, forming a distinctive red band in the otherwise gray lake clays. The chronology of these various ice advances is not well controlled by radiocarbon dates, but all of them are probably younger than the Nickerson phase at the point of the Superior Lobe.

DES MOINES LOBE

The great advance of the Des Moines Lobe[1] from Manitoba to central Iowa, with its eastern offshoot the Grantsburg Sublobe, left such an impressive and well-defined record (Fig. 4) that it was identified in its major features by Upham (1880) and Winchell (1877) in the very early days of glacial mapping. The interpretations of these pioneer field workers were modified in important respects by Sardeson (1916) and Leverett (1932) and by subsequent workers, especially with respect to the Grantsburg Sublobe. Many important details in drift distribution and in chronolo-

---

[1]Chamberlin (1883, p. 388) named this ice lobe the Minnesota Valley Glacier. Upham (1895, p. 130) called it the Minnesota Lobe. The name Des Moines Lobe first appears in a report of the Iowa Geological Survey (Bain, 1897, p. 437). Leverett (1932, p. 56-57) used the names Minnesota-Iowa Lobe and Des Moines Lobe for the same feature and proposed that different names, such as Minnesota Valley Lobe and Red River Lobe, be applied to it at particular positions during its shrinkage (Leverett, 1932, p. 90).

gy remain in doubt, however, and certain problems will be solved only by detailed mapping and the discovery of materials suitable for radiocarbon dating.

For years the deposits of the Des Moines Lobe were termed the Young Gray Drift, to distinguish them from the Old Gray Drift of presumed pre-Wisconsin glaciations and from the Red Drift of eastern ice lobes. This terminology has long been abandoned, however, in favor of more specific names based on type localities. The drift of the Des Moines Lobe is actually gray only in its unoxidized condition—rarely observed except in deep exposures or in situations of poor drainage where oxygenated waters have penetrated only to shallow depths. Elsewhere it is light brown, even yellowish-brown or olive-brown. It characteristically is dominated by pebbles of limestone and dolomite from the Paleozoic rocks of the Winnipeg Lowland. Fragments of granitic rocks are also abundant. These derive from the widespread Precambrian terrane of western Minnesota and from Canada, but they are not particularly diagnostic as to source. Of greater significance is the content of fragments of siliceous shale eroded from the Upper Cretaceous Pierre Formation of eastern North and South Dakota and southern Manitoba. The distinctive gray-green clasts of this material constitute a major fraction of the coarse sand size in the drift—in fact, their abundance in this size range is more reliably measured than in the pebble size because the larger fragments shatter on exposure and cannot be collected easily. Ground-up shale probably contributes to the relatively fine-grained texture of the matrix of the Des Moines Lobe till, which classifies as clay loam or loam (Fig. 5).

Whereas Upham (1880) and Leverett (1932) assigned all of the Young Gray Drift to the single Des Moines Lobe and its offshoots, subsequent work has shown that two drifts of different lithology and provenance are involved, and thus two distinct ice lobes. The first clues for the relationships came from work in the Wadena Drumlin Field of west-central Minnesota, which Leverett had identified but attributed to the Des Moines Lobe expanding from the west and southwest. Studies of the fan-shaped form of the drumlin field as well as the stone fabric of the till show that the ice came from the northeast rather than the southwest (Wright, 1957, 1962), and that the drumlin till, although rich in pebbles of carbonate and granite, lacks fragments of Cretaceous shale (Arneman and Wright, 1959). The Wadena Drumlin Field and associated features are now therefore assigned to a separate ice lobe, which had its source in the carbonate terrane of southern Manitoba but which deployed eastward into a shallow bedrock lowland in north-central Minnesota rather than southward into the Red River Lowland. Its diversion to the southwest to form the drumlin field is believed to have resulted from blockage by an eastern ice lobe (Wright, 1962).

The Wadena Lobe drift is now identified broadly throughout western and central Minnesota, not only on the surface in the Wadena Drumlin

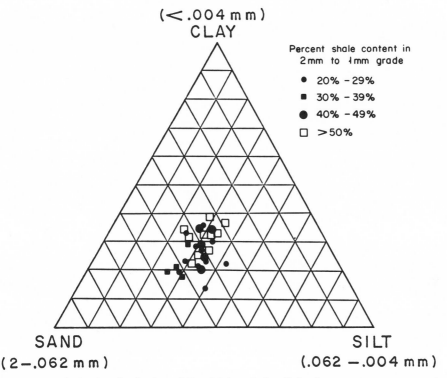

Figure 5. Grain-size distribution of Des Moines Lobe till in the map area represented by Figure 6.

Field but beneath the Des Moines Lobe drift on the west and south and beneath the Superior Lobe drift and other deposits on the east. Its main advance is believed to be of Wisconsin date, perhaps about 34,000 yrs B.P. in west-central Minnesota (Matsch, 1971), but the record of a later advance, the Itasca Moraine, is clearly contemporaneous with the St. Croix Moraine of the Superior Lobe.

Till of the Des Moines Lobe in many exposures along the Minnesota River valley and its tributaries in southwestern Minnesota commonly is separated from the underlying Wadena Lobe till by a stone line (Matsch, 1971). The stone lines are outcrops of a boulder pavement, generally one stone thick, that is a mosaic of predominantly coarse-crystalline igneous rocks ranging in diameter from more than a meter to just a few centimeters. The plane of contact between the upper till and the boulder pavement is commonly a polished and striated horizontal facet on the boulders.

The boulder pavement is almost certainly a lag deposit concentrated by subaerial erosion, especially slope wash, and the concentration of stones implies that several meters of finer material must have been removed. The pavement differs from some other lag deposits, especially in Iowa, because no loess is associated with it. Smoothed and striated boulder pavements separating two tills have been reported in South Dakota (Flint, 1955, p. 58), North Dakota (Clayton, 1966), west-central Saskatchewan (Christiansen, 1968), and west-central Manitoba (Klassen, 1967). Striated boulder pavements have also been described from in front of retreating glacial margins on north-central Baffin Island (Andrews, 1965) and in the Thule area in Greenland. Concerning the latter, Swinzow (1962, p. 225) noted:

> These rounded boulders have been observed in many places directly in front of the ice sheet; their most characteristic feature is the presence of glacial striations on the flat top surface. This is considered good evidence of ice sliding over permafrost.

The extensively developed faceted and striated boulder pavement along the axis of the Des Moines Lobe in Minnesota is evidence that the ice slid along this part of its course across a bouldery interface, but that it accomplished very little erosion.

The age of the pavement is not known, because the drifts above and below are not well dated. Lack of weathering implies a date younger than the last interglacial interval, although application of this criterion to the stone line on top of the "Iowan till" of eastern Iowa is now considered to have led to incorrect conclusions (Ruhe, 1969). The alternative is an early or middle Wisconsin date.

An interesting lithologic feature of the Des Moines Lobe till sheet in Minnesota is the systematic variability in content of Cretaceous shale, as measured in the size fraction 1 to 2 mm (Matsch, 1971). The content is highest (>50 percent) in drift along the axis of the lobe along the Minnesota River valley, and it grades to <20 percent on the southwest flank and to <30 percent in the northeast (Fig. 6). The remarkable constancy of the shale content along axial flow paths that exceed 150 km in length suggests that very little comminution took place as a result of glacial transport. Progressive destruction of shale fragments along flow lines flanking the glacier axis is not reflected in textural fining (Fig. 5).

The observed distribution of shale is considered to be a result of differential subglacial erosion in this stretch of the Des Moines Lobe. This explanation assumes that the englacial sediment had a constantly high shale content after the glacier had crossed the broad area underlain by siliceous Cretaceous shale (Fig. 6). The initially broad ice sheet became lobate and constricted within the long topographic trough on the northeastern side of the Coteau des Prairies; the new geometry was similar to that of a valley glacier. The ice stream advanced across the boulder-paved and graded landscape that had developed on the older drift surface. Almost no erosion occurred in the axial part, as indicated by the preservation there of the boulder pavement beneath the drift. In that part the ice was moving down a regional slope. Along the shallower margins of the ice tongue, however, where the ice was expanding outward against the flanks of the Minnesota River Lowland, erosion of the underlying

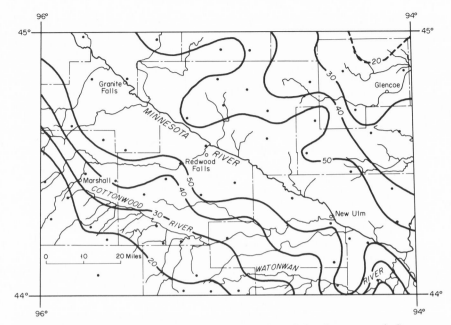

Figure 6. Map showing lines of equal content of sand-size fragments of Cretaceous shale in the Des Moines Lobe drift of the Minnesota Valley. Lower values toward both sides of the valley are interpreted as a result of dilution by drift eroded from beneath by the ice as it flowed toward its lateral margins.

till sheet was fairly active. The boulder pavement was destroyed and the original shale-rich load of the glacier became diluted with Wadena Lobe till, which is rich in limestone and granite but poor in shale. The shale content thus became systematically lower with increasing dilution. In this model, the ice lobe moved by different flow mechanisms in its different parts. The thicker part, perhaps at its pressure melting point, may have moved by sliding on a layer of water, although the friction was enough to permit the boulders to be faceted and striated. Along the thinner margins, however, where flow was compressive and the ice colder, the underlying material was being actively sheared into the glacier.

Alternatively, the field relationships described above may illustrate a mechanism for differential subglacial erosion proposed by Weertman (1961) that involves regelation rather than shearing as the important debris-entraining process. In his model three different subglacial thermal environments are encountered progressively inward from the edge of a cold ice sheet: (1) a shallow margin frozen to the bed because the 0° C isotherm lies deeper than the interface between ice and bedrock; (2) an intermediate zone where the bottom of the ice is at the melting point, but where the temperature gradient is such that refreezing takes place; and (3) an interior zone at 0° C where the ice is melting, but where no refreezing takes place. As the boundaries between these three subglacial environments shift under nonsteady-state conditions, debris is eroded along an intermediate swath between margin and interior by the progress of regelation. Shear deformation above the glacier floor could effectively disperse the newly entrained debris.

The Des Moines Lobe originated as a protuberance from the Laurentide Ice Sheet somewhere in southern Manitoba. Southward it became channeled along the Red River Valley and the Minnesota River valley between the Coteau des Prairies on the west and the Alexandria Moraine Complex on the east. The Prairie Coteau is a wedge-shaped plateau in eastern South Dakota and southwestern Minnesota that is capped on the eastern portion by rugged moraines of the Des Moines Lobe and on the western by moraines of the James Lobe (Fig. 4). The central part is a topographic sag that is drained by the Big Sioux River. Although Cretaceous sediment and Sioux Quartzite may core much of the plateau, the drift cover has a thickness up to 200 m, and few outcrops of bedrock are found along the bounding scarps on east and west. The scarps themselves are unusually sharp and straight—that on the east rises 200 m over a distance of only 5 km. Although the outermost lateral moraine (Bemis) of the Des Moines Lobe lies on the east crest of the plateau itself, successively younger ones are present on the scarp, so that this western lateral margin of the ice lobe must have looked like the side of a broad valley glacier.

The northeastern margin of the Des Moines Lobe was not so well defined. It was basically delimited by the high ridges of the Alexandria Moraine Complex, which is believed to have been formed as a major end moraine of the Wadena Lobe at an earlier phase of Wisconsin glaciation.

The Wadena Lobe had earlier advanced farther into southeastern Minnesota, for its distinctive drift underlies the stone pavement beneath the Des Moines Lobe cover in the Minnesota Valley (Matsch, 1971). But the Alexandria Moraine Complex is a very prominent feature reflecting a long still-stand of the ice, during which time the Wadena Drumlin Field on the east and northeast had formed.

The Des Moines Lobe is believed to have expanded eastward over the massive Alexandria Moraine Complex, which at that time was probably cored with stagnant Wadena Lobe ice and was thus even higher than today. Much of the moraine consists of ice-contact stratified deposits and ice-stagnation landforms, covered locally with a veneer of Des Moines Lobe drift.

The main flow of the Des Moines Lobe continued southeastward as a tongue 150 km wide along the Minnesota River Lowland. This broad regional topographic sag had been developed across Precambrian rocks previous to the later Cretaceous weathering that produced a thick kaolinitic regolith on the igneous and metamorphic complex. Later the trough was filled with Cretaceous sediments, which were subsequently largely removed by stream erosion so that only patches remain; these consist of sand, clay, and lignite rather than the siliceous marine shale that now underlies the eastern Dakotas.

The Minnesota Valley takes an abrupt turn to the northeast near Mankato and continues to the Minneapolis area as a lowland cut along relatively soft Cambrian sandstone—the so-called Minneapolis Lowland, which projects northeastward almost to Lake Superior (Fig. 1). When the Des Moines Lobe reached this great bend near Mankato, it divided into two arms. One arm moved northeastward up the Minneapolis Lowland as the Grantsburg Sublobe, eventually reaching as far as western Wisconsin. The second arm turned south, crossed a low divide over early Paleozoic carbonate beds, and moved down the shallow Des Moines River valley to south-central Iowa.

The west margin of the Des Moines Lobe continued as a straight line all the way to the Iowa border, even as the bounding scarp of the Minnesota River Lowland became less distinct. The outermost moraine on this side of the lobe (Bemis Moraine), as followed from its interlobate junction with the moraines of the James Lobe at the north end of the coteau southward to the rounded end of the Des Moines Lobe in Iowa, is about 450 km long. It decreases in elevation from 600 m to 330 m along this distance at a fairly regular gradient. The lobe was 150 km broad in Iowa.

The Bemis Moraine along the northeastern edge of the Prairie Coteau contrasts sharply in morphology with the Altamont and successively younger moraines constructed by the Des Moines Lobe. In southwestern Minnesota it is a well-dissected belt of drift as much as 6 km wide, with a relatively sharp crest that forms the drainage divide between the Big Sioux and Minnesota Rivers. Along its course in eastern South Dakota and southwestern Minnesota, the moraine is cut transversely at intervals by narrow gorges eroded by the outlet streams of long, narrow, proglacial lakes pooled between the moraine and the retreating Des Moines Lobe. The Altamont Moraine in the same region, on the other hand, consists of a poorly drained complex of hummocky terrain, whose broad crest lies as much as 60 m lower on the flanks of the Prairie Coteau. The region between is a complex of stagnant-ice features, including perched lake plains, ice-disintegration ridges, and other distinctive landforms.

Although the degree of dissection may imply that the Bemis Moraine may be appreciably older than the Altamont Moraine, the drift lithologies in the two moraines are similar along the southwestern side of the lobe. The Altamont Moraine, which lies inside and essentially parallel to the Bemis Moraine, marks an important recessional position of the Des Moines Lobe. Earlier the name had been applied to the outermost moraine by Chamberlin (1883), but Leverett (1932) revised the designation after a reinterpretation of the type area in eastern South Dakota. Traced northward from western Iowa, it converges to crowd closely against the Bemis Moraine just below the crest of the east side of the Prairie Coteau.

The next moraine in the sequence at the end of the lobe in Iowa is the Humboldt; then comes the Algona, which is broad and low. Narrow features on the slopes of the Prairie Coteau inside the Altamont Moraine were considered by Leverett as still younger moraines, recording progressive narrowing of the Des Moines Lobe. Actually, they are largely crevasse fillings (for example, the Antelope Moraine) or the scarps of longitudinal meltwater channels (for example, the Marshall Moraine). Even though these linear features are not moraines, they have the same significance—they are probably features that mark the marginal positions of the wasting Des Moines Lobe.

Whereas the west margin of the Des Moines Lobe was straight and sharp against the escarpment of the Prairie Coteau, the east margin in Minnesota is not marked continuously by a moraine. The east edge of the advancing ice lobe was only temporarily confined by the Alexandria Moraine Complex. The moraine still has summits above 500 m, and at the time it was overrun it may have been appreciably higher because of the stagnant Wadena Lobe ice that it is believed to have contained. The ice spread eastward beyond the moraine and covered the outer portion of the Wadena Drumlin Field with silty shale-bearing drift. Meanwhile, subglacial streams cut great valleys eastward through the Alexandria Moraine Complex and spread their outwash in vast plains to the southeast and northeast, further obscuring the drumlins.

The eastward overflow of the Alexandria Moraine by the Des Moines Lobe reached grand proportions farther south, where the Minnesota River Lowland and the Minneapolis Lowland join near Mankato. The ice moved northeastward up this lowland and overflowed the St. Croix Moraine, which had earlier been formed by the Superior Lobe moving from the opposite direction. It completely buried the St. Croix Moraine in the segment from Albany to Minneapolis. Perhaps this area was along the axis of the Minneapolis Lowland and was thus lower than adjacent segments, but the St. Croix Moraine in this segment could have been cut by several tunnel valleys, which produced gaps that allowed the ice from the west to leak through the otherwise massive moraine. At least the Superior Lobe tunnel valleys mostly trend in the direction of the area in question before they are obscured by younger drift, and their mode of origin requires that they terminate at the ice front. Once the ice broke through the barrier in several gaps, it could overwhelm the moraine and move on northeastward along the Minneapolis Lowland, thus forming the Grantsburg Sublobe.

An outstanding exposure showing the stratigraphic and geomorphic relations at the edge of the overlap area is presented by a deep roadcut through an esker 2 km east of St. Rosa in southwestern Morrison County (Swanville quadrangle). The esker itself is made of calcareous sand and gravel of the Wadena Lobe, and it probably formed in a tunnel valley during wastage of this ice mass from its terminus at the Alexandria Moraine Complex (Fig. 7). The gravel is overlain by interlayered red

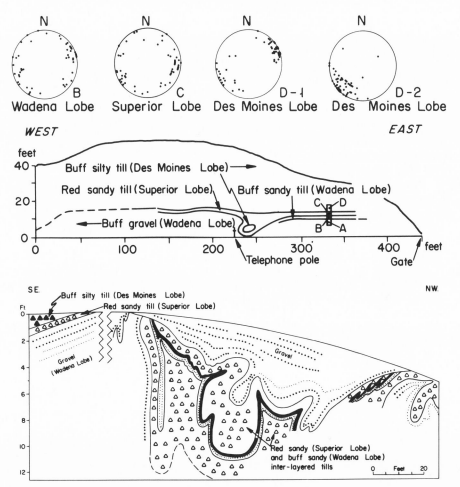

Figure 7. Above: Section along roadcut through esker 4 km east of St. Rosa, Todd County. Circles above show lower hemisphere stereographic projections of long axes of 50 pebbles of tills taken from points B, C, and D-1. Fabric for D-2 comes from same drift 125 m northeast along the esker on the southeast flank. For lithology of pebbles at A-D, see Table 1. Below: Section of St. Rosa Esker formerly exposed in gravel pit 0.5 km southwest of the above roadcut, showing isoclinal folds in esker gravel and its sandy till cap. The additional cap of Des Moines Lobe till (left end) is not deformed.

and light-brown sandy tills. This represents the overriding of the esker by the Superior Lobe, which had picked up light-brown calcareous Wadena Lobe drift on its way to its terminus at the St. Croix Moraine. Capping the interlayered till is a section of characteristic light-brown calcareous silty till as much as 10 m thick, bearing the fragments of Cretaceous shale that are diagnostic of the Des Moines Lobe. In a gravel pit south of the roadcut the esker gravel and the overlying interlayered till are thrown into vertical isoclinal folds, and the complex is truncated and overlain by till of the Des Moines Lobe that caused the deformation.

TABLE 1. LITHOLOGY OF 100 PEBBLES IN EACH OF FOUR DRIFTS EXPOSED IN THE
ROADCUT THROUGH THE ST. ROSA ESKER*

	Wadena Gravel (A)	Wadena Till (B)	Superior Till (C)	Des Moines Till (D)
Northeastern provenance				
Gabbro	5	6	20	5
Red syenite	1	4	8	—
Felsite	9	1	8	1
Red sandstone	1	3	—	—
Western provenance				
Limestone and dolomite	29	31	2	65
Cretaceous shale	—	—	—	3
Undiagnostic				
Granite	28	19	19	11
Basalt	23	33	38	12
Slate, schist, quartzite	4	3	5	3

*Points A to D in Figure 7.

Stone counts and stone fabrics for the drifts at the St. Rosa Esker are shown in Table 1 and Figure 7.

Grantsburg Sublobe

The north margin of the Grantsburg Sublobe drift, east of the Mississippi River (Fig. 4), is relatively distinct, although not marked by an end moraine. It truncates obliquely the well-defined Superior Lobe drumlins in Morrison and Benton Counties, and it crosses but does not completely obscure the Superior Lobe tunnel valleys and eskers between the Mississippi and St. Croix Rivers. This northern margin of the Grantsburg Sublobe was limited merely by the slightly higher ground on the relatively smooth and gentle northwestern flank of the Minneapolis Lowland, along which the sublobe moved northeastward. A small end moraine was formed near the east end of the sublobe near Pine City and Grantsburg.

The south margin of the Grantsburg Sublobe was limited in the segment from Grantsburg southwest to Minneapolis by the inner flank of the St. Croix Moraine of the Superior Lobe. The St. Croix Moraine is one of the largest and most distinct in the Great Lakes region, and at the time it was invaded by the Grantsburg Sublobe it may have been even more prominent as a topographic barrier because of the persistence of stagnant ice, for which independent evidence of several types exists (Florin and Wright, 1969). The Grantsburg Sublobe nowhere passed southward across this barrier, although its outwash probably did.

Where the Grantsburg Sublobe drift overlaps the St. Croix Moraine, especially in the Minneapolis–St. Paul area, the contact is generally marked by interlayering of the two drifts. Individual layers range from a few centimeters thick down to a millimeter. The makeup of the layers closely matches that of the main till masses: red noncalcareous sandy till with stone types from the Lake Superior district, and yellowish-brown calcareous silty till with fragments of limestone, shale, and other rocks from northwestern Minnesota. In places, some of the layers are made of sand rather than of till.

Although the interlayering has been attributed to primary deposition by two adjacent ice lobes, perhaps as proglacial mud flows rather than as basal till (Wright, 1953), the interlayering is now considered to be basically metamorphic structure rather than sedimentary, in that it results from the incorporation and redeposition of red drift by the overriding Grantsburg Sublobe. Critical in this interpretation is the finding that the platy and linear pebbles and sand grains of both the red and the yellow tills show parallel orientation, implying deposition in the same stress environment. The thin layers of sand show no primary sedimentary structures—they also are believed to represent masses of drift that have been stretched out and redeposited.

Although the mechanism of incorporation and deposition cannot be reconstructed in detail from the available evidence, it appears likely that the Grantsburg Sublobe, in overriding the irregular terrain of the St. Croix Moraine, picked up masses of red drift and incorporated them into its load. The moraine at this time was still filled with stagnant ice, and the presence of this ice may have facilitated erosion and remobilization of the drift. As the Grantsburg Sublobe continued to flow forward, the included till masses were strung out into layers like schlieren in metamorphic rocks. The red masses could be projected upward from the bottom where the ice passes over an obstruction, in the manner described by Boulton (1970) for basal ice flow in Spitsbergen. Meanwhile, the rest of the basal ice was carrying the normal debris from the source area. Slow basal melting ultimately brought on the deposition of all basal till, with faithful preservation of the glacial structure.

Glacial Lake Grantsburg

The Grantsburg Sublobe dammed the normal southward drainage of much of north-central Minnesota. The upper Mississippi River had established its modern course by that time in the 80-km segment between Brainerd and St. Cloud. With its major headwaters being the Crow Wing River system, it drained the east edge of the Des Moines Lobe and passed through the St. Croix Moraine via Pillager Gap. Drainage was also contributed to the Mississippi River at that time from northeastern Minnesota, for the Superior Lobe completely filled the Lake Superior Basin and thus diverted much of the St. Louis River system into the Mississippi drainage.

The dammed drainage produced glacial Lake Grantsburg along the north edge of the Grantsburg Sublobe (Fig. 8). The limits of the lake were traced by Cooper (1935), not from shorelines (for none could be identified), but from patches of lake clay discontinuously distributed on the till plain north of the areas of Grantsburg Sublobe till. The

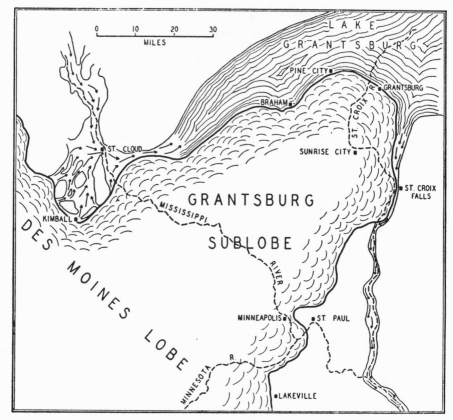

Figure 8. Development of glacial Lake Grantsburg and the Anoka Sandplain (from Cooper, 1935). The eastern part of Lake Grantsburg received red clay from the Superior Lobe via the Grindstone, Kettle, and St. Croix Rivers, but most of the Minnesota portion of

gaps in the distribution may be in part a result of nondeposition caused by wave action, but much of it may result from the occurrence of stagnant ice left over from the Superior Lobe coverage of the area. The outlet of Lake Grantsburg was postulated by Cooper to be the margin of the ice near its terminus. However, various possible proglacial outlets occur in Wisconsin, now occupied by underfit streams—for example, the Apple River.

The deposits of glacial Lake Grantsburg within the Minnesota portion west of Pine City consist almost entirely of thin patches of gray calcareous silty clay, locally varved. The sections are generally less than a meter thick—in fact, so thin that the varving, the carbonate content, and the textural characteristics have been modified by soil-forming processes. However, the limits as mapped by Cooper (1935), even without adequate topographic-map control, are reasonable.

The gray calcareous nature of the Lake Grantsburg sediments within Minnesota indicates a source within the adjacent Grantsburg Sublobe itself, although certainly some contribution must have been added by

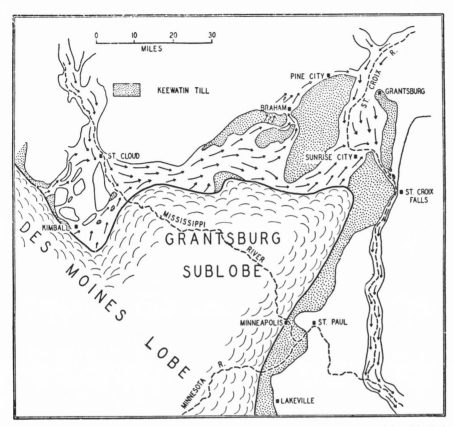

the lake held only gray clay. Arrows indicate stream flow on the Anoka Sandplain. Stippled pattern ("Keewatin till") shows the distribution of Grantsburg Sublobe till.

the Mississippi River, which was dammed by the sublobe. Much of the Mississippi River drainage at that time carried the same kind of gray calcareous sediment—the drainage from the east margin of the Des Moines Lobe via Pillager Gap. The fine-grained sediment from the eastern tributaries of the Mississippi River, fed in part by the much diminished Superior Lobe, may have been largely screened out by proglacial lakes farther north.

The absence of red clays in most of the Lake Grantsburg deposits provides an important clue to the position of the Superior Lobe at the time of maximum advance of the Grantsburg Sublobe. It means that the Superior Lobe must have been tightly confined to the Lake Superior Basin and that it did not cross the divide into the Minneapolis Basin. If it had crossed the divide, then its meltwater would have brought red clays to Lake Grantsburg via the Snake River, or via other streams that may have followed the old tunnel valleys. As it was, the only stream contributing red clays to Lake Grantsburg was the Grindstone River, which came from the tip end of the Superior Lobe (at that time near Finlayson) and entered Lake Grantsburg east of Hinckley. The diversion eastward from Hinckley was a consequence of a large block of stagnant Superior Lobe ice that occupied the area south of Hinckley. The Grindstone outwash stream formed an excellent outwash fan at Hinckley, which eastward became a delta into Lake Grantsburg at a point now deeply dissected by the St. Croix River.

The Lake Grantsburg deposits exposed along the bluffs of the St. Croix River valley (for example, at St. Croix State Park) consist of red clays that are varved in part. As described by Berkey (1905), they reach a thickness of 12 m, according to well records. These sediments came not only from the Grindstone River but also from the upper St. Croix River, which drained the east flank of the Superior Lobe. If the sedimentation rate represented by exposed varves is representative of the entire section, then a total of 1,700 yrs may be represented by the deposit. The lake there lasted much longer than it did farther west. The significance of the chronology is discussed below, after a consideration of the Anoka Sandplain.

Anoka Sandplain

The Anoka Sandplain was formed as the Grantsburg Sublobe wasted (Fig. 8). The sandplain occupies the same area as the sublobe, except the morainic hills were not covered by sand. For the most part, the deposits of Lake Grantsburg were not covered by those of the Anoka Sandplain. The lake deposits were proglacial (at least in the Minnesota portion), whereas the sand covered the area bared by the wasting ice. The main exception is the St. Croix Valley, where the red varved clays have a cover of red sand graded to the Lake Grantsburg level. Braided patterns on air photographs show that the streams forming the sandplain in eastern Minnesota occupied different courses, successively cross cutting. Apparently as the ice wasted, slightly lower courses of the river systems were established, in places across dead-ice blocks that are now represented by lakes. Much of the drainage at that time continued to come from the Grantsburg Sublobe itself, for the sands of the Anoka Sandplain commonly contain lenses of gravel close to islands of Grantsburg Sublobe till—presumably the islands represent late-lasting masses of dead ice that continued to supply meltwater and coarse outwash to the sandplain (Farnham, 1956).

The Anoka Sandplain is by no means a featureless surface. In addition to the gentle swells and swales of the river braids, the surface is interrupted by low hills of red or yellow till, representing slightly higher areas of ground moraine of the Superior or Grantsburg Lobes that were not inundated by the outwash. Sand dunes also occur as positive features, the sand having been blown from the sandplain surface either during late-glacial time when the floodplain was dry or during the mid-postglacial dry period. Some of the dune areas show enough structure to infer that the dominant strong summer winds were southwesterly. A patch of dunes northwest of the town of Elk River is a particularly good example (Cooper, 1935).

Negative topographic features that interrupt the Anoka Sandplain include principally the hundreds of lakes and marshes that represent blocks of stagnant glacial ice buried by the sand. Many of these abruptly break the intricate braided pattern of the sandplain surface. The most conspicuous lakes occur in strings that mark the positions of tunnel valleys or former courses of the St. Croix and Mississippi Rivers. These valleys had been plugged with stagnant ice and then evenly buried by the sandplain deposits. Subsequent melting of the ice produced the lines of lake depressions.

As the Grantsburg Sublobe slowly wasted back to the west, the drainage shifted gradually southward. Eventually the waters were ponded directly north of the St. Croix Moraine in a series of small lakes that drained southward through the moraine (Stone, 1966). By this time the outlet of the sandplain via the St. Croix River had been abandoned, but a much diminished Lake Grantsburg may still have existed in the St. Croix Valley. Finally, the Mississippi River took up its course through Minneapolis and St. Paul, leaving the Anoka Sandplain proper as a terrace.

St. Louis Sublobe

In addition to the Grantsburg Sublobe, the Des Moines Lobe had a second offshoot to the east—the St. Louis Sublobe, which occupied the Red Lakes Lowland of north-central Minnesota (Fig. 3). This lowland had previously been filled by the Wadena Lobe, which, however, encountered the Rainy Lobe and was diverted southwestward to form the Wadena Drumlin Field and Alexandria Moraine Complex and, on retreat, the Itasca Moraine, as recounted above. This time the ice advancing from the west encountered no such opposition, and it filled the lowland and spread almost across the state to within 35 km of Lake Superior.

The till of the St. Louis Sublobe in the Red Lakes Lowland can be identified by its high content of silt and of fragments of Cretaceous shale, whereas the Wadena Lobe till on which it rests is sandy and devoid of shale. Eastward, however, the local substratum was eroded by the ice and the lithology of the drift changes. This substratum for the eastern part of the sublobe consisted of drift derived from northern and eastern sources. As the sublobe reached its extremities it overrode the basins of glacial Lakes Aitkin and Upham. These lakes had originally been formed in the Automba phase, when the Superior Lobe stood at the Highland–Mille Lacs Moraine and dammed the southward drainage of the Mississippi and St. Louis Rivers. Meltwater from the Superior Lobe at that time had deposited red silt and clay in the lake, and meltwater from the Rainy Lobe at the Vermilion Moraine brought in fine brown sediments. The reddish-brown silty till that resulted when the St. Louis Sublobe advanced over the lake beds was deposited around the margins of the lake basins.

The St. Louis Sublobe at its extremities was probably quite thin, for it adopted a form that followed the irregular course set by such low boundaries as previously existing moraines. Thus, on the south it came as far as the Mille Lacs Moraine, on which it deposited a thin cap up to an elevation of 410 m. On the east it was bounded in part by the Highland Moraine, and in part by the Toimi drumlins. On the north the ice spread onto the flank of the Mesabi iron range, where its typical red-brown silty till is exposed as the surface drift in many iron mines (Winter, 1971). There it reached an elevation of 470 m (1,550 ft), and it even protruded at one place through a gap in the ridge. The higher elevation limit on the north may merely reflect the extent of postglacial crustal tilting.

When the St. Louis Sublobe stood at its maximum position, its melt-waters were drained eastward by the St. Louis River toward Lake Superior. But the Lake Superior Basin at that time was filled by ice of the Nickerson phase, so the meltwater was diverted southward to the Moose River channels, and thence to the St. Croix River. Retreat of the St. Louis Sublobe brought into being glacial Lakes Upham II and Aitkin II, which were interconnected and drained out the St. Louis River by way of the series of diversion channels previously described. Eventually the ice withdrew beyond the Red Lakes Lowland, and the meltwater was trapped in the east arm of Lake Agassiz. Lake Aitkin then dried up, as the Mississippi River flowed directly into Lake Upham, and a soil and organic layer formed on the sediment surface near Aitkin (Farnham and others, 1964). Subsequent southward tilting of the crust brought the Mississippi River back into the Lake Aitkin plain, burying the soil with more lake sediments and ultimately with peat as the outlet stream cut a channel through the bounding moraine and drained the lake.

Chronology of the Des Moines Lobe and Its Sublobes

Radiocarbon dates in drift of the Des Moines Lobe and its sublobes, although not abundant, are enough to provide a general chronology of events, especially because supplemental dates can be used from certain outwash features of the Superior Lobe that can be connected geomorphically with features of the Grantsburg and St. Louis Sublobes. This chronology differs in important respects from that previously followed (Wright and Ruhe, 1965).

The oldest event recorded is the advance of the Grantsburg Sublobe to its maximum at the Pine City Moraine and the formation of glacial Lake Grantsburg. The controlling C-14 dates are from basal sediments of kettle lakes on outwash deposits of the Split Rock phase of the Superior Lobe. At least some of this outwash drained down the Grindstone

River, forming a delta in glacial Lake Grantsburg, which is, therefore, at least 16,000 yrs old.

Lake Grantsburg was first confined to the St. Croix Valley, and it lasted there from 16,000 to 14,000 yrs ago while receiving red clay from the Superior Lobe in its Split Rock phase. As the Grantsburg Sublobe expanded, the lake level rose and the long west arm of Lake Grantsburg developed across eastern Minnesota. The gray clays deposited in this part of the lake were locally overridden by the ice at its maximum (Cooper, 1935). This part of the lake was short-lived, because the outlet stream around the point of the lobe eroded its channel and lowered the level of the lake.

Up to this time the main flow of the Des Moines Lobe along the Minnesota River valley had all been directed around the Mankato bend and into the Minneapolis Lowland as far as the terminus in Wisconsin. This was the downhill route, including the break through the Albany-to-Minneapolis segment of the St. Croix Moraine. The floor elevation of the centerline of the lobe decreased from about 330 m above sea level at the divide against the Red River Valley, to 300 m at the Mankato bend, to 240 m at Minneapolis, then rose to only 300 m at the terminus at Grantsburg. Much of this long ice tongue was contained at this time between the Prairie Coteau on the west and the Alexandria Moraine on the east.

As the ice tongue rose on its northeastern flank it spread over the ice-cored Alexandria Moraine and covered the margin of the Wadena Drumlin Field. On its southwestern flank it rose to its highest moraine (Bemis) on the crest of the Prairie Coteau, and it spilled southward at the Mankato bend and flowed over the 330-m divide that leads south to the Des Moines River valley in Iowa. At the maximum the ice flow was thus diverted in part from its northeastward course (Grantsburg Sublobe) to a southerly course into central Iowa, where it reached its terminus about 14,000 yrs ago (Ruhe, 1969), according to several C-14 dates of wood beneath the till and from basal bog sediments on top of the till.

The Grantsburg Sublobe, thus partially beheaded, wasted by stagnation, and the Anoka Sandplain developed in its place. By that time (14,000 yrs ago) glacial Lake Grantsburg had drained completely by wastage of the ice dam. The oldest radiocarbon date on basal lake sediments in kettle holes on the Anoka Sandplain is 13,530 ± 240 yrs B.P. (Y-1478) for Horseshoe Lake—completely consistent with the chronology.

Wastage of the main Des Moines Lobe proceeded rapidly from its maximum position at the Bemis Moraine. The Algona Moraine is dated at about 13,000 yrs ago (Ruhe, 1969). Madelia, a lake site near Mankato, was uncovered before 12,650 ± 350 yrs ago (W-824).

During all the time of Des Moines Lobe development in the south, the St. Louis Sublobe formed as a protrusion into the Red Lakes Lowland of north-central Minnesota. It reached its maximum at some time after the Split Rock phase of the Superior Lobe (dated as 16,000 to 14,000 yrs ago, according to the chronology presented above). The evidence for this is that some of the Split Rock outwash was directed west up the St. Louis River and thence into the Mississippi drainage, which therefore could not have been blocked by the St. Louis Sublobe.

The only other controlling dates are from the upper sediments of glacial Lake Aitkin II, which was formed as the St. Louis Sublobe retreated. These average about 11,650 yrs B.P. The oldest date for glacial Lake Agassiz, 11,740 ± 200 yrs B.P. (Y-1327) from the Herman Beach, is also pertinent, because at the Herman phase, Lake Agassiz filled the Red Lakes Lowland that had previously contained the St. Louis Sublobe. Accordingly, an approximate minimal date of 12,000 yrs ago is used for the St. Louis Sublobe maximum.

Glacial Lake Agassiz

Eventual retreat of the Des Moines Lobe and its sublobes into the Red River Lowland resulted in the formation of glacial Lake Agassiz, dammed on the south by the Big Stone Moraine in westernmost Minnesota (Fig. 4). Some small preliminary lakes at the south edge of the wasting ice lobes had different outlets across the moraine, but as the small lakes grew and joined they were released by a single outlet, whose channel became deepened until it was paved with large granite boulders washed out of the moraine (Matsch and Wright, 1966). The lake was thus stabilized at the Herman level (326 m at the outlet), which is recorded by a beach ridge traceable around the margins of the Red River Valley as far as the Canadian border and also is found eastward around the Red Lakes Lowland of north-central Minnesota. The lake at that time was about 60 km wide in its southern part and was at least 30 m deep. In the shallow water the waves smoothed out some of the primary irregularities of the ground moraine that was flooded by the lake, but little sediment was deposited. Till is exposed at the surface, in some cases still with its washboard topography. Where the water was more than about 15 m deep, sand was deposited. Silt and clay covered the bottom where the lake was deeper than about 60 m.

As the ice continued to recede, the lake grew larger, and the volume of water passing through the outlet increased enough to erode through the boulder-paved channel, resulting in a lowering of the lake level. This event occurred about 11,700 yrs ago, according to a radiocarbon date on the basal sediment of a beach pond. After three such steplike lowerings, each recorded by a strandline extending progressively farther to the north, the outlet became stabilized, perhaps on the granitic bedrock, and the Campbell Strandline was formed, with an elevation of 302 m at the outlet (Matsch and Wright, 1966).

This was the last lake level to serve a southern outlet. By that time the lake was very large, and the outlet stream, the glacial River Warren, cut a deep channel through thick drift up to 60 m and into bedrock along the Minnesota River valley. The effect continued far downstream to the Mississippi River junction and beyond. In the Minneapolis area the River Warren thus cut deeply below the level of the outwash terraces that had previously formed when the Anoka Sandplain was drained. At St. Paul, it cut 80 m below the terrace surface into the Paleozoic bedrock.

Further retreat of the ice, including lobes east of the Des Moines Lobe, uncovered lower outlets to the east to northern Lake Superior (Zoltai, 1965). The Campbell Strandline and the southern outlet of the lake were abandoned, and the lake temporarily fell to a lower level about 10,000 yrs ago. Minor readvance of the ice, along with southward tilt of the land caused by crustal upwarp, again closed the eastern outlets, and the lake rose to the Campbell level. Final retreat of the ice about 9,200 yrs ago led gradually to reoccupation of outlets to the east, and then ultimately to the north, to Hudson Bay, about 8,000 yrs ago.

REFERENCES CITED

Andrews, J. T., 1965, Glacial geomorphological studies on north-central Baffin Island, Northwest Territories, Canada (Ph.D. thesis): Nottingham, England, University of Nottingham, 476 p.

Arneman, H. F., and Wright, H. E., Jr., 1959, Petrography of some Minnesota tills: Jour. Sed. Petrology, v. 29, p. 540–554.

Bain, H. F., 1897, Relations of the Wisconsin and Kansan drift sheets in central Iowa, and related phenomena: Iowa Geol. Survey Ann. Rept., v. 6, p. 429–476.

Berkey, C. P., 1905, Laminated interglacial clays of Grantsburg, Wisconsin: Jour. Geology, v. 13, p. 35–44.

Black, R. F., 1969, Valderan glaciation in western Upper Michigan: Internat. Assoc. Great Lakes Research, Proc. 12th Conf., Ann Arbor, p. 116–123.

Boulton, G. S., 1970, On the origin and transport of englacial debris in Svalbard glaciers: Jour. Glaciology, v. 9, p. 213–230.

Chamberlin, T. C., 1883, Preliminary paper on the terminal moraine of the second glacial epoch: U.S. Geol. Survey Ann. Rept., v. 3, p. 291–402.

Christiansen, E. A., 1968, A thin till in west-central Saskatchewan, Canada: Canadian Jour. Earth Sci., v. 5, p. 329–336.

Clayton, Lee, 1966, Notes on Pleistocene stratigraphy of North Dakota: North Dakota Geol. Survey Rept. Inv. 44, 24 p.

Cooper, W. S., 1935, The history of the upper Mississippi River in Late Wisconsin and postglacial time: Minnesota Geol. Survey Bull 26, p. 116.

Farnham, R. S., 1956, Geology of the Anoka sandplain, in Glacial geology, eastern Minnesota: Geol. Soc. America Guidebook, Ann. Mtg., Minneapolis, p. 53–64.

Farnham, R. S., McAndrews, J. H., and Wright, H. E., Jr., 1964, A Late Wisconsin buried soil near Aitkin, Minnesota, and its paleobotanical settings: Am. Jour. Sci., v. 262, p. 393–412.

Farrand, W. R., 1969, The Quaternary history of Lake Superior: Internat. Assoc.

Great Lakes Research, Proc. 12th Conf., Ann Arbor, p. 181–197.

Flint, R. F., 1955, Pleistocene geology of eastern South Dakota: U.S. Geol. Survey Prof. Paper 262, 173 p.

Florin, Maj-Britt, and Wright, H. E., Jr., 1969, Diatom evidence for the persistence of stagnant glacial ice in Minnesota: Geol. Soc. America Bull., v. 80, p. 695–704.

Hack, J. T., 1965, Postglacial drainage evolution and stream geometry in the Ontonagon area, Michigan: U.S. Geol. Survey Prof. Paper 504 B, 40 p.

Klassen, R. W., 1967, Surficial geology of the Waterhen–Grand Rapids area, Manitoba: Canada Dept. Energy, Mines, and Resources Paper 66-36, 6 p.

Leverett, Frank, 1932, Surficial geology of Minnesota and parts of adjacent areas: U.S. Geol. Survey Prof. Paper 161, 149 p.

Matsch, C. L., 1971, Pleistocene stratigraphy of the New Ulm region, southwestern Minnesota (Ph.D. thesis): Madison, Wisconsin Univ., 78 p.

Matsch, C. L., and Wright, H. E., Jr., 1966, The southern outlet of Lake Agassiz, in Mayer-Oakes, W. J., ed., Life, land, and water: Conf. on Environmental Studies of the Glacial Lake Agassiz Region Proc.: Winnipeg, Manitoba, Univ. Manitoba Press, p. 121–140.

Ruhe, R. V., 1969, Quaternary landscapes in Iowa: Ames, Iowa State Univ. Press, 255 p.

Sardeson, F. W., 1916, Minneapolis–St. Paul folio: U.S. Geol. Survey Geol. Atlas, Folio 201.

Schneider, A. F., 1961, Pleistocene geology of the Randall region, central Minnesota: Minnesota Geol. Survey Bull. 40, 151 p.

Stone, J. E., 1966, Surficial geology of the New Brighton quadrangle: Minnesota Geol. Survey Geol. Map Ser., 6-M2, 39 p.

Swinzow, G. K., 1962, Investigation of shear zones in the ice sheet margin, Thule area, Greenland: Jour. Glaciology, v. 4, p. 215–229.

Upham, Warren, 1880, Preliminary report on the geology of central and western Minnesota: Minnesota Geol. Nat. Hist. Survey 8th Ann. Rept. (1879), p. 70–125.

———— 1895, The glacial Lake Agassiz: U.S. Geol. Survey Mon. 25, 658 p.

Wasylikowa, Krystyna, and Wright, H. E., Jr., 1970, Late-glacial plant succession on an abandoned drainageway, northeastern Minnesota: Acta Palaeobotanica, v. 11, no. 1, p. 23–43.

Weertman, J., 1961, Mechanism for the formation of inner moraines found near the edge of cold ice caps and ice sheets: Jour. Glaciology, v. 3, p. 965–978.

Winchell, N. H., 1877, Geology of Hennepin County, Minnesota: Minnesota Geol. Survey, 5th Ann. Rept. (1876), p. 131–201.

Winter, T. C., 1971, Sequence of glaciation in the Mesabi-Vermilion iron range area, northeastern Minnesota: U.S. Geol. Survey Prof. Paper 750-C, p. 82–88.

Wright, H. E., Jr., 1953, Interbedded Cary drifts near Minneapolis, Minnesota: Jour. Geology, v. 61, p. 465–471.

———— 1957, Wadena drumlin field, Minnesota: Geog. Annaler, v. 39, p. 19–31.

———— 1962, Role of the Wadena Lobe in the Wisconsin glaciation of Minnesota: Geol. Soc. America Bull., v. 73, p. 73–100.

———— 1973, Tunnel valleys, glacial surges, and subglacial hydrology of the Superior Lobe, Minnesota: Geol. Soc. America Mem. 136, p. 251–276.

Wright, H. E., Jr., Mattson, L. A., and Thomas, J. A., 1970, Geology of the Cloquet quadrangle, Carlton County, Minnesota: Minnesota Geol. Survey Geol. Map Ser. GM-3, 30 p.

Wright, H. E., Jr., and Ruhe, R. V., 1965, Glaciation of Minnesota and Iowa, *in* Wright, H. E., Jr., and Frey, D. G., eds., The Quaternary of the United States: Princeton, N.J., Princeton Univ. Press, p. 29–41.

Wright, H. E., Jr., and Watts, W. A., 1969 (with contributions by Jelgersma, Saskia, Waddington, J. C. B., Winter, T. C., and Ogawa, Junko), Glacial and vegetational history of northeastern Minnesota: Minnesota Geol. Survey Spec. Pub. 11, 59 p.

Zoltai, S. C., 1965, Glacial features of the Quetico-Nipigon area: Canadian Jour. Earth Sci., v. 2, p. 247–269.

MANUSCRIPT RECEIVED BY THE SOCIETY DECEMBER 27, 1971

Printed in the United States of America

TOPICAL

GEOLOGICAL SOCIETY OF AMERICA
MEMOIR 136
© 1973

Glacial Dispersal of Rocks, Minerals, and Trace Elements in Wisconsinan Till, Southeastern Quebec, Canada

W. W. Shilts

Geological Survey of Canada, Ottawa, Ontario, Canada

ABSTRACT

Lennoxville Till of the Lac-Mégantic region is homogeneous with respect to fabric, texture, clay-mineral composition, and color. The homogeneity resulted from erosion and transportation by a glacier that moved southeastward over bedrock that had little petrologic variation. Lennoxville Till is, however, also characterized by southeast-trending bands with high concentrations of certain rocks, minerals, and chemical components derived from scattered igneous bodies.

Chromium, nickel, and magnetite are dispersed in ribbonlike bands at least 50 km southeast of their principal source areas in the ultrabasic-basic rock complex at Thetford Mines. Concentrations of surface erratics derived from these same sources have dispersal patterns similar to those of Cr, Ni, and magnetite.

Plagioclase grains and granodioritic erratics are dispersed in bands southeast of the granodiorite stock of the Little Megantic Mountains. Granodioritic erratics that occur on the surface are thought to have been let down onto lodgment facies of Lennoxville Till during melting of the Lennoxville glacier, because they mantle the surface of its granodiorite-poor clay-till facies. By extension, most of the boulder mantle that characteristically rests on Lennoxville lodgment facies may be an ablation deposit with dispersal characteristics locally reflecting glacier deflections and lobations that occurred during deglaciation.

Dispersal data strongly confirm ice-flow patterns inferred independently

from striation, fabric, and geomorphic data. Most till components discussed show evidence of long-distance transport, except where topographic prominences have blocked or deflected ice-transported sediment.

No evidence was found to support the concept of glacier flow into
Quebec from late-glacial highland centers of outflow located south or
east of the area.

Texture of till has an important influence on observed concentrations
of trace elements. Zr is concentrated preferentially in the silt fraction
of silt + clay; Ti, Cu, V, Zn, Pb, Cr, and Ni are concentrated preferentially
in the clay fraction. Concentrations of Cr, Ni, and Ti are thought to
be texture-independent in samples located down-ice from igneous sources
rich in these components.

INTRODUCTION

Objective

The objective of this study is to define dispersal trends of selected
rocks, minerals, and trace elements for a single till sheet in an area
of moderate relief (800 m). It is thought that this information leads
to a clearer understanding of (1) ice-flow directions, inferred for surface
and older tills from fabric and striation data; (2) the effect of irregular
topography upon patterns of dispersal; (3) the relationship between the
continental glacier and possible late-glacial ice caps in the New England
highlands; (4) the relative distances of transport and rate of dilution
among diverse till components; and (5) the origin of the bouldery mantle
that commonly overlies surface till in the region.

Study Area

This study was conducted near Lac-Mégantic, Quebec (henceforth called
the Megantic area) in the Appalachian Mountains (Fig. 1). Most samples
are located in that part of Canada east of long 71° W. and south of
lat 46° N., an area of approximately 3,000 km².

The principal advantages of studying dispersal trends in the Megantic
area are: (1) a large number of exposures offering access to relatively
unweathered till; (2) presence of several lithologically distinct source
areas surrounded by slate and sandstone that vary little in composition;
(3) presence of boulder piles constructed by farmers while clearing fields;
(4) a well-known history of ice movement established independently by
striations, till fabrics, and moraine orientations; and (5) contrasting physiographic sections—relatively flat areas with interspersed hilly sections.

Physiography

The Megantic area has been divided into several physiographic sections

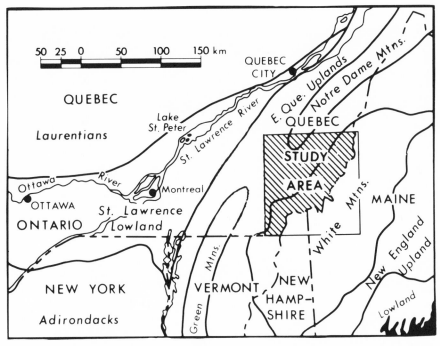

Figure 1. Location map showing generalized physiographic subdivisions and study area.

(Fig. 2). The surface of most of the region is gently rolling and is called the Chaudière Hills. Glacial deposits are relatively thick over this section, and glaciers have leveled the terrain by planing off prominences and filling depressions with up to 75 m of glacial sediments. Altitudes range between 270 m and 490 m.

The Lac-aux-Araignées Basin is a flat, low area developed on a granodiorite pluton. The section is generally thickly drift-covered. Altitudes within the section vary between 400 m and 500 m.

The Portage Uplands section has a high, relatively level surface, commonly corrugated by northeast-trending strike ridges. The surface is thinly mantled with till and is bounded at its north- and northwest-facing sides by an escarpment 100 m to 300 m high. Altitudes vary between 520 m and 640 m.

The Boundary Mountains are the northeastern extension of the White Mountains of New Hampshire. This section is characterized by high peaks (900–1,175 m), steep slopes, and thin till cover.

The Little Megantic Mountains form a 10 km-long, northeast-trending ridge which stands 100 m to 500 m above the Chaudière Hills. These mountains played a significant role in development of dispersal patterns by blocking and diverting debris transported from sources northwest of them, by protecting glacial deposits laid down in their lee, and by being the source for distinctive indicators eroded from their granodiorite core and its surrounding contact-metamorphic zone.

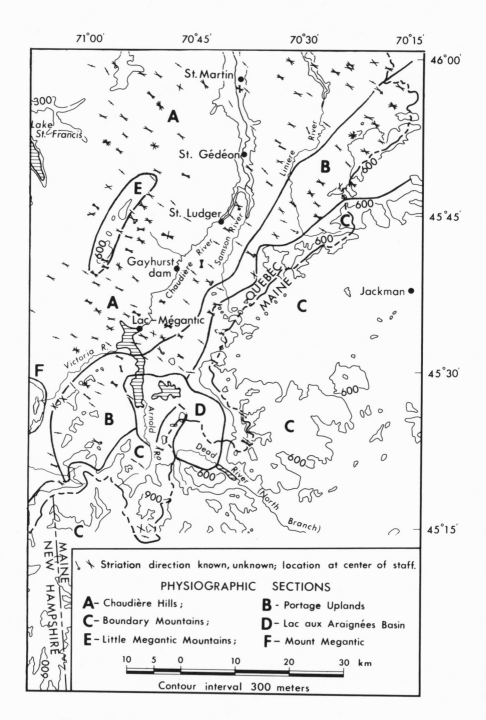

Figure 2. Physiographic sections, topography, and striations—Megantic area.

Mount Megantic (1,088 m) is the most easterly of the distinctive Monteregian Hills. It consists of an outer syenitic rim standing 500 m to 600 m above the surrounding hills, and an inner gabbroic and granitic core.

Method of Investigation

Till and its mantle of cobbles and boulders were examined to determine regional variations of lithologic components. Samples of till were collected largely from unoxidized, unleached exposures at least 2 m below the ground surface; about 25 percent of the till samples came from the oxidized portion of the C horizon. Because of compositional differences among various size grades of till, certain grades were selected for analysis of specific components.

Cobbles, Boulders, and Pebbles. Frequencies of granodioritic, gabbroic, and ultrabasic clasts of cobble and boulder size (15 cm to 30 cm maximum diameter) are calculated at over 160 stations. At each station the calculation was made by identifying 100 to 200 stones from a farmer's boulder pile or, rarely, from a stream bed.

Lithologic frequencies are also calculated for pebbles collected from the till matrix (matrix is defined as the portion of till finer than 2 mm). The most important indicator lithologies (granodioritic and ultrabasic pebbles) are largely decomposed in the upper 2 m of till exposures. Therefore, the number and distribution of sample sites was limited by the availability of deep exposures, and frequency data are presented for granodioritic pebbles only. Sample points were not random enough to study the dispersal of other types of pebbles systematically.

Fine Sand. Fine sand portions (0.125 mm to 0.250 mm) of till were separated into heavy minerals (SG > 2.85) and light minerals (SG < 2.85) with bromoform in separatory funnels. Magnetite was separated from heavy minerals by hand magnet, and its concentration was tabulated as weight percent of the total heavy concentrate.

Light minerals were mounted in Canada balsam, and ground to a thickness of less than 0.125 mm. Mounts were stained for plagioclase and potassium feldspar according to the method of Boone and Weaver (1968). Initial HF etching of 27 sec and Amaranth immersion time of 4 sec at 23° C were modifications of the published method.

Plagioclase and potassium feldspar percentages are calculated from 500 to 1,000 grains of the light-mineral concentrate. Percentages reported exclude rock fragments, serpentine, and calcite which average, respectively, 40 percent, less than 1 percent, and less than 2 percent of the total concentrate.

Trace Elements. Concentrations of copper, chromium, nickel, zirconium, vanadium, and titanium were determined in ppm of the silt + clay (<63 μ) fraction of till. Analyses were made by emission spectrography. Concentrations of zinc and lead were determined in ppm of silt + clay by atomic absorption after hot, $HCl-HNO_3$ leach. The <63 μ fraction was obtained by sieving through a stainless steel screen after gently disaggregating the dry till sample in an agate mortar and pestle.

Previous Investigations

Many studies of glacial dispersal have been undertaken in the past hundred years and a voluminous literature exists. Indicator studies have commonly dealt with boulder and pebble distribution, but since about 1950, distribution patterns of sand-size minerals, clay minerals, and trace elements have received much attention. Dreimanis (1956) gives an excellent discussion and reference list of publications on boulder tracing; Potter and Pettijohn (1963) also present a discussion of till provenance studies. Hawkes and Webb (1962) and several authors in Kvalheim (1967) discuss the development of geochemical prospecting techniques in glacial or related deposits.

Recent publications by Lee (1963), Sitler (1963), Willman and others (1963, 1966), Gillberg (1964, 1967), Dreimanis and Vagners (1969), and White and others (1969) are examples of systematic modern approaches to regional mineral and rock dispersal studies for till.

Although much information has been gathered on the geochemical properties of unconsolidated sediments in glaciated regions, relatively little attention has been given to regional dispersal patterns of trace elements in unweathered till. Among those who have investigated this phenomenon specifically for till are Kauranne (1957, 1959), Lee (1963), Pawluk and Bayrock (1969), and Donovan and James (1967). Most of these authors show an awareness of long-distance transport of certain till components—particularly those occurring in the sand-size or finer grades. This concept conflicts with the formerly widespread belief, based largely on similarity of cobble-size and larger clasts to local bedrock, that till is composed mostly of locally derived detritus. Donovan and James (1967), however, although demonstrating areal trace-element anomalies caused by transport down-ice from an ore body, concluded that the bulk of the till matrix was transported only a few hundred meters and that only the larger erratics were far-travelled.

In southeastern Quebec, McDonald (1966, 1967a, 1967b) mapped indicator trains trailing southeast from several distinct source areas. He interpreted erratics found west of their source areas as having been transported southwest across regional strike during a southwest flow phase of the penultimate glaciation. Cooke (1937b) and Duquette (1960b) interpreted a similar occurrence of erratics as reflecting north or northwestward flow of ice into Quebec from highlands in New England (for more detailed discussion, *see* McDonald and Shilts, 1971).

Bedrock Lithology

Metasedimentary Rocks. Bedrock of the Lac-Mégantic region consists mainly of slightly metamorphosed (chlorite grade), northeast-striking, nearly vertical interbeds of sandstone, slate, and calcareous mudstone (Fig. 3). The metasedimentary units are interbedded with basic metavolcanic rocks east and south of the town of Lac-Mégantic and are interbedded with conglomerate, limestone, and acid and basic metavolcanic rocks northwest of the Little Megantic Mountains. The metasediments are exten-

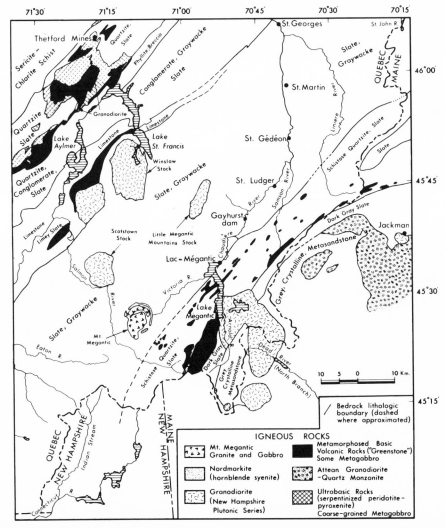

Figure 3. Generalized bedrock map of Megantic area, *after* Cooke (1937a, 1950), Lord (1938), Gorman (1955), Duquette (1960a, 1960b, 1961), Reid (1960), St. Julien (1963-1965), Doyle (1967), and Marleau (1968).

sively cut by quartz, calcite, and epidote veins, and by felsic dikes less than 5 m wide.

The metasedimentary rocks, because of their extensive areas of outcrop and general lack of compositional variation, are considered to produce background erratic concentrations upon which are superimposed anomalous concentrations of detritus eroded from the igneous suites that intrude them. Most till samples collected north of St. Gédéon can be considered to have background compositions with respect to the till components studied.

Granodiorite. Several stocks of New Hampshire Plutonic Series granodiorite (Doyle, 1967; Marleau, 1968, p. 37–38) intrude the metasedimentary and metavolcanic rocks (Fig. 3). Each stock is surrounded by a zone, generally less than 1 km wide, of cordierite-, hornblende-, and andalusite-rich contact hornfels. Glacial erosion of these stocks produced distinctive, light gray (rarely pink), medium- to coarse-grained, roughly equidimensional granodioritic clasts. Granodioritic erratics are generally unaltered below 3 m; at the surface they commonly retain their glacially polished and striated surfaces.

Other Acid Intrusions. Attean granite or quartz monzonite crops out in several areas east of the International Boundary. It tends to be coarser grained than rocks from the New Hampshire Series stocks and is distinguished from them by its texture, pinkish color, and chloritized mafic minerals.

Mount Megantic consists of an outer ring dike of nordmarkite that surrounds a granite and gabbro complex.

Ultrabasic and Basic Intrusions. Serpentinized peridotite and pyroxenite comprise the bulk of ultrabasic rocks cropping out in the Thetford Mines region. Associated with the ultrabasic rocks are distinctive, coarse-grained metagabbros that crop out most commonly near and southwest of Thetford Mines. Ultrabasic erratics weather rapidly in the zone of oxidation of till but, like granodioritic erratics, they retain their glacially polished surfaces above ground.

Ultrabasic outcrops also occur near the International Boundary south and south-southeast of Mount Megantic.

GLACIAL HISTORY AND SEDIMENTS

Stratigraphy

The Megantic region was glaciated twice in the late Wisconsinan Stage, once in the early Wisconsinan and at least one earlier time (Table 1). Tills of the last three glaciations have been identified (McDonald, 1967a; Shilts, 1969, 1970; McDonald and Shilts, 1971). The oldest (Johnville) and youngest (Lennoxville) tills were deposited by glaciers with regional southeast movement; an intermediate till (Chaudière) was deposited by a glacier flowing first southwestward and later southeastward. During

TABLE 1. GENERALIZED STRATIGRAPHIC COLUMN OF PLEISTOCENE DEPOSITS; SOUTHERN QUEBEC

Age	Formation	Inferred Ice-Flow Direction and Glacial Events
Wisconsinan	Lennoxville Till	Ice flow southeast to east-southeast into New England at maximum; lobe in Chaudière Valley flowed south-southwest at beginning and end of glaciation; Drolet Lentil deposited by early lobe.
	Drolet Lentil	
	Gayhurst Formation	Ice-filled St. Lawrence Lowlands, blocking drainage for 3,000 to 4,000 yrs (based on estimation of number of varves); lakes formed in Chaudière Valley and deposited predominantly fine-grained sediments of Gayhurst Formation.
	Chaudière Till	Ice flow toward west-southwest or southwest during most of Chaudière glaciation; flow gradually shifted toward southeast at end of glaciation; Chaudière glacier retreated only to St. Lawrence Lowlands.
	Massawippi Formation	Fluvial and lacustrine sediments of nonglacial interval; correlated with St. Pierre Beds of St. Lawrence Lowlands; some weathering and leaching of till.
	Johnville Till	Ice flow southeast.
Pre-Wisconsinan (?)	Pre-Johnville sediments	Fluvial and lacustrine sediments of nonglacial interval; fluvial gravels locally oxidized; gravel contains fragments of Laurentian gneiss thought to have been transported south during an earlier glaciation.
	Paleozoic bedrock	

each of these glaciations, northward-flowing drainage was ponded during advance and retreat, with the result that tills are commonly separated by lake sediment.

Lennoxville Till

The Lennoxville glacier entered the Megantic area from the northwest and north, but a broad lobe was deflected south up the Chaudière Valley, around the north end of the Little Megantic Mountains (Fig. 4). This lobe rode over thick lake sediment of the Gayhurst Formation below 430 m altitude and incorporated the laminated silty clay into a very clayey till, here informally named the Drolet Lentil (Table 1; Fig. 4). As the glacier grew thick enough to cover the mountains, it flowed S. 20° E. to S. 40° E. over the whole area, removing the Drolet Lentil north and south of the obstructing ridge, but apparently shearing over

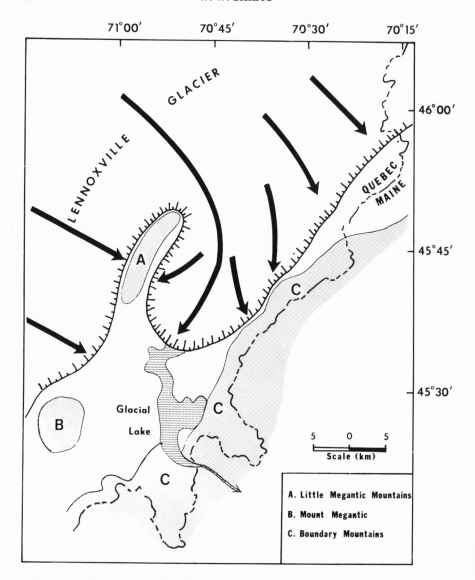

Figure 4A. Generalized ice-flow directions during deposition of Drolet Lentil Till at onset of Lennoxville glaciation; flow patterns were similar during deglaciation.

the ice of the advance lobe in the Chaudière Valley. For this reason, a portion of the Drolet Lentil was preserved.

Till fabric and striation measurements leave little doubt that the Lennoxville Till was deposited by a glacier flowing southeastward during its maximum development (Fig. 2). During deglaciation, the retreating glacier front became lobate, and ice was again deflected up the Chaudière Valley around the Little Megantic Mountains.

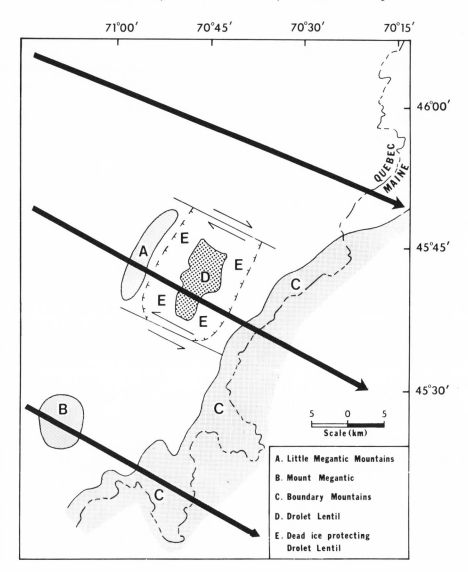

Figure 4B. Generalized ice-flow directions during maximum extent of Lennoxville glacier; note location of Drolet Lentil and inferred position of stagnant ice that protected it from erosion.

Lennoxville Till Petrology

The matrix of Lennoxville Till typically contains subequal amounts of sand, silt, and clay, in which cobbles and pebbles but few boulders occur. The Drolet Lentil commonly contains over 70 percent clay-size particles. The <2 mm fraction of Lennoxville Till has a mean grain size ($\bar{M}_z$, Folk and Ward, 1957) of 6.0 Φ (σ = 0.9 Φ), a sorting coefficient

($\bar{\sigma}_I$) of 3.7 Φ (σ = 0.4 Φ), and a kurtosis ($\bar{K}_G$) of 1.0 (σ = 0.1) (data based on 40 samples). For the Drolet Lentil, $\bar{M}_Z$ = 8.8 Φ (σ = 0.7 Φ), $\bar{\sigma}_I$ = 3.0 Φ (σ = 0.6 Φ), and $\bar{K}_G$ = 0.9 (σ = 0.1) (based on 11 samples). Lennoxville Till is olive-black to medium gray where unoxidized. It reacts vigorously with HCl below 2 m, but it generally contains less than 4 percent carbonate in the fine sand and finer fractions and less than 20 percent calcareous pebbles. The <4 μ fraction of unoxidized Lennoxville Till is composed of well-crystallized chlorite and 10 Å mica with minor amounts of quartz, feldspar, amphibole, calcite, dolomite, and serpentine. Most till is noticeably rich in pyrite derived from pyritiferous slates.

Where oxidized, Lennoxville Till is moderate yellowish-brown and noncalcareous to weakly calcareous. Chlorite in the <4 μ fraction shows variable degrees of alteration to expansible 14 Å mineral phases. In extremely oxidized samples, alteration of chlorite produces a broad diffraction band between 10 Å and 14 Å. Pyrite is altered to a rusty color or destroyed altogether.

Other mineralogical, chemical, and lithological properties of Lennoxville Till vary areally, and this predictable variation serves as a basis for the following discussion.

DISPERSAL

Glacial dispersal is discussed for eight till components for which unique sources can be defined. In addition to those components, several others were studied but are discussed only briefly. Other components include many of the pebble and cobble types listed in Table 2, rock fragments and potassium feldspars in fine sand, heavy minerals in fine sand, clay minerals, and trace elements (Cu, V, Zr, Ag, Zn, Pb) in silt and clay. Dispersal patterns of these components are not figured, either because too few sample points exist to draw meaningful dispersal maps, or because

TABLE 2. LITHOLOGIC FREQUENCIES OF PEBBLES AND SURFACE ERRATICS

Rock Type	Surface Cobbles (percent)	In-matrix Pebbles (percent)
1. Sandstone, slate, calcareous mudstone	10–95	—
2. Sandstone	—	30–70 (median = 40)
3. Slate	—	6–50 (median = 40)
4. Calcareous mudstone	—	0–30 (median = 14)
5. Granite gneiss (Canadian Shield)	1 ± 0.5	0–1
6. Vein quartz	1–10	1–12 (median = 5)
7. Chlorite schist	—	0–1
8. Conglomerate	0–1	—
9. Basic volcanic rock	0–3	0–3
10. Felsic dike rock	0–5 (rare)	0–1
11. Contact hornfels	0–14	0–5
12. Granodiorite	0–95 (see text)	0–49 (see text)
13. Gabbro	0–10 (see text)	0–3
14. Serpentinized peridotite, pyroxenite	0–50 (see text)	0–38 (median = 3)

frequency variations are too slight to be interpreted as relating to specific sources in or near the study area (maps showing concentrations of components not figured in this text are available from the author on request).

Table 2 lists the types of rock most frequently encountered in samples of clasts associated with Lennoxville Till. Sandstone, slate, calcareous mudstone, granite gneiss, and vein quartz are common to all counts. Basic volcanic rocks, conglomerates, felsic dike rocks, and contact hornfels occur locally; contact hornfels has a predictable distribution that varies directly with the frequency of granodiorite.

Frequencies of ultrabasic plus gabbroic clasts and of granodioritic clasts are summarized by isopleth maps (Figs. 5, 6, and 7). Deviations from straight-line, constant-gradient dispersal have been accounted for by consideration of one or more of the following variables: (1) sampling error (misidentification, removal from a sample site of certain rock sizes or types for construction purposes, dilution of the sample by human addition of broken local bedrock, preferential removal of certain rock types from the sample pile by the investigator, and differential weathering of erratics; these errors are considered unimportant because of regional consistency of data and demonstrated reproducibility of counts by different investigators); (2) blocking or deflection of debris by high areas; (3) southward deflection of flow up the Chaudière Valley during early and late Lennoxville glaciation; and (4) concentrated glacial erosion of certain parts of the source areas or shift in flow of basal ice that cannot be accounted for by the data presently available.

Dispersal of Ultrabasic and Gabbroic Rocks

Figure 5 depicts the frequency distribution of ultrabasic plus gabbroic cobbles collected from the Lennoxville Till surface. Source areas for these erratics are outcrops of gabbro, serpentinized periodotite, and pyroxenite that extend southwestward from Thetford Mines. The northeastern part of the ultrabasic belt is characterized by predominance of ultrabasic over gabbroic rocks; the southwestern part is characterized by the reverse.

The ultrabasic plus gabbroic dispersal pattern comprises two southeast-trending ribbons that pass north and south of the Little Megantic Mountains. Ultrabasic rocks commonly comprise 100 percent of the ultrabasic-gabbroic component in the northern ribbon, but gabbro makes up 30 to 100 percent of this component in the southern ribbon. Figure 5 illustrates ice-flow divergence around the Little Megantic Mountains and blocking of southeast-transported debris as reflected by low concentrations of ultrabasic and gabbroic indicators on the southeastern side of the mountains.

Frequencies of ultrabasic rocks decrease from greater than 50 percent near their source to less than 4 percent near the north end of the Little Megantic Mountains; the decrease is followed down-ice by an

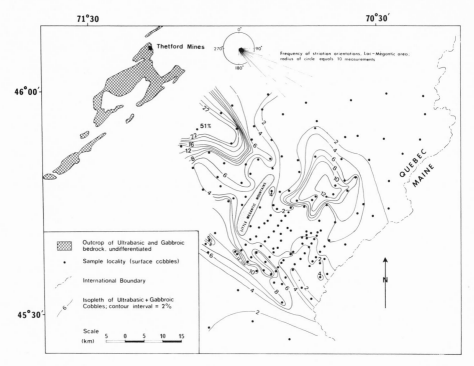

Figure 5. Dispersal pattern of ultrabasic + gabbroic cobbles collected from the surface.

abrupt increase to about 13 percent near the junction of the Chaudière and Samson Rivers. The explanation of the anomalous increase is not clear from available data. Shilts (1969) has suggested the presence of an unmapped ultrabasic source northwest of the anomalous zone, but no ultrabasic outcrops were found on detailed traverses at its northwestern edge. The common occurrence of pyroxenite cobbles within the zone also suggests that all or most of the ultrabasic erratics were derived from the Thetford Mines area where pyroxenite is common. A second explanation of the anomaly may be that the train or ribbon is so narrow that sampling density was not great enough to bring it out between the areas of high concentrations. The area between the two concentration zones is heavily wooded, and suitable sampling sites are difficult to locate.

If the separation of areas of high concentrations is not caused by either of the above possibilities, some glacial process must have enriched ultrabasic erratics in the southeastern zone. The increase may be related to local blocking by the northeastern extension of the Little Megantic Mountains of ultrabasic erratics transported in the lower portions of the glacier. Alternatively, the increase could be related to concentrations of erratics in the terminal zone of the Lennoxville glacier when it was thin and resting against the Portage Uplands. Such a "conveyor-belt" concentration of far-travelled erratics has been observed in morainal belts in Ohio (Goldthwait and others, 1961).

Other features apparent in the ultrabasic plus gabbroic dispersal pattern are: (1) bowing of concentration lines up the Chaudière Valley, probably due to late-glacial lobate flow; and (2) ribbonlike shape of the principal ultrabasic plus gabbroic dispersal trends. There is no evidence of development of fans of erratics such as those depicted for the classic boulder trains of New England (Flint, 1957, p. 126).

Dispersal of Granodioritic Cobbles and Boulders

The dispersal pattern of granodioritic cobbles collected from the surface (Fig. 6) is similar in aspect to the patterns for ultrabasic plus gabbroic rocks. The dispersal pattern has two principal, southeast-trending ribbons. The northernmost ribbon trails away from the south end of the Little Megantic Mountains. The resulting dispersal trend has two anomalously high zones separated by the Chaudière Valley. The gap between them may have been caused by late-glacial, southward deflection of the ribbon of englacial debris, or it may be the result of deposition of englacial debris on topographic prominences flanking the valley. The southern ribbon represents material eroded from granodiorite stocks west of the Little Megantic Mountains.

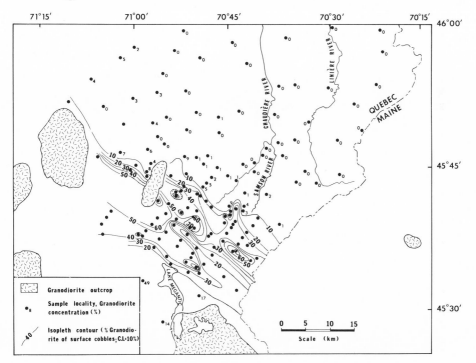

Figure 6. Dispersal pattern of granodioritic cobbles collected from the surface.

Dispersal of Granodioritic Pebbles

Although the dispersal pattern of granodioritic pebbles collected from within Lennoxville Till matrix generally reflects southeastward movement of the glacier (Fig. 7), samples collected from the Drolet Lentil consistently contain few or no granodioritic clasts. Low frequencies in Drolet Lentil samples cause the tongue-shaped depression in the northern boundary of the dispersal pattern of granodioritic pebbles.

The paucity of granodioritic pebbles in the Drolet Lentil can only have resulted from deposition from ice that did not erode any significant outcrops of granodiorite. This conclusion supports the concept that the Drolet Lentil was deposited by an ice lobe that swung *around* the Little Megantic Mountains and flowed *up* the Chaudière Valley. The Drolet Lentil is, however, mantled by a dense cover of boulders and cobbles that contains high percentages of granodioritic clasts. Therefore, it is concluded that in the lee of the Little Megantic Mountains, dispersal of granodioritic clasts proceeded in the following sequence, as depicted schematically in Figure 8 (*see also* Fig. 4):

1. Clay till of the Drolet Lentil was deposited along the axis of the Chaudière Valley by a lobe of the advancing Lennoxville glacier.

2. The Lennoxville glacier overrode the Little Megantic Mountains

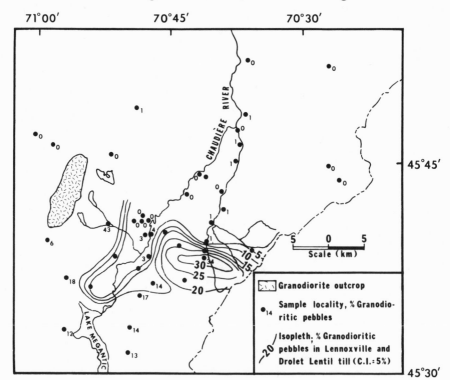

Figure 7. Dispersal pattern of granodioritic pebbles collected from subsurface exposures of Lennoxville Till and Drolet Lentil Till.

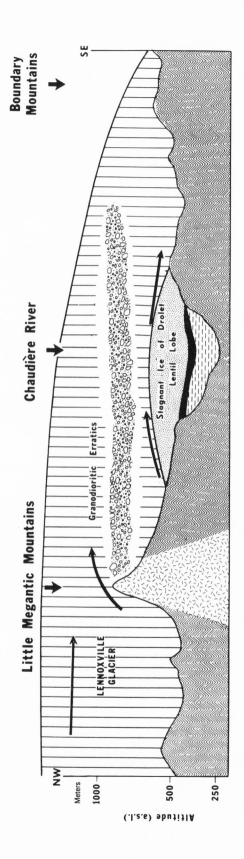

Figure 8. Diagrammatic cross section of Lennoxville glacier showing probable relationship of englacial debris that formed ablation mantle to Drolet Lentil clay till (*see also* Figs. 4, 6, and 7).

and sheared over the stagnant remnants of the lobe in the Chaudière Valley.

3. The Lennoxville glacier flowed southeastward, across the Boundary Mountains and transported granodioritic clasts, derived from the Little Megantic Mountains, across the Chaudière Valley into Maine.

4. As the Lennoxville glacier retreated, granodioritic clasts were let down onto the surface of the Drolet Lentil, forming an ablation mantle. The ablation mantle is visualized by the author as a thin, matrix-free equivalent of ablation till as defined by Flint (1957, p. 120-122).

Ablation Mantle

A thin mantle of boulders and cobbles occurs on the surface of Lennoxville Till everywhere in the Megantic region. Similar mantles have been described by Flint (1957, p. 118-120) as lag concentrates formed by sheet wash or deflation of a till surface. Frost jacking also causes development of boulder mantles in the highlands of New England, and fields cleared of boulders during one season become mantled by boulders jacked to the surface in succeeding years.

Examination of numerous vertical exposures of Lennoxville and Drolet Lentil Tills indicates that, except in the mountains near the border, these units contain few clasts larger than cobbles. Therefore, it is not reasonable to assume that the large clasts that form a dense mantle over the youngest tills could have been concentrated from till by postdepositional processes. It is concluded that the boulder mantle that covers surface till of the Megantic area was deposited during the ablation phase of Lennoxville glaciation. It is suggested here that ablation mantle may be more widespread than previously thought and should be considered as an alternate interpretation of surface concentrations of boulders previously considered to be lag deposits.

Magnetite Dispersal

Obvious sources of anomalous concentrations of magnetite are ultrabasic rocks in the gabbroic-ultrabasic belt that passes through Thetford Mines (Fig. 3). McDonald (1966, 1967a) has already demonstrated magnetite dispersal trends southeast of ultrabasic terranes in the Sutton (Green) Mountains and southeast of the nordmarkite and gabbro at Mount Megantic.

Figure 9 shows high magnetite contents in belts trending southeast from the main mass of ultrabasic rocks in the Thetford Mines region. High magnetite values also extend southeastward from the southern end of the Little Megantic Mountains, but the high magnetite concentrations may partially reflect transport of ultrabasic debris around the obstructing ridge.

Samples from the Drolet Lentil and a few samples of Lennoxville

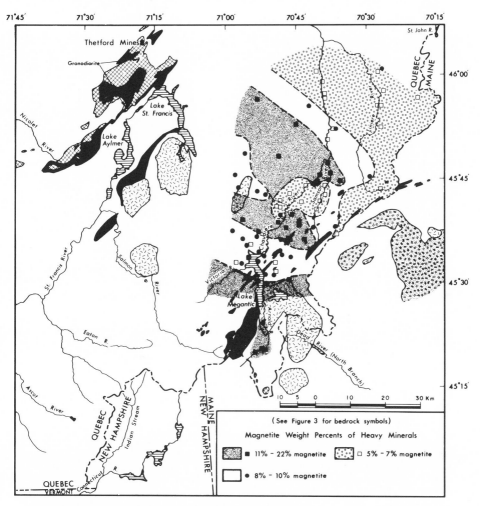

Figure 9. Dispersal pattern of magnetite concentrated from heavy-mineral separates from the fine-sand fraction of Lennoxville and Drolet Lentil Tills.

Till in the Samson River Valley have an abnormally low magnetite content, suggesting affinity with Lennoxville Till occurring north of the magnetite dispersal shadow.

Feldspar Dispersal

Granite, granodiorite, syenite, and gabbro are potential sources of plagioclase and potassium feldspar. Both feldspars together comprise only about 20 to 25 percent of sand-size detritus in the metasediments that underlie most of the area (Marleau, 1968).

Figure 10 shows the distribution of feldspar percentages. Samples located north of a line extending to the southeast from the northern end

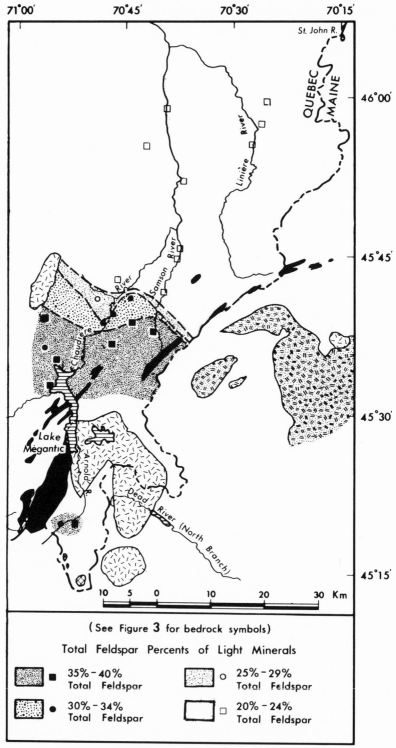

Figure 10A. Dispersal pattern of total feldspar (Ca−Na+K) in fine-sand fraction of Lennoxville and Drolet Lentil Tills.

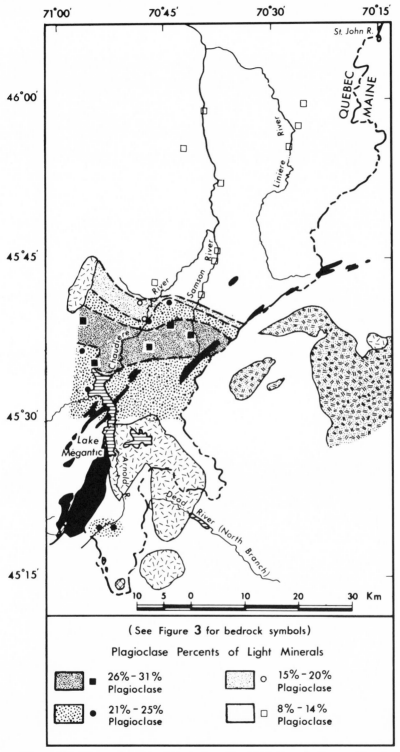

Figure 10B. Dispersal pattern of plagioclase (Ca−Na) feldspar in fine-sand fraction of Lennoxville and Drolet Lentil Tills.

of the Little Megantic Mountains contain 20 to 24 percent total feldspar (Fig. 10A)—a range that agrees well with the average content of underlying metasediments. Samples located southeast of the granodiorite that forms the core of the Little Megantic Mountains contain 1.5 to 2.0 times as much feldspar as samples located farther north. The increase in feldspar percentage is largely due to an increase in plagioclase derived from granodiorite (Fig. 10B).

Feldspar contents of the samples plotted on Figure 10 indicate that the Lennoxville glacier transported feldspar grains at least 16 km southeast of their source in the Little Megantic Mountains.

Trace-Element Dispersal

Concentrations of Cr and Ni (Figs. 11 and 12) from the <63 μ (silt + clay) fraction of Lennoxville Till are clearly related to source areas

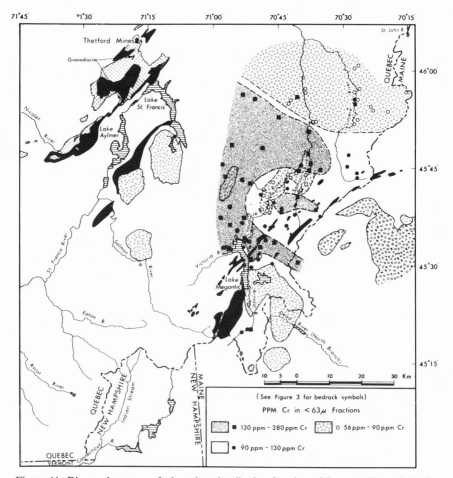

Figure 11. Dispersal pattern of chromium in silt-clay fraction of Lennoxville and Drolet Lentil Tills.

northwest of the study area, most logically the Thetford Mines ultrabasic complex. Dispersal patterns for both of these elements are similar to those of ultrabasic surface cobbles and magnetite. Chromium dispersal (Fig. 11) shows the "blocking" effect of the Little Megantic Mountains. Till with high chromium content generally has high amounts of ultrabasic or basic volcanic, gabbroic, and ultrabasic pebbles. The chromium dispersal pattern also shows the lack of reworking of the Drolet Lentil by later (southeast) Lennoxville glacier flow.

The dispersal pattern for nickel (Fig. 12) is similar to that of Cr. Ni concentrations in samples located 55 km from the presumed Ni source area are as high or higher than Ni concentrations determined by Kauranne (1959, p. 3) in till only 0.5 to 1.0 km down-ice from a major nickel ore body in Finland (Ni ≈ 100 ppm).

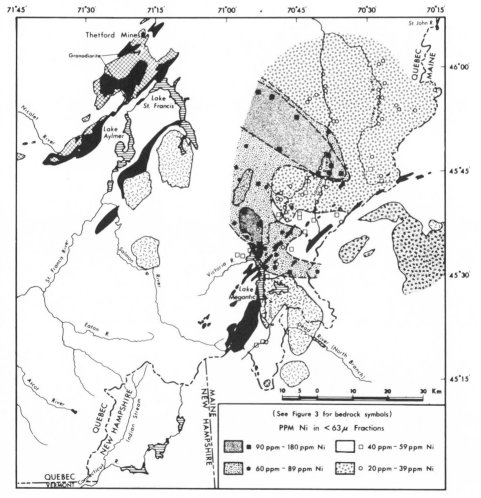

Figure 12. Dispersal pattern of nickel in silt-clay fraction of Lennoxville and Drolet Lentil Tills.

Cu, V, Ti, Zr, Zn, and Pb were also analyzed for all samples from the study area. Ti (Fig. 13) seems to be concentrated in the region of the ultrabasic fans and in the lee of the syenitic intrusion at Mount Megantic. Other trace elements show little variation and cannot be grouped in any meaningful way on a dispersal map.

Textural Control of Trace-Element Concentration

Texture plays an important role in controlling apparent trace-element concentration; the clay fraction tends to adsorb cations and to release them readily during artificial or natural leaching. In effect, clay is the chemically active part of a sediment or soil, and the amount of clay

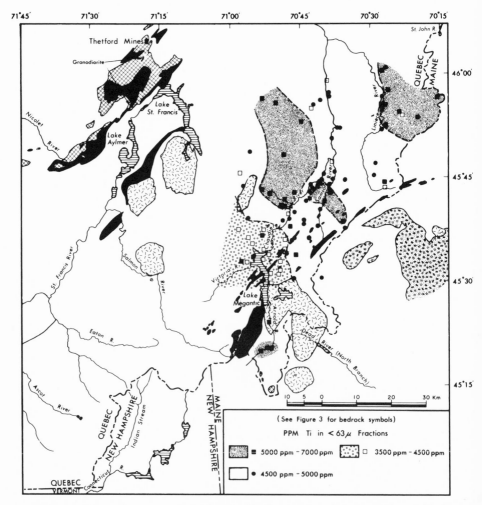

Figure 13. Dispersal pattern of titanium in silt-clay fraction of Lennoxville and Drolet Lentil Tills.

in a silt-clay mixture strongly controls apparent cation concentration either determined spectrographically or by atomic absorption.

Figure 14 illustrates textural control on apparent trace-element concentration. All elements except Zr show a tendency to be concentrated preferentially in the clay-size fraction. Zr is concentrated in the silt fraction. Secondary trends on the Cr and Ni diagrams include samples

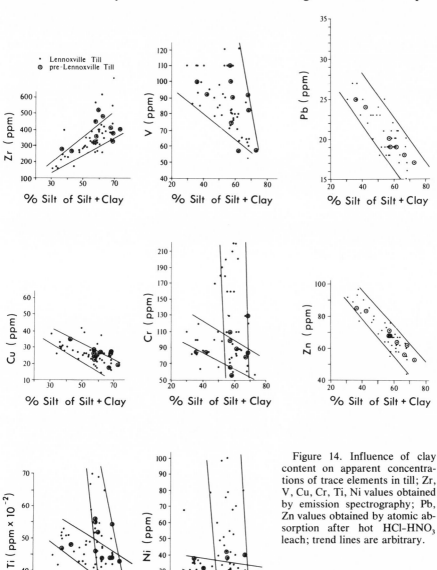

Figure 14. Influence of clay content on apparent concentrations of trace elements in till; Zr, V, Cu, Cr, Ti, Ni values obtained by emission spectrography; Pb, Zn values obtained by atomic absorption after hot HCl-HNO$_3$ leach; trend lines are arbitrary.

from the dispersal fans of ultrabasic debris (Figs. 11 and 12). Most of the samples in the secondary trend of Ti are located either southeast of the ultrabasic outcrops at Thetford Mines or southeast of the syenitic rock at Mount Megantic. Cr, Ni, and Ti concentrations are texture independent for samples from within the dispersal trends of these elements. The lack of clear secondary trends for other elements suggests that no sources for these elements were detected by this sampling pattern, although several small sulfide ore bodies are known to exist in the region.

It is suggested that regional trace-element concentrations for till should be interpreted with caution and that apparent variations in concentration be rationalized with respect to texture (and clay mineralogy) before interpretations are made and background or threshold levels established.

DISCUSSION

Ice-Movement Directions

Lithologic, mineralogic, and elemental components that could be related to unique source areas reflect southeastward flow of the Lennoxville glacier. Many of the components studied could have been used to substantiate striation and fabric data or to establish ice-flow directions in lieu of these data.

The significance of the Drolet Lentil has been clarified by comparison of its petrologic characteristics to those of Lennoxville Till. The southwestward trend of its outcrop area, its southwestern fabric, and southwestward striations on bedrock directly under it only indicate that it was deposited by an early or late ice lobe flowing to the southwest along the axis of the depression between the Little Megantic Mountains and the Boundary Mountains. Mineralogic and geochemical similarity of the Drolet Lentil to till deposited outside the principal granodiorite and ultrabasic dispersal trends permits further interpretation of the Drolet Lentil as an early Lennoxville glacial deposit.

The presence of high percentages of granodioritic erratics on the surface of the Drolet Lentil confirms that the Lennoxville glacier flowed southeastward over it. Furthermore, granodioritic debris must have been englacial or supraglacial to have been let down onto and not reworked into the clay till.

Topographic Influence on Dispersal Patterns

Topographic characteristics of the study area influenced dispersal patterns of various components studied. The most obvious manifestation of topographic influence is the blocking effect by prominent ridges or hills. These features tended to catch ice-transported debris on their stoss sides, causing anomalously low concentrations of indicator components in their lee.

The Little Megantic Mountains have blocked components originating in the ultrabasic-basic belt to the northwest. The dispersal patterns of ultrabasic and gabbroic rocks, Ni, Cr, and magnetite all show leeside gaps resulting from blocking by the Little Megantic Mountains (Figs. 5, 9, 11, and 12).

Late Glacial Highland Centers of Outflow

Till components studied show no evidence of northward or westward deflections from the highlands of western Maine and such distinctive indicators as Attean granite were not identified in Quebec. Among several authors, Flint (1951), Duquette (1960b), and Borns and Hagar (1965) have suggested that active, late-glacial ice caps in the Maine and New Hampshire highlands may have flowed into Quebec. No evidence to support these ideas was found during this study; the evidence presented indicates that movement in the study area during the last glaciation was southeastward with early and late southwest deflection by valleys and ridges. Lamarche (1971) cites good evidence of local northward movement in the Thetford Mines area. Late northward movement of a remnant ice mass left between the Green–Notre Dame Mountains and Boundary Mountains may have occurred in a broad east-west band north of the study area.

Stratigraphic Implications

In the Megantic region comparative studies of the petrographic properties of surface and older tills have been particularly useful. The penultimate (Chaudière) glaciation was characterized by ice flow from the north or northeast with the result that, for most of the region, the till of the older glaciation bears few of the components derived from ultrabasic or granodioritic sources (Table 3).

Sections near the mouth of the Samson River illustrate the compositional disparity between Chaudière Till, largely deposited by ice that flowed southward or southwestward, and Lennoxville Till, deposited by ice that flowed southeastward (Table 3). Sections B and C are located in the ultrabasic dispersal fan for Lennoxville Till; Chaudière Till in these sections has none of the ultrabasic components.

CONCLUSIONS

Major conclusions of this investigation are:

1. Most till components studied—granodioritic and ultrabasic plus gabbroic cobbles and pebbles, magnetite, plagioclase, and trace elements

TABLE 3. PETROLOGIC VARIATIONS OF TILL IN SELECTED STRATIGRAPHIC SECTIONS

Location	Sample Number/ Stratigraphic Unit/ Depth below Surface (m)	K	P	H	M	Cu	Ti	Cr	Ni	V	Zr	Pb	Zn	Ag
Chaudière River, St. Martin 70°39.4'W. 45°59.4'N.	9/Lennoxville Till/4.0	3	17	2.2	5	42	4300	110	29	84	280	23	68	0.8
	65/Lennoxville Till/9.0	9	11	1.8	5	25	4500	66	30	98	290	24	69	0.7
	8/Chaudière Till/40.0	13	15	1.3	4	27	4400	89	24	57	480	19	64	0.7
Samson River* 70°37.8'W. 45°46.3'N.	0130/Lennoxville Till/1.0	9	11	2.8	17	38	4400	190	110	84	340	23	66	0.6
	0117/Lennoxville Till/4.0	—	—	—	—	32	4200	110	88	77	380	18	57	0.6
	0109/Lennoxville Till/6.0	5	16	2.4	7	35	4400	130	67	86	400	23	68	0.8
	095/Lennoxville Till/8.5	—	—	—	—	33	5100	83	64	91	370	20	65	0.8
	77/Chaudière Till/23.5	10	18	2.5	4	19	4300	57	21	57	400	17	53	0.7
	118/Chaudière Till/24.5	11	16	3.9	3	22	4600	65	23	74	360	19	68	0.7
Samson River* 70°38'W. 45°45.3'N.	27B/Lennoxville Till/12.0	4	8	3.0	13	25	4900	180	99	90	350	25	64	0.8
	27C/Chaudière Till/22.0	11	16	2.0	4	35	4800	84	32	92	270	24	83	0.7
Samson River 70°40'W. 45°42'N.	25A/Lennoxville Till/8.0	9	13	1.8	4	38	5100	82	34	78	620	19	58	0.7
	25B/Lennoxville Till/10.0	12	12	2.1	3	19	5300	76	37	73	570	19	55	0.6
Eugénie River 70°46.5'W. 45°43.5'N.	30/Drolet Lentil/3.0	10	12	1.7	5	29	5200	82	36	93	250	24	82	0.9
	29A/Chaudière Till/15.0	7	12	2.3	7	27	4400	84	40	82	330	20	62	0.7
Gayhurst Dam 70°48'W. 45°39.9'N.	45A/Drolet Lentil/3.0	11	20	1.9	9	26	4200	82	31	86	160	27	89	1.0
	S24/Chaudière Till/98.0	10	10	4.2	5	19	4300	92	20	72	330	17	62	0.6
	S26/Chaudière Till/105.5	9	10	4.7	5	30	5100	110	34	100	270	20	76	0.7
Arnold River 70°52.5'W. 45°20.6'N.	53/Ablation facies/5.0	23	23	7.5	18	32	5500	60	42	94	390	22	67	0.7
	75/Lennoxville Till/13.5	13	30	12.7	21	23	7000	110	46	110	390	19	66	0.9
	74/Chaudière Till/14.5	20	22	8.2	12	28	5600	110	42	110	320	20	71	0.8
	73/Chaudière Till/18.0	20	19	8.5	13	26	5500	99	38	100	450	20	68	0.8
Northwest Lac-Mégantic* 70°54.7'W. 45°35.6'N.	S30/Lennoxville Till/7.0	—	—	—	—	75	3900	260	89	73	260	18	80	1.2
	S32/Lennoxville Till/10.0	—	—	—	—	27	4400	210	82	82	280	19	71	1.0
	S35/Lennoxville Till/15.0	—	—	—	—	22	4300	270	77	76	200	19	68	1.0
	S38/Lennoxville Till/21.0	—	—	5.5	7	24	3800	230	68	71	300	19	58	1.0
	S48/Lennoxville Till/41.0	11	19	4.3	5	17	4000	220	55	77	270	20	61	0.8
Route 24 70°46.6'W. 45°41'N.	134/Drolet Lentil/2.0	—	—	2.0	5	30	4900	75	38	99	240	23	86	0.8
	134A/Lennoxville Till/2.0	—	—	3.2	13	26	4500	190	66	89	380	24	75	0.7

K-percentage of potassium feldspar in light-mineral separate of fine sand, exclusive of rock fragments; P-plagioclase percentage, same qualification as "K"; H-heavy minerals, percentage in weight of fine sand; M-magnetite, percentage in weight of heavy minerals from fine sand; Cu, Ti, Cr, Ni, V, Zr, Pb, Zn, Ag-concentrations of elements in silt and clay fraction (ppm); heavy line separating stratigraphic units represents lake sediments; *stratigraphic sections located in dispersal fan of ultrabasic rocks and minerals for Lennoxville Till.

(Cr, Ni, Ti)—indicate southeastward movement of the Lennoxville glacier. Lack of evidence of northward or westward displacement of components eliminates the possibility of significant late-glacial flow from highland ice caps in Maine, New Hampshire, or Quebec. Unfortunately, sample points were not located far enough north or west to shed light on the northward movement described by Lamarche (1971). The Drolet Lentil was deposited while a Lennoxville glacier lobe was advancing up the Chaudière Valley at the onset of Lennoxville glaciation. The glacier lobe that deposited the Drolet Lentil was overridden by later Lennoxville ice and protected the lentil in the lee of the Little Megantic Mountains.

Late-glacial, southwesterly flow up the Chaudière Valley displaced englacial or supraglacial erratics southward, resulting in up-valley bowing of their concentration lines.

2. The Lennoxville glacier carried debris in zones above its base. During melting, this coarse debris was let down onto the surface of Lennoxville and Drolet Lentil lodgment till, forming a thin ablation mantle.

3. Topographic prominences blocked both basal and englacial sediment. Components derived from the Thetford Mines ultrabasic belt were blocked by the Little Megantic Mountains.

4. Chromium, nickel, and magnetite have anomalous concentrations in till at least 50 km down-ice from their source areas.

5. Trace elements can present as consistent and reliable a picture of ice-flow history in the Lac-Mégantic region as can boulders or mineral components. Although trace elements are easy to sample for and their concentrations are generally not susceptible to subjective interpretation, their distribution should be evaluated in the light of other petrographic and stratigraphic data.

6. The concentrations of trace elements are at least partially influenced by the texture of samples studied. Zr is preferentially concentrated in the silt fractions (4 μ to 63 μ) and Cu, V, Zn, Pb, Ti, Cr, and Ni occur preferentially in the $< 4 \mu$ fraction, but concentrations of Cr, Ni, and Ti are texture-independent in samples from dispersal fans of those elements. Texture is an important variable that should be evaluated when establishing background values and interpreting trace-element anomalies in till.

7. Dispersal trends in the Lac-Mégantic area are ribbon-shaped in plan view; although the maximum limits of distribution of indicators from a source area may be fan shaped, the principal concentrations of most components examined lie along narrow, subparallel ribbons or bands.

ACKNOWLEDGMENTS

This paper is partially abstracted from a Ph.D. thesis presented at Syracuse University. Research was supported by the Geological Survey of Canada and by a postdoctoral grant from the National Research Council of Canada.

Doctors G. M. Boone, B. C. McDonald, E. H. Muller, J. H. Hartshorn, G. W. White, and H. E. Wright, Jr., provided helpful criticism and suggestions for improvements of the manuscript, but the author assumes full responsibility for conclusions drawn. Marcel Ouelett and Robert Bélanger competently assisted the author in field determinations of clast frequencies.

Trace elements were analyzed by the Analytical Chemistry Section of the Geological Survey of Canada and by Bondar-Clegg and Company, Ltd.

REFERENCES CITED

Boone, G. M., and Weaver, E. P., 1968, Staining for cordierite and feldspars in thin section: Am. Mineralogist, v. 53, p. 327-331.

Borns, H. W., Jr., and Hagar, D. J., 1965, Late-glacial stratigraphy of a northern part of the Kennebec River valley, western Maine: Geol. Soc. America Bull., v. 76, p. 1233-1250.

Cooke, H. C., 1937a, Thetford, Disraeli, and eastern half of Warwick map-areas, Quebec: Canada Geol. Survey Mem. 211, 160 p.

_____ 1937b, Further note on northward moving ice: Am. Jour. Sci., v. 34, p. 221.

_____ 1950, Geology of a southwestern part of the Eastern Townships of Quebec: Canada Geol. Survey Mem. 257, 142 p.

Donovan, P. R., and James, C. H., 1967, Geochemical dispersion in glacial overburden over the Tynagh (Northgate) base metal deposits, west-central Eire: Canada Geol. Survey Paper 66-54, p. 80-110.

Doyle, R. C. (compiler and editor), 1967, Preliminary geologic map of Maine: Maine Geol. Survey.

Dreimanis, A., 1956, Steep Rock iron ore boulder train: Geol. Assoc. Canada Proc., v. 8, pt. 1, p. 27-70.

Dreimanis, A., and Vagners, J., 1969, Lithologic relation of till to bedrock: Canada Natl. Acad. Sci. Pub. 1701, p. 93-98.

Duquette, G., 1960a, Preliminary report on Weedon area: Quebec Dept. Mines Prelim. Rept. 416, 9 p.

_____ 1960b, Preliminary report on Gould area: Quebec Dept. Mines Prelim. Rept. 432, 10 p.

_____ 1961, Preliminary report on Lake Alymer area: Quebec Dept. Mines Prelim. Rept. 457, 13 p.

Flint, R. F., 1951, Highland centers of former glacial outflow in northeastern North America: Geol. Soc. America Bull., v. 62, p. 21-38.

_____ 1957, Glacial and Pleistocene geology: New York, John Wiley & Sons, Inc., 553 p.

Folk, R. L., and Ward, W. C., 1957, Brazos River bar: a study in the significance of grain size parameters: Jour. Sed. Petrology, v. 27, p. 3-26.

Gillberg, G., 1964, Till distribution and ice movements on the northern slopes of the South Swedish highlands: Geol. Fören. Stockholm Förh., v. 86, p. 433-483.

_____ 1967, Further discussion on the lithological homogeneity of till: Geol. Fören. Stockholm Förh., v. 89, p. 29-49.

Goldthwait, R. P., White, G. W., and Forsyth, J. L., 1961, Glacial map of Ohio: U.S. Geol. Survey Misc. Geol. Inv. Map I-316.

Gorman, W. A., 1955, Preliminary report on St. Georges-St. Zacharie area: Quebec Dept. Mines Prelim. Rept. 314, 5 p.

Hawkes, H. E., and Webb, J. S., 1962, Geochemistry in mineral exploration: New York, Harper and Row, 415 p.

Kauranne, L. K., 1957, On prospecting for molybdenum on the basis of its dispersion in glacial till: Bull. Comm. Géol. Finlande, no. 180, p. 31-43.

_____ 1959, Pedogeochemical prospecting in glaciated terrain: Bull. Comm. Géol. Finlande, no. 184, p. 1-9.

Kvalheim, A., 1967, Geochemical prospecting in Fennoscandia: New York, Interscience, 349 p.

Lamarche, R. Y. 1971, Northward moving ice in the Thetford Mines area of southern Quebec: Am. Jour. Sci., v. 271, p. 383–388.

Lee, Hulbert A., 1963, Glacial fans in till from the Kirkland Lake fault: Canada Geol. Survey Paper 63-45, 36 p.

Lord, C. S., 1938, Megantic sheet (west half), Frontenac County, Quebec: Canada Geol. Survey Map 379A.

Marleau, R. A., 1968, Woburn-East Megantic-Armstrong area, Frontenac and Beauce Counties: Quebec Dept. Nat. Resources Geol. Rept. 131, 55 p.

McDonald, B. C., 1966, Auriferous till in the Eastern Townships, southeastern Quebec: Canada Geol. Survey Paper 66-2, p. 51–54.

——— 1967a, Pleistocene events and chronology in the Appalachian region of southeastern Quebec, Canada [Ph.D. dissert.]: New Haven, Yale Univ., 161 p.

——— 1967b, Wisconsin stratigraphy and ice-movement directions in southeastern Quebec, Canada [abs.]: Geol. Soc. America, Special Paper 115, p. 277.

McDonald, B. C., and Shilts, W. W., 1971, Quaternary stratigraphy and events in southeastern Quebec: Geol. Soc. America Bull., v. 76, p. 683–697.

Pawluk, S., and Bayrock, L. A., 1969, Some characteristics and physical properties of Alberta tills: Research Council Alberta Bull. 26, 72 p.

Potter, P. E., and Pettijohn, F. J., 1963, Paleocurrents and basin analysis: New York, Academic Press Inc., 296 p.

Reid, A. M., 1960, Preliminary report on the geology of Mount Megantic: Quebec Dept. Mines Prelim. Rept. 533, 6 p.

Shilts, W. W., 1969, Quaternary geology of the upper Chaudière River drainage basin, Quebec: Canada Geol. Survey Paper 69-1A, p. 218–220.

——— 1970, Pleistocene geology of the Lac-Mégantic region, southeastern Quebec, Canada [Ph.D. dissert.]: Syracuse, N.Y., Syracuse Univ., 154 p.

Sitler, R. F., 1963, Petrology of till from northeastern Ohio and northwestern Pennsylvania: Jour. Sed. Petrology, v. 33, p. 365–379.

St. Julien, P., 1963-1965, Unpublished maps of Thetford Mines-Disraeli area: on file, Quebec Dept. Nat. Resources, Quebec.

White, G. W., Totten, S. M., and Gross, D. L., 1969, Pleistocene stratigraphy of northwestern Pennsylvania: Pennsylvania Geol. Survey Gen. Geol. Rept. G55, 88 p.

Willman, H. B., Glass, H. D., and Frye, J. C., 1963, Mineralogy of glacial tills and their weathering profiles in Illinois, pt. I, Glacial tills: Illinois Geol. Survey Circ. 347, 55 p.

——— 1966, Mineralogy of glacial tills and their weathering profiles in Illinois, pt. II, Weathering profiles: Illinois Geol. Survey Circ. 400, 76 p.

MANUSCRIPT RECEIVED BY THE SOCIETY DECEMBER 27, 1971

Printed in the United States of America

GEOLOGICAL SOCIETY OF AMERICA
MEMOIR 136
© 1973

Differentiation of Glacial Tills in Southern Ontario, Canada, Based on Their Cu, Zn, Cr, and Ni Geochemistry

R. W. MAY

A. DREIMANIS

*Department of Geology, University of Western Ontario,
London, Ontario, Canada*

ABSTRACT

Geochemical investigations were performed on 109 samples of late Wisconsin glacial till from southern Ontario. The -0.037 mm size fraction was analyzed for Cu, Zn, and Cr by atomic absorption spectrophotometry and for Ni by colorimetry. Means and standard deviations (in ppm) for 109 samples (group 1) are: Cu, $\bar{x} = 23$, $\sigma = 6$; Zn, $\bar{x} = 62$, $\sigma = 14$; Cr, $\bar{x} = 51$, $\sigma = 15$; and Ni, $\bar{x} = 28$, $\sigma = 8$. The statistics were also computed for the following groups of samples: (2) oxidized till (67 samples); (3) unoxidized till (42 samples); (4) Erie Lobe till (22 samples); (5) Ontario Lobe till (13 samples); (6) Huron Lobe till (24 samples); and (7) Georgian Bay Lobe till (11 samples). Cr and Ni are slightly higher in group 3 than in group 2. Results of t and F tests performed on groups 4, 5, 6, and 7 suggest that the samples are derived from two populations; one characterized by the Erie Lobe samples (population E) and the other by samples from the Georgian Bay and Huron Lobes (population N). Population E has a higher content of all four elements than population N.

Results of a discriminant analysis suggest that the Ontario Lobe samples are derived from both populations. Three samples from the northwestern end of the Ontario Lobe show definite affinities with population N, whereas the others belong to population E. This difference suggests that the glacier which deposited the former till may have incorporated drift material from a northern source.

This preliminary investigation suggests that the application of multi-variate statistical techniques to geochemical analysis of selected fractions of glacial deposits could provide another basis for differentiation of glacial tills.

INTRODUCTION

Numerous criteria have been used to characterize, distinguish, or corre-late individual tills. Many deal with the lithologic composition, often by investigation of selected size fractions. Pebbles, sand, combined silt and clay, and clay have all been used. Chemical composition, except for quantitative carbonate determinations, has seldom been utilized. Analysis of till for selected elements has been done for specific purposes: (1) geochemical prospecting for ore deposits (Kvalheim, 1967); (2) determi-nation of plant nutrients and other significant constituents of till as a parent material of soils (Bear, 1964); and (3) determination of toxic elements (Warren and others, 1967).

One of the most extensive regional investigations of tills has been done recently in Alberta by Pawluk and Bayrock (1969), who determined 10 chemical parameters in 475 till samples from an area of 170,000 sq mi (an average of one sample per 360 sq mi). They concluded that "the minor elements iron, boron, cobalt, copper, zinc, and molybdenum, present in tills of Alberta appear to have originated from a common source, and are principally associated with the clay-size fraction." No attempt was made to relate geochemical differences of tills to different glacial movements, nor were tills differentiated on the basis of the chemi-cal parameters. Forslev (1957) analyzed the silt-clay fraction of unleached samples of late Wisconsin tills deposited by the Des Moines, Lake Michigan, Saginaw, and Erie Lobes. Des Moines Lobe tills contain more Zr, Ba, and Mn, and less Sc than the others; Lake Michigan Lobe tills contain more Ti and Al and less Sr; Saginaw and Erie Lobe tills have more Ca and Sr; and Erie Lobe tills more V. It was concluded that the chemical composition of the fine fraction of tills could prove useful in distinguishing tills having different source areas.

This study attempts to test this conclusion in southern Ontario where previous till investigations by Dreimanis and others (1957), Dreimanis (1965), Karrow (1963, 1967), and other workers have indicated the pres-ence of lithologically different tills. As the chemical composition of till is a function of its lithology, it was hoped that a fast and simple method for differentiation of tills could be established by selecting those trace elements which show the greatest differences between tills deposited by different lobes, and by analyzing that size fraction where these elements are most abundant (their terminal grades, according to the usage of Dreimanis, 1969). Multivariate statistical techniques are utilized to evaluate the results of the analyses.

Composition of till depends greatly upon the composition of the rocks and sediments that were overridden by the glacier. Bedrock (Fig. 1) in southern Ontario ranges in age from Precambrian to Mississippian. The Paleozoic bedrock is dominantly limestone and dolostone. Shale is abundant in an area east of the Niagara escarpment and in part of southwestern Ontario.

Late Wisconsin tills in southern Ontario were deposited by glacial lobes which flowed down four major bedrock depressions—Ontario, Erie, Huron, and Georgian Bay—and outward onto the surrounding areas. Although smaller glacial lobes existed, this study is concerned with deposits from the four major lobes (Dreimanis and Goldthwait, this volume, p. 71-105).

SAMPLING AND LABORATORY PROCEDURES

Sampling locations are shown in Figure 2. Samples of both oxidized and unoxidized tills were taken generally at least 0.5 to 1.0 m below the B horizon of the soil. Secondary carbonates and iron and manganese oxide coatings were cleaned from the fracture surfaces where necessary.

Atomic absorption spectrophotometry (Angino and Billings, 1967) was used to analyze the −0.037 mm size fraction for Cu, Zn, and Cr. Ni analyses were done colorimetrically using a modification of a method given by Stanton (1966). Analytical error was found to be less than 10 percent of the amount present. The choice of the −0.037 mm size fraction was made after a preliminary semiquantitative investigation of 13 till samples which were wet-sieved into 2.0 to 1.0, 1.0 to 0.500, 0.500 to 0.250, 0.250 to 0.125, 0.125 to 0.063, 0.063 to 0.037, and −0.037 mm fractions, and analyzed for Cu, Zn, and Ni. The −0.037 mm fraction was found to contain a detectable amount of each element and yielded consistent results.

Choice of elements was made after examination of published data on trace-element content of the possible source rocks for the tills (Warren and Delavault, 1961; Shaw and others, 1967; Fahrig and Eade, 1968).

ANALYTICAL RESULTS

Table 1 gives a summary of the means and standard deviations of the seven groups considered. The only apparent consequence of oxidation is in the Cr analyses and possibly the Ni analyses. However, fewer unoxidized samples than oxidized ones were analyzed, and the unoxidized samples include 13 from the Erie Lobe, which tend to have a higher Cr and Ni content than the others. For these reasons, discussion of the four lobes includes analyses of both oxidized and unoxidized tills.

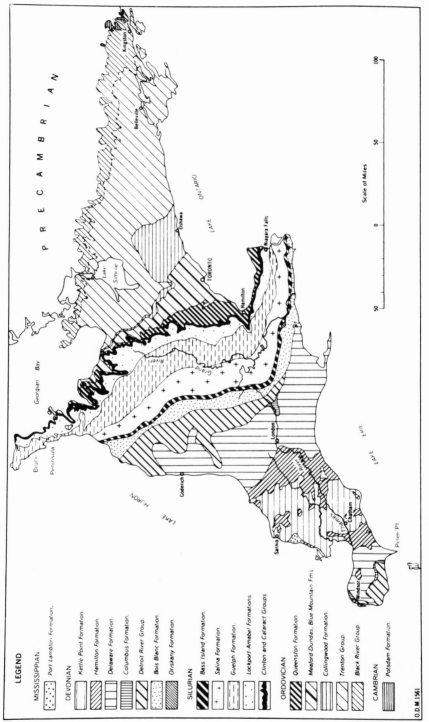

Figure 1. Bedrock geology of southern Ontario (after Hewitt, 1964).

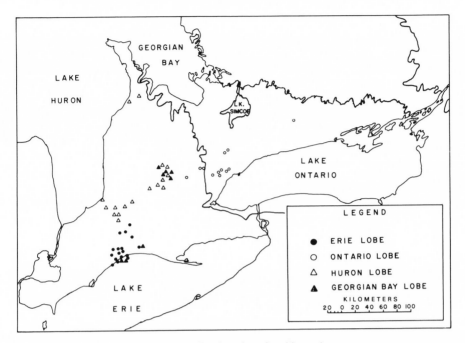

Figure 2. Sampling locations for this study.

In general, the Erie Lobe tills contain more Ni, Cu, Zn, and Cr (in order of less difference) than the Huron–Georgian Bay Lobe tills. The Ontario Lobe samples are similar to the Erie Lobe samples but have a relatively larger standard deviation.

STATISTICAL ANALYSIS

F and t tests (Neville and Kennedy, 1964) on each variable indicate the existence of two populations: one characterized by samples from the Erie Lobe and the other characterized by samples from the Georgian Bay and Huron Lobes. Visual inspection of the results from the Ontario Lobe suggests that these samples may be a mixture of both populations. This hypothesis would account for the large standard deviations encountered in Ontario Lobe results. Multivariate discriminant analysis (Klovan and Billings, 1967) was used to quantitatively test this hypothesis.

Discriminant analysis considers all variables simultaneously, unlike F and t tests which operate on one variable at a time. It is thus possible to test further the validity of the assumption of two different groups, Erie and Huron–Georgian Bay, considering all four variables simultaneously. This assumption was found to be valid. The next step in the procedure is one of decision making. Using the chemical analyses from the two known groups, the Erie and the Huron–Georgian Bay Lobes, a decision function or discriminant function is derived. This function,

$$D = 0.36 \text{ Cu} + 0.03 \text{ Zn} - 0.01 \text{ Cr} + 0.93 \text{ Ni},$$

TABLE 1. SUMMARY OF GROUP STATISTICS

Group	(No. of samples)		Cu	Zn	Cr	Ni
1	All samples	$\bar{x}$	23	62	51	28
	(109)	σ	3	14	15	9
2	Oxidized	$\bar{x}$	22	61	46	27
	(67)	σ	7	15	14	9
3	Unoxidized	$\bar{x}$	24	63	59	31
	(42)	σ	5	12	16	7
4	Erie Lobe	$\bar{x}$	25	66	63	38
	(22)	σ	3	6	17	5
5	Huron Lobe	$\bar{x}$	21	58	46	24
	(24)	σ	3	10	10	5
6	Georgian Bay Lobe	$\bar{x}$	19	54	52	24
	(11)	σ	4	8	22	5
7	Ontario Lobe	$\bar{x}$	30	75	56	31
	(13)	σ	10	15	15	13

Groups 1, 2, and 3 include some samples not in groups 4 to 7 because sufficient stratigraphic information was not available for a definitive lobal assignment.

is a linear combination of the four variables studied. When values for a sample are substituted into the function, the value D obtained can be used to classify that particular sample into one or other of the two populations, depending on whether it is above or below a particular critical value D_0. This parameter is obtained by substituting into the discriminant function the mean values of the variables for the combined groups of samples. In this study D_0 is 37. Calculation of D values for the known groups of samples (Fig. 3) yields the following decision rule: if D is less than 37, then the sample is classifed as Huron–Georgian Bay (population N); if D is greater than 37, then the sample is classified as Erie (population E). If this rule is applied to the known groups, then four samples are misclassified. This fact gives a measure of the probability of misclassification when the rule is applied to unknown samples. In this case the probability would be 7 percent. If this rule is now applied to the Ontario Lobe group (group 0), three of the samples have definite N affinities whereas the other ten samples are classified in population E.

The results of the discriminant analysis lend support to the original hypothesis, namely that the Ontario Lobe group is a mixture of two populations.

These results suggest that in the northwestern part of the Ontario Lobe area, the glacier must have overridden and incorporated pre-existing material derived from a northern source. Previous investigations by Vagners (1966), and field observations by the authors lend further support to this conclusion.

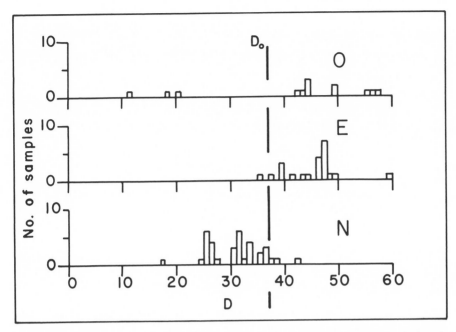

Figure 3. Distribution of discriminant scores for populations E and N and Ontario Lobe (0) samples.

SUMMARY

Chemical analyses of the -0.037-mm size fraction of 109 till samples show that the Erie-Ontario Lobe tills may be distinguished from the Huron–Georgian Bay Lobe tills by a higher content of Ni, Cu, Zn, and Cr. The best distinction is achieved by application of multivariate statistical techniques to the analytical data on all four elements combined.

Where one glacial lobe incorporated sediments from another lobe, the resulting tills have a mixed composition. Along its northwestern edge the Ontario Lobe incorporated older drift derived from a northern source. This preliminary investigation suggests that the application of multivariate statistical techniques to geochemical analysis of selected fractions of glacial deposits could provide an additional criterion for differentiation.

ACKNOWLEDGMENTS

Financial support for this study came from the National Research Council of Canada Grant A4215. Geochemical laboratory facilities at the University of Western Ontario were used during this study. Dr. M. Fleet (U.W.O.) is thanked for advice on geochemical matters and Dr. P. Sutterlin (U.W.O.) for discussion of the statistical procedures.

REFERENCES CITED

Angino, E. E., and Billings, G. K., 1967, Atomic absorption spectrometry in geology, *in* Methods in geochemistry and geophysics No. 7: Amsterdam, Elsevier, 144 p.

Bear, F. E., 1964, Chemistry of the soil: Am. Chem. Soc. Mon., Ser. 160, 2d ed., New York, Reinhold Pub. Co., 515 p.

Dreimanis, A., 1965, Tills of Southern Ontario, *in* Legget, R. F., ed., Soils in Canada: Royal Soc. Canada Spec. Pub. 3 (revised ed.), Toronto, Canada, Univ. Toronto Press, p. 80–96.

―――― 1969, Selection of genetically significant parameters for investigations of tills: Poznan, Zesz, Nauk. UAM Geografia 8, p. 15–29.

Dreimanis, A., and Goldthwait, R. P., 1973, Wisconsin glaciation in the Huron, Erie, and Ontario Lobes: Geol. Soc. America Mem. 136, p. 71–105.

Dreimanis, A., Reavely, G. H., Cook, R. J. B., Knox, K. S., and Moretti, F. J., 1957, Heavy mineral studies in tills of Ontario and adjacent areas: Jour. Sed. Petrology, v. 27, no. 2, p. 148–161.

Fahrig, W. F., and Eade, K. E., 1968, The chemical evolution of the Canadian Shield: Canadian Jour. Earth Sci., v. 5, p. 1247–1252.

Forslev, A., 1957, Geochemical study of some late Wisconsin tills: Geol. Soc. America Bull., v. 68, p. 1727–1728.

Hewitt, D. F., 1964, Rocks and minerals of Ontario: Ontario Dept. Mines Geol. Circ. 13, 108 p.

Karrow, P. F., 1963, Pleistocene geology of the Hamilton-Galt area: Ontario Dept. Mines Geol. Rept. 16, 68 p.

―――― 1967, Pleistocene geology of the Scarborough area: Ontario Dept. Mines Geol. Rept. 46, 108 p.

Klovan, J. E., and Billings, G. K., 1967, Classification of geological samples by discriminant-function analysis: Bull. Canadian Petroleum Geology, v. 15, p. 313–330.

Kvalheim, A., ed., 1967, Geochemical prospecting in Fennoscandia: New York, Interscience, 350 p.

Neville, A. M., and Kennedy, J. B., 1964, Basic statistical methods for engineers and scientists: Scranton, Pa., International Textbook Co., 325 p.

Pawluk, S., and Bayrock, L. A., 1969, Some characteristics and physical properties of Alberta tills: Research Council Alberta Bull. 26.

Shaw, D. M., Reilly, G. A., Magnuson, J. R., Pattendon, G. E., and Campbell, F. E., 1967, An estimate of the chemical composition of the Canadian Precambrian Shield: Canadian Jour. Earth Sci., v. 4, p. 829–853.

Stanton, R. E., 1966, Rapid methods of trace analysis for geochemical applications: London, Edward Arnold, Ltd., 96 p.

Vagners, U. J., 1966, Lithologic relationships of till to carbonate bedrock in southern Ontario (M.Sc. thesis): London, Ont., Univ. Western Ontario, 154 p.

Warren, H. V., and Delavault, R. E., 1961, The lead, copper, zinc, and molybdenum content of some limestone and related rocks of southern Ontario: Econ. Geology, v. 56, p. 1265–1272.

Warren, H. V., Delavault, R. E., and Cross, C. H., 1967, Possible correlations between geology and some disease patterns: New York Acad. Sci. Annals, v. 136, p. 657–710.

MANUSCRIPT RECEIVED BY THE SOCIETY DECEMBER 27, 1971

GEOLOGICAL SOCIETY OF AMERICA
MEMOIR 136
© 1973

DeKalb Mounds: A Possible Pleistocene (Woodfordian) Pingo Field in North-Central Illinois

RONALD C. FLEMAL

Department of Geology, Northern Illinois University, DeKalb, Illinois 60115

KENNETH C. HINKLEY

United States Soil Conservation Service, Batavia, Illinois 60510

JAMES L. HESLER

Department of Geography, University of Iowa, Iowa City, Iowa 52240

ABSTRACT

More than 500 circular to elliptical mounds occur in the late Pleistocene (Woodfordian) deposits of north-central Illinois. The mounds rise from 1 to 5 m above the general ground level and are either flat-topped or have slightly depressed centers. They range in diameter from 30 m to approximately 1 km; the smaller mounds are most abundant. The mounds consist of a core of lacustrine silt and clay surrounded by a sandy rim. The lacustrine sediments overlie Woodfordian till and outwash, and are in turn overlain by Woodfordian loess.

The morphologic and stratigraphic characteristics of the mounds suggest that they are deposits formed within the lakes of pingo craters. These characteristics include (1) a high degree of symmetry, (2) overlapping and superpositional relationships, (3) occurrence over both till and outwash, (4) surrounding annular depression, (5) confinement to a low-relief inter-morainic area, (6) mineralogical identity with the underlying materials, and (7) the association of the mounds with indicators of permafrost. The pingo lake hypothesis is also consistent with implied groundwater conditions and ice margin locations during the Woodfordian.

INTRODUCTION

Small circular mounds or ridges formed under glacial or periglacial conditions have been noted commonly in the literature (for example, Gravenor, 1955; Gravenor and Kupsch, 1959; Svensson, 1964, 1969; Wiegand, 1965; Bik, 1967, 1969; Clayton and Cherry, 1967; Pissart, 1968; Quinn, 1968). These features have had a long list of names applied to them and have been ascribed to an equally long list of origins. Recent studies of aerial photographs combined with detailed soil mapping have indicated the existence of a field of more than 500 small, highly symmetrical, low-relief mounds in the glacial terrain near DeKalb, Illinois (Fig. 1). An excellent opportunity to study these mounds in detail has been provided by a soil mapping project of the Soil Conservation Service in the DeKalb area. In addition, the associated glacial deposits have been extensively studied and mapped by the Illinois State Geological Survey (Kempton, 1963; Anderson, 1964; Kempton and Hackett, 1968; Gross, 1970; Willman and Frye, 1970), and thus the stratigraphic relationships of the mounds are well established.

The authors conclude that the mounds represent the remnants of a sizable Pleistocene pingo field. If this interpretation is correct, it has strong implications regarding the climate in the DeKalb region during the time of formation of the mounds. It also reflects on the possible origin of similar mounds in other regions.

GEOGRAPHIC AND GEOLOGIC SETTING

The DeKalb mound field covers an area of approximately 300 sq km. The city of DeKalb, located 100 km directly west of Chicago, Illinois, is centrally located within the field (Fig. 1). The mounds are confined to a topographic depression between two morainic complexes; the lower parts of the depression are from 15 to 45 m below the present crests of the moraines. The depression is partially drained by the south branch of the Kishwaukee River and its tributaries. Surface drainage, however, is poor, and several large areas within the mound field lack surface streams. Except for the mounds themselves, relief within the depression is generally very low, commonly less than 3 m per sq km.

The two morainic complexes, both of which were developed during the Woodfordian substage (Frye and Willman, 1960), are the morphostratigraphic expression of two major Woodfordian ice advances. The drift

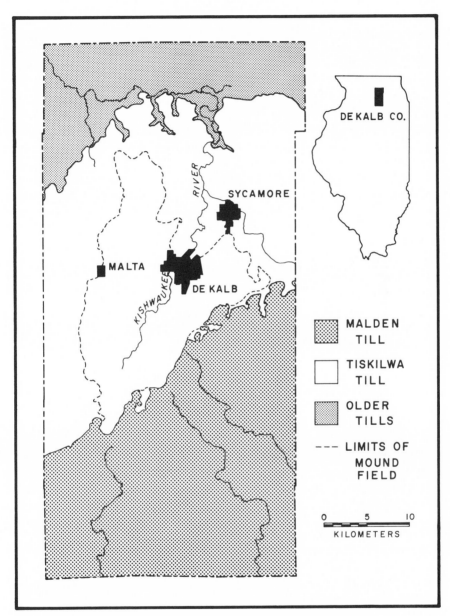

Figure 1. Location of mound field in northern Illinois.

of the Bloomington Morainic System, which is also the drift that floors most of the depression in which the mounds occur, is the Tiskilwa Till Member of the Wedron Formation (Willman and Frye, 1970). The till is readily identifiable on the basis of its reddish-gray color in oxidized exposures and its maroon color at depth. The drift of the Arlington and Shabbona Moraines, which border the mound field on the southeast, is the Malden Till Member of the Wedron Formation (Willman and Frye,

1970). It is a distinct yellowish-gray till. Both till members may be further distinguished on the basis of their mineralogy (Kempton and Gross, 1970) and their related soil characteristics (McComas and others, 1969).

Within the general intermorainal depression, the Tiskilwa Till is locally covered by proglacial outwash derived from both the Bloomington and Arlington glaciers. Valley-train deposits from the Arlington ice occur in the valley of the south branch of the Kishwaukee River and its major tributaries. Confined to the area of the Tiskilwa Till are other small outwash plains and valley trains that apparently formed as the Bloomington ice melted.

The drift thickness in the area of the mounds ranges from about 15 m to more than 125 m (Piskin and Bergstrom, 1967). The drift consists of numerous interbedded tills and stratified deposits, possibly as old as pre-Illinoian, and includes definite Illinoian, Altonian, and Woodfordian units (Kempton, 1963).

MORPHOLOGY

The mounds all have low relief compared to their horizontal dimensions. The majority of definitely identified mounds range from several tens of meters to approximately 100 m across. The most striking, however, are about 25 "giants" ranging up to as much as 400 to 1,200 m in longest dimension (Fig. 2). The tops of all mounds stand from 1 m to 5 m above the adjacent ground surface; in general, those with the largest horizontal dimensions also exhibit the greatest relief. Because the mounds are comparatively low with respect to their horizontal dimensions, they are most readily identified on aerial photographs where slight variations in soil color between the mound flanks and the adjacent lows stand in sharp contrast (Figs. 2 and 3). The contrast in soil color is due not only to variation in moisture content but to differences in slope and in soil parent material. Soil mapping, consequently, has been highly successful in recognizing the lower and smaller mounds. Nevertheless, because of the difficulty in recognizing the smaller mounds, particularly where they lie close to surface drainage lines and have been partly eroded, it is likely that we have underestimated the total number.

The mounds tend to occur in clusters (Fig. 2), and within the clusters are very closely spaced and commonly are separated by a depression only a few meters wide. In many cases mounds actually overlap one another (Fig. 3), and the top of the smaller satellite mound is at a lower elevation than the top of the main mound. The smaller one appears to abut against the main mound and thus disrupts its symmetrical form. In the extreme case, satellite mounds are located completely on top of larger mounds (Fig. 2). Most giant mounds have one or more satellite mounds on their crests.

In plan view the mounds display a high degree of symmetry. With the exception of demonstrably overlapping mounds, the forms range from

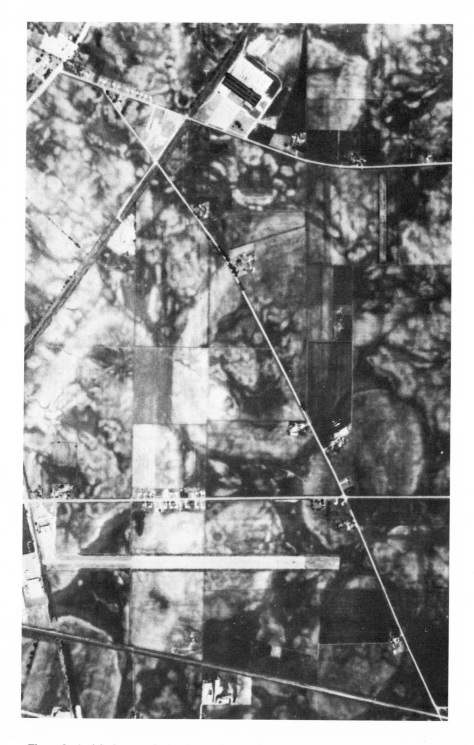

Figure 2. Aerial photograph showing a cluster of large mounds east of DeKalb, Illinois. Long dimension of largest mound at center of photograph is 1 km. Several small mounds are superimposed on large mounds (U.S. Department of Agriculture aerial photograph 3EE-127, 6-3-64, DeKalb County, Illinois).

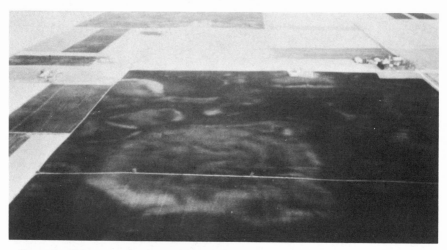

Figure 3. Oblique aerial photograph showing main mound and two overlapping satellite mounds (lower right and upper right) south of Malta, Illinois. Diameter of main mound is 350 m. Line across mound is a fence line.

nearly perfect circles to ovals or ellipses. The larger mounds tend to exhibit maximum elongation, with length-width ratios approaching 2:1 (Fig. 2). Even these mounds, however, possess symmetry across the long axis. There is some suggestion of a local preferred orientation of the major axes of the elongated mounds (Fig. 2). However, the preferred orientation is apparent only where the mounds are closely spaced or partially overlapping, and it is not consistent from one mound cluster to another.

An attempt has been made to correlate the suggested preferred orientation with some physical characteristic of the mound area. Three general relationships exist. The long axes are generally parallel to the trend of the adjacent moraine crests, and perpendicular both to the regional slope within the intermorainic depression and to the dominant Pleistocene wind direction (west-northwest) inferred from nearby sand dune orientation and paha alignment (Vail, 1969). It is not known, however, whether any of these relationships reflect significantly on the origin of the mounds.

One of the most salient characteristics of the larger mounds is the distinctive flatness of their tops (Fig. 3). Relief on the crests of these mounds commonly is as little as 0.5 to 1 m over a distance of 0.5 km. Most of the crest relief occurs near the margins of the mounds where slightly raised rims are locally evident, but seldom extend completely around a mound. Relief on the smaller mounds is generally of the same order of magnitude as on the larger ones, but because of the shorter horizontal dimensions the slopes are greater. Where a slightly elevated rim occurs on the smaller mounds, dark, partly colluvial soils develop in the center of the mound and are seen on aerial photographs as a dark circular area surrounded by a light rim (Fig. 2). In a few instances breaches occur in the rims of mounds. These breaches lead into currently ephemeral drainage channels (Fig. 4) and suggest that water at one time flowed from the center of the mound outward through the breaches.

In regions remote from surface drainage lines, most mounds are completely enclosed by an annular depression. These depressions, when filled with water, as commonly occurs during the spring thaw, appear as moats surrounding the mounds. Those mounds near drainage lines tend also to have a depression surrounding them, but the depression is not closed.

INTERNAL CHARACTERISTICS

The internal characteristics of the mounds have been investigated by numerous borings by the Soil Conservation Service field team. All of the mounds exhibit a similar composition and stratigraphic sequence. One of the larger mounds (Fig. 5) was singled out for detailed coring. A total of 51 cores of 1.5 in. (3.8 cm) diameter was taken from and adjacent to this mound, 41 on a north-south line and 10 on an east-west line. The lines were located off-center of the mound because of greater accessibility. Cores were taken to a maximum depth of 7.1 m. Coring beyond this depth was prevented by the presence of an exceedingly wet, fluid sand, which continuously flowed into the holes and prevented

Figure 4. Two small mounds with breached rims (center and center left) leading into ephemeral drainage channels. Modern drainage is in artificial channel at bottom of photograph. Largest mound is approximately 100 m in diameter.

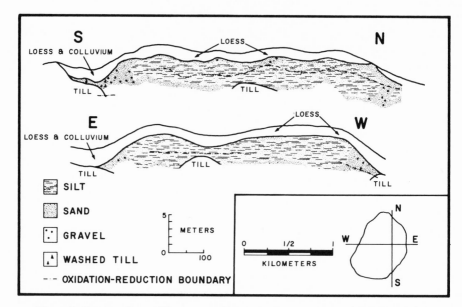

Figure 5. Cross sections of large mound located 3 km southwest of DeKalb, Illinois.

further core retrieval. In a few holes a screw auger was used to attempt to probe through the sand but without success.

Approximately the upper meter in each core (both those located on the mounds and those adjacent to the mounds) consists of a leached, mottled-tan silt loam passing upward into a well-developed A-horizon, rich in organic matter. The upper layer is thickest on the western flanks of the mounds where it appears to be embanked against the mounds. There is also a slight thickening of the upper layer in the depression surrounding the mounds, although part of this thickening is probably the result of colluvial and alluvial deposition. X-ray analysis of the clay-size fraction in this upper layer indicates a high montmorillonite content. In contrast, the remaining portions of the cores either lack montmorillonite or contain it in markedly lesser quantities. The loesses in northern Illinois are montmorillonite-rich (Frye and others, 1962), whereas the ice-contact or proglacial deposits are dominantly illitic. On the basis of composition, areal distribution, and grain size, the upper layer is correlated with the Richland Loess (Frye and Willman, 1960).

Immediately beneath the loess is a thin, discontinuous band of sand and gravelly sand. It marks the major stratigraphic break within the mounds, separating what is apparently the original material from the overlying loess mantle. This coarse layer is probably a lag concentrate formed by the partial deflating of the original surface by wind prior to loess deposition.

The main mass of the mounds consists of finely laminated plastic silt and clay that is rich in organic matter and contains occasional sand stringers (Fig. 5). Grain size increases downward as the sand stringers increase in both thickness and abundance. Both in illitic composition and in reddish color, the silt and clay of the mounds are identical to the fine-grained fraction of the adjacent and subjacent Tiskilwa Till. Toward the margins of the mounds, the silt and clay interfinger with and grade into sand that contains occasional pebbles but otherwise is very well sorted. The marginal sands are thickest where raised rims exist on the surface. Although no detailed work has been done on the organic materials within the silts and clays, cursory study indicates the presence of spruce pollen.

The finely laminated sediments within the central portion of the mounds show little or no evidence of either postdepositional disruption or disturbance. A few cores contain very local nonhorizontal bedding, but it is difficult to ascertain whether this is original or whether it was caused by plastic deformation during coring. In the coarser sediments on the flanks, bedding is mostly massive, but locally, cross-beds dip gently inward toward the center of the mounds.

The silts, sands, and clays were deposited in the quiet waters of lakes or ponds. The thin laminations are suggestive of varves, although there is no direct evidence that supports a seasonal deposition. The thickness of the lacustrine sediments, which extend well beneath the adjacent till surface (Fig. 5), suggests that the lakes may have at one time exceeded 6 m in depth. Both the gentle angles of dip of the few marginal cross beds (which may have been postdepositionally increased by relatively greater compaction of the core areas) and the horizontal structure of the upper silts and clays suggest, however, that the water was probably not deep nor the bottom topography irregular when deposition of the sediments ceased.

A noteworthy characteristic of the mounds is the relatively shallow depth of the oxidation-reduction boundary (Fig. 5). In the adjacent till this boundary occurs at depths between 3 m and 7 m, depending upon the surface topography and drainage. In the mounds, in spite of their elevation, this boundary occurs at depths from 2 m to 4 m. It is likely that the oxidation-reduction boundary closely coincides with the long-term minimum elevation of the water table, and therefore implies a long-term water table nearer the surface beneath the elevated mounds than in the adjacent lower till areas. This is the reverse of the normal relationship of the water table in hills.

STRATIGRAPHIC RELATIONS

In the mound studied in most detail (Fig. 5) only 4 of the 37 holes bored into the mound proper passed completely through the lacustrine sequence. These four holes bottomed in normal Tiskilwa Till. The till was also encountered outside the mound and was traced on three ends of the cross section drilled to the margins of the mound. At these points the till surface dips beneath the lacustrine materials. Although it is not certain what the lacustrine materials lie on everywhere beneath this mound, they at least locally overlie the Tiskilwa Till and therefore had to be formed after the till was deposited.

Several mounds occur in areas where they are completely surrounded by outwash. Because the outwash materials differ little in texture from the sandy materials within the mounds, it is difficult to determine stratigraphic relationships. No embankment of outwash abuts the mounds in the direction from which the outwash was derived nor do outwash bars occur on their downstream side. Furthermore, the mounds surrounded by outwash show no evidence of having been erosionally modified as they might have been had they existed in the middle of an outwash-carrying river. Neither do they appear to be partly buried, because they have the same relief as mounds of the same size in till areas. Therefore, it is probable that these mounds were formed after the outwash was deposited.

TIME OF ORIGIN

The time of origin of the mounds is established by their stratigraphic position between the underlying till and overlying loess. The underlying Tiskilwa Till was deposited in mid-Woodfordian time by the Bloomington glacier (Willman and Frye, 1970). The absolute age of the Tiskilwa Till has not been definitely established, but it is estimated to be about 16,500 radiocarbon yrs B.P. (Frye and others, 1965). This, therefore, is the maximum age of the mounds.

The loess which overlies the mounds is the Richland Loess (Frye and Willman, 1960). Although Richland Loess deposition in northeastern Illinois may have occurred locally as late as Valderan time, the main episode of deposition occurred during Woodfordian time, subsequent to the withdrawal of the successive Woodfordian glaciers (Willman and Frye, 1970). The appreciable thickness of the Richland Loess in the mound area suggests that the loess deposition began early, probably soon after the Bloomington glacier had withdrawn from the area.

The occurrence of mounds in areas of Tiskilwa Till and not in areas of Malden Till suggests that the mounds probably were formed before the Arlington glacier retreated to expose the Malden Till. The absolute age of the Malden Till is unknown, but a major retreat and readvance before deposition of the Malden Till probably required 1,000 to 1,500 yrs (Willman and Frye, 1970), and the mounds were probably formed during this interval in mid-Woodfordian time, in the range of 15,000 to 16,500 radiocarbon yrs B.P.

HYPOTHESES OF ORIGIN

It is very unlikely that the mounds represent the erosional remnants of sediments formed in a large lake that might have covered the area in the past. The high degree of symmetry, their occurrence in some regions that lack present surface drainage, the regular pattern of coarsening of the sediments toward the edge of each mound, the overlapping and superpositioning of some onto others, the variation in surface elevation, and the occurrence of some mounds over lows in the till topography indicate that they are constructional features only slightly modified by erosion and deposition.

This raises the major question regarding their origin. How could a large number of circular lakes develop and be maintained in such a manner that after the lakes disappeared their sediment surfaces were above the level of the adjacent landscape? The lakes must have had at one time a confining wall which has since disappeared. Two types of lakes require such a confining wall—ice-walled lakes and pingo lakes.

ICE-WALLED LAKE HYPOTHESIS

The term "ice-walled lake" is applied to lakes formed in depressions in a stagnating glacier. The landforms resulting from sediment deposited in these lakes have been variously called "ice-walled lake plains," "perched lacustrine plains," "moraine plateaus," "moraine-lake plateaus," and "impounded glacial lake plains" (Clayton and Cherry, 1967). The formation of ice-walled lakes has been described by Clayton (1962) and Clayton and Cherry (1967). Ice-walled lakes have been recognized throughout large areas of the Missouri Coteau in North Dakota and in adjacent Saskatchewan and Alberta (Winters, 1960; Christiansen, 1961; Clayton, 1962, 1967; Clayton and Cherry, 1967; Royse and Callender, 1967), where both their morphological characteristics and associated stagnant-ice features substantiate their origin.

It has been suggested by some of our correspondents familiar with deposits of ice-walled lakes that, because of the similarity in internal stratigraphy, the DeKalb mounds are ice-walled lake deposits. Numerous characteristics of the mounds, however, other than their internal composition, argue against their having this origin. The two most important are the overlapping and superpositioning, and the almost certain occurrence of some mounds on outwash.

It is difficult to imagine how stagnant-ice walls could account for the common occurrence of smaller mounds on top of larger mounds. This would necessitate a highly untenable uniform expansion of the ice over the existing broader lacustrine deposits to form the walls of the later, smaller lake. It is equally difficult to imagine how overlapping mounds that have surfaces at different elevations (Fig. 3) could be formed by ice-walled lakes. It is certainly unlikely that overlapping mounds were formed in contemporaneous broadly connected lakes because of their disparate surface elevations, and there is otherwise no place for a stagnant-ice wall to have existed between them. If the overlapping mounds are not to be considered contemporaneous, then the stagnant-ice wall must have moved during the formation of the mounds, with a rapidity that is highly unlikely.

Even the more common occurrence of closely clustered but not overlapping mounds is difficult to explain by the stagnant-ice hypothesis. Very narrow and high walls of ice would have had to exist long enough for the lakes to fill with sediments. Such a narrow wall could hardly be of sufficient size to contribute the 7 m of sediment that have been found in some of these mounds. It is also significant in this regard that mounds in the center of a cluster contain neither less sediment nor finer grained sediment than either solitary mounds or mounds at the margin of a cluster.

Some mounds probably occur over glacial outwash, which indicates that the ice must have completely melted from the site of these mounds before the outwash and, later, the mounds were formed. These mounds could not, therefore, have been surrounded by stagnant glacial ice. It is implausible that the morphologically and stratigraphically identical mounds that overlie till had a different origin.

Aside from the major difficulties with an ice-walled lake origin, there are other objections to this hypothesis. Both in morphologic and stratigraphic characteristics, the mounds bear little resemblance to those more definitely formed by deposits of ice-walled lakes. The mounds are generally much smaller and much more symmetrical in outline than those attributed to ice-walled lakes in the northern Great Plains region. They are essentially without internal collapse structures, which are common in mounds formed by ice-walled lakes. The sediments in the mounds lack most of the coarse fraction of the adjacent tills. This suggests that the source of the sediment was not till that had slid into the lake and was later sorted. Except possibly for the mounds themselves, there are few other features that suggest ice stagnation in the vicinity. Such stagnation features as kames and eskers are remarkably scarce. Frye and others (1965) have, in fact, called attention to what appears to be an unusually high level of continuous glacier activity in the entire Woodfordian of Illinois.

The probable occurrence of the mounds over glacial outwash not only rules out the ice-walled lake origin, but any other ice-contact origin as well. However, there was glacial ice in the region at the time the mounds formed because the presence of the loess mantle over them indicates that they formed during Woodfordian time. Although the glacial ice left the DeKalb region for the last time following the deposition of the Tiskilwa Till, it remained only a short distance to the east throughout the remainder of the Woodfordian, as is shown by the thick sequence of later Woodfordian deposits that occur in the region immediately east of the mound field (Willman and Frye, 1970). Therefore, although the mounds did not form in contact with the ice, they did form in close proximity to the ice margin.

PINGO LAKE HYPOTHESIS

The hypothesis that some small-scale circular mounds or ridges may represent the remnants of former pingos is common in European literature (for example, Cailleux, 1956; Picard, 1961; Pissart, 1963, 1968; Svensson, 1964, 1969; Mullenberg and Gullentops, 1969), but less so in North America (Wolfe, 1953; Bik, 1967, 1969). In part this may be due to unfamiliarity with the characteristics and associations of modern pingos, even though much study has been given them recently (for example, Müller, 1963; Mackay, 1963; Maarleveld, 1965; Krinsley, 1965; Holmes and others, 1966).

Pingos are dome-shaped mounds developed in permafrost that result "from the uplifting of a layer of frozen ground by the pressure of water freezing in the substratum to form large ice lenses" (Müller, 1968, p. 845). There are two major varieties of pingos: open-system and closed-system (Müller, 1968). Inasmuch as the DeKalb mounds appear to be derived from open-system pingos, the discussion hereafter is confined to this variety.

Open-system pingos form when sufficient hydrostatic pressure exists under permafrost to push up the overlying impervious frozen ground into a dome. The hydrostatic pressure may originate in ground water beneath permafrost under the impetus of a hydraulic head (Mackay, 1963; Müller, 1968) or when permafrost is locally aggrading in such a manner that pore pressure is built up in saturated ground confined by impervious permafrost or other substratum. In certain cases the hydrostatic pressure has been shown to be reinforced by the pressure of gas expansion (Müller, 1963). The basal diameters of the domes thus produced are generally tens of meters across, but domes as much as 500 m across have been observed (Krinsley, 1965). Characteristically, they are circular or slightly ellipsoidal in plan. The most favorable localities for the development of open-system pingos are low, sloping surfaces underlain by an aquifer that has access to surface waters (Holmes and others, 1966).

The formation of a pingo dome appears to involve not simply an arching of the permafrost, but also the segregation within the permafrost of a lens of pure ice. This conclusion is borne out by the observation that all modern pingos have a solid core of pure ice (Müller, 1963). Furthermore, the raising of a pingo dome also apparently involves marginal subsidence, since pingos are commonly surrounded by an annular depression (Krinsley, 1965).

As a pingo is raised, severe tensile stresses arise in the rigid frozen ground on the growing crest of the expanding dome, thus producing fractures which may extend downward into the aqueous center, forming springs. Later, gravity movement of material off the top of the pingo may further expose the ice core to the atmosphere. The core then begins to melt and the pingo begins to degenerate. Open-system pingos in various stages of degeneration, but still retaining a partial ice core, have been described by Müller (1963), Krinsley (1965), and Holmes and others (1966). Svensson (1964, 1969) has described pingos from which the ice core has completely melted.

Partially degenerated pingos resemble small volcanic craters, which commonly contain circular lakes (Porsild, 1938; Müller, 1963, 1968; Svensson, 1964, 1969; Krinsley, 1965; Holmes and others, 1966; Birot, 1968). The size of the lakes gradually increases as the ice core continues to melt until eventually a lake surrounded by a high wall may exist (Fig. 6). Muller (1963) has reported several pingo crater lakes in east Greenland that are more than 100 m in diameter and surrounded by debris-mantled ice walls that rise more than 10 m above the surface of the lakes. The level of the water surface in these lakes is generally above that of the ground outside the pingo wall, commonly by as much as 2 to 7 m.

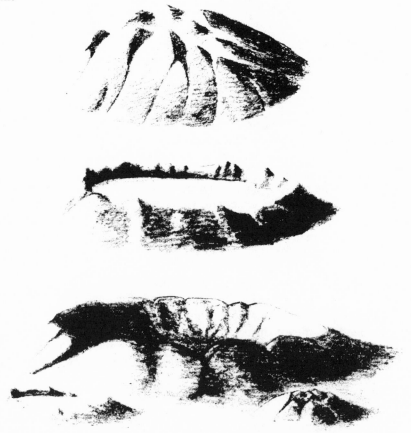

Figure 6. Characteristics of pingos and pingo craters at various stages of degeneration. Top: newly formed pingo. Middle: pingo crater containing lake. Bottom: multiple stage pingos (redrawn from photographs in Müller, 1963).

The lakes formed within open-system pingo craters may be partially maintained by the discharge of ground water into them (Müller, 1963). Müller (1963) observed one such discharge to be nearly 2 l per sec. When hydrostatic pressure is sufficient, mud, sand, and even gravel may be "thrown up to the surface" or into the lakes (Rosenkrantz, 1942). This phenomenon is apparently related to the "suffusion bursts" described by Patterson (1940), and has been taken by Müller (1963) to be characteristic of open-system pingos.

With sufficient time, the material carried into the lakes by the springs, in addition to whatever material is washed in from the pingo walls or blown in by wind, will tend to fill up the lake. If filling is sufficient to raise the bottom of the lake above the level of the adjacent terrain, a circular lacustrine mound will result when the ice wall completely melts.

Unfortunately, in the best documented examples of modern open-system pingo lakes (Müller, 1963), most of the pingos occur in areas of either very coarse detrital or bedrock substrata. Therefore, they are not markedly infilled. Nevertheless, Müller does indicate that some infilling occurs in all the lakes and that he has measured lacustrine fills in excess of 2 m. Because the sediments within modern pingo lakes are always similar to the materials adjacent to and apparently underlying them, the turbid springs are evidently the source of most of the material within the lakes. The amount of infilling that will occur is evidently determined by the characteristics of the underlying materials and the amount of water discharge. Where the underlying materials are thick tills and sands, the lake infilling will likely be great.

One of the most significant characteristics of open-system pingos is that they regularly occur in clusters, commonly with several subsidiary pingos surrounding a main pingo (Fig. 6), or with a later stage pingo inside the remnants of a former pingo (Müller, 1963, 1968). Such pingo clusters are produced by repeated degeneration and formation of new pingos at a single site. Müller (1963) suggested that multiple pingo generation is to be expected, because of the manner in which ground ice continues to form and melt in an open-system pingo field.

The type of landform that results from the complete collapse and melting of a pingo is determined by local conditions, including the type and mobility of underlying materials, history of ice-melting, amount and source of water, proximity of other pingos, and form, type, and size of the original pingo. The landform may also have been modified by later erosion or deposition. Some of the characteristics that may vary are: (1) plan form, ranging from circular to ellipsoidal; (2) dimensions, including width, length, and height; (3) characteristics of deposits in the rim and core, including the amount of lacustrine infilling; (4) presence or absence of rim breaches; and (5) distribution and density of resultant landforms. Variations in these characteristics are found in modern pingos (Müller, 1963) and in remains of Pleistocene pingos (Pissart, 1968).

The many morphological and stratigraphic similarities between the De-Kalb mounds and the deposits of a field of completely degenerate open-system pingos are striking. There are eight major points of similarity: (1) regular circular and ellipsoidal form; (2) surrounding annular depression; (3) overlapping and superpositioning; (4) occurrence in areas of loosely consolidated substratum; (5) confinement to areas of low slope; (6) over-all size; (7) lacustrine composition; and (8) mineralogical similarity with underlying materials. Of particular significance is the overlapping and superpositioning relationship which, although it is a phenomenon to be expected with pingo craters, is also unique to them, and therefore strongly supports the pingo-lake origin of the DeKalb mounds.

Pissart (1968) suggested that there are two basic forms of "pingo-scars"—simple closed depressions and closed depressions surrounded by a rampart. Studies of the deposits of modern pingo lakes, however, suggest that there is a third form in which sediments completely (or nearly completely) fill the lake up to a level of the rampart and thereby produce flat-topped mounds, as exemplified by the DeKalb mounds. Although these three basic forms are part of a continuum, and each may be present in any one field, generally one form predominates at a single locality. The apparent reason for this is that hydrologic, climatic, and stratigraphic conditions tend to be fairly uniform at a given locality, and these exert the major controls on the morphologic character of the resultant landforms. The occurrence of the DeKalb mounds in an area underlain by thick glacial deposits is probably the significant factor that accounts for the mound-type pingo remnants. It is significant that the only other mounds which have been interpreted as of pingo origin (Bik, 1967, 1969) also occur in an area of relatively thick glacial drift.[1]

Although the morphology and stratigraphy suggest that the DeKalb mounds represent the deposits of former pingo lakes, two conditions are necessary for open-system pingo development—permafrost and ground water under hydraulic pressure beneath the permafrost. Both of these conditions must have occurred in the DeKalb region during the stratigraphically defined time of origin of the mounds.

Many independent observations confirm the fairly widespread presence of permafrost along the Woodfordian ice margin in northern Illinois and adjacent regions. Smith (1949, 1962) and Black (1964, 1965, 1969a) have recorded a variety of features of permafrost origin, including ice-wedge casts, blockfields, solifluction phenomena, and patterned ground in southwestern Wisconsin. Involutions have been reported within northern Illinois by Sharp (1942) and by Frye and Willman (1958). Horberg (1949) has described a possible ice-wedge cast. Wayne (1967) found patterned ground and a possible pingo remnant in Indiana. Thus, it is fairly certain that permafrost did exist in the mound region at the appropriate time. Further-

---

[1] A third example has recently been reported by Edward Watson in Nature, v. 236, no. 5346, p. 343-344.

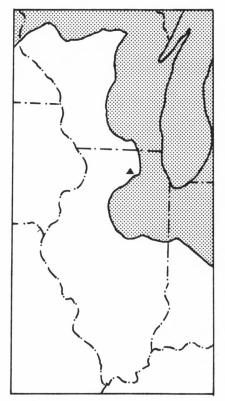

Figure 7. Location of mound field (triangle) with respect to generalized boundary of middle Woodfordian ice mass (stippled).

more, the fact that the mound field occurs within a general re-entrant in the ice sheet (Fig. 7) suggests that if permafrost were present anywhere along the ice margin, it would logically have occurred in the climatically protected region of the re-entrant.

The existence of ground water under hydraulic pressure beneath the former permafrost is less easily proven. However, the most probable explanation for the present high water table within the mounds is that it results from a hydraulic head produced by the small elevational differences between the adjacent moraines and the topographic depression in which the mounds occur. The high permeability of the sands within the mounds precludes the presence of a perched water table. The higher level of the ground water under the mounds than in the surrounding till implies a greater ease of passage of the ground water into the mounds than into the adjacent till areas, and, therefore, the existence of some vertical conduit for water movement beneath the mounds.

Because no agent now operating in the region could produce such conduits, they must have been produced in the past under different conditions. The disruption of the otherwise impervious till by first a rising ice block and then rising ground water, such as occurs in open-system pingos, provides a plausible mechanism by which these conduits could have formed.

McGinnis (1968) showed, on the basis of theoretical considerations, that normally ground water will flow outward from beneath an ice sheet. Confirmation of this theory may exist in the occurrence of ground-water discharges along the margins of modern ice sheets, as has been noted by Black and others (1965) and Black (1969b). In theory, the amount of discharged ground water is a function of both the thickness of the glacier and its rate of bottom melting. For a large temperate glacier, such as that which occupied the region of the mounds, the volume of ground-water discharge must have been large (McGinnis, 1968). The water would have been forced away from the ice through both the bedrock and the thick surficial glacial deposits. Downward movement would have been inhibited by the presence of normal ground water and upward movement by the frozen crust of permafrost. Under these conditions large pressures, depending upon the head provided by the glacier, could have developed beneath the permafrost and eventually caused it to buckle into pingo domes. Therefore, it is not only morphologically and stratigraphically plausible to suggest that the DeKalb mounds are remnants of pingos, but with the favorable permafrost and ground-water conditions the formation of pingos would be a natural result.

CONCLUSIONS AND INTERPRETATION OF EVENTS

The pingo lake hypothesis of the origin of the DeKalb mounds is the most likely explanation for two reasons: (1) it explains all of the salient morphological and stratigraphic characteristics of the mounds; and (2) it is consistent with the inferred climate and ground-water movement at the time the mounds formed.

The authors suggest the following sequence of events leading to the formation of the mounds: (1) During middle Woodfordian time the Wisconsinan glacier retreated from the position of the Bloomington Morainic System, uncovering an area of low relief ground moraine and outwash topography before it readvanced to the position of the Shabbona and Arlington Moraines. (2) Climatic conditions, in part promoted by the isolation of the area with a re-entrant of the ice front, caused development of permafrost. (3) Concurrent with the permafrost development, ground water was forced from beneath the ice to a position beneath the impervious permafrost. (4) Locally, ground-water pressure built up sufficiently to cause pingos to develop. (5) The pingos began to degenerate, for a time each forming a crater in which a lake developed. (6) Materials from the underlying glacial deposits were carried into the lake by upwelling ground water, and, perhaps in part augmented by eolian deposition and slopewash from the pingo walls, tended to fill the lake. (7) In some craters water overflowed the rims, perhaps shortly before the crater wall completely melted, eroding breaches in the rims of some of the lake sediments. (8) As individual pingos completely decayed, their sedimentary fill was left as a mound, the height of which depended upon the size of the pingo and the amount of infilling. (9) Successive generations of pingos continued to develop at favorable sites, producing overlapping and superimposed mounds. (10) Eventually either climatic conditions sufficiently warmed so that the permafrost melted or the glacier retreated so as to reduce the local hydraulic head. At any rate, pingo formation ceased, and mounds composed of lacustrine deposits remained. (11) Loess was deposited as a mantle over the mounds.

The remnants of pingos in areas of former periglacial environments are probably more abundant than current awareness indicates, as has been suggested by Pissart (1968). There is no reason to believe that pingos, which are so abundant in some parts of the modern periglacial environment, were less common in similar environments in the past.

Former pingos have not been identified in many areas due, in part, to the lack of attention paid to the small-scale topographic features of most of the landscapes in which they are likely to occur. Perhaps also they have not been widely recognized because of the scarcity of examples of former pingos to serve as models of what a Pleistocene-age pingo would look like at present. Thus, it is hoped that the DeKalb mounds, as probable pingo remnants, will serve as a model in the future search for other remains of these highly intriguing landforms.

ACKNOWLEDGMENTS

The authors express their gratitude to the many individuals who have participated in one manner or another in this study. We in particular wish to thank J. P. Kempton, D. L. Gross, and A. M. Jacobs, all of the Illinois State Geological Survey, R. C. Anderson of Augustana College, and R. F. Black of the University of Connecticut, who shared their ideas with us in the field. We also thank J. R. Mackay of the University of British Columbia, W. J. Wayne of the University of Nebraska, and M. J. J. Bik of the Geological Survey of Canada, who provided us with stimulating ideas through correspondence.

REFERENCES CITED

Anderson, R. C., 1964, Sand and gravel resources of DeKalb County: Illinois Geol. Survey Circ. 367, 16 p.

Bik, M.J.J., 1967, On the periglacial origin of prairie mounds, *in* Clayton, L., and Freers, T. F., eds., Glacial geology of the Missouri Coteau and adjacent areas: North Dakota Geol. Survey Misc. Ser. 30, p. 83–94.

——— 1969, The origin and age of the prairie mounds of southern Alberta, Canada: Biul. Peryglacjalny, no. 19, p. 85–130.

Birot, P., 1968, The cycle of erosion in different climates (C. I. Jackson and K. M. Clayton, trans.): Berkeley, Calif., Univ. California Press.

Black, R. F., 1964, Periglacial phenomena of Wisconsin, north-central United States: VI Internat. Cong. Quaternary Rept., v. 4, p. 21–28.

——— 1965, Ice-wedge casts of Wisconsin: Wisconsin Acad. Sci., Arts and Letters Trans., v. 54, p. 187–222.

——— 1969a, Climatically significant fossil periglacial phenomena in north-central United States: Biul. Peryglacjalny, no. 20, p. 225–238.

——— 1969b, Saline discharges from Taylor Glacier, Victoria Land, Antarctica: Antarctic Jour. U.S., v. 4, p. 89–90.

Black, R. F., Jackson, M. L., and Berg, T. E., 1965, Saline discharge from Taylor Glacier, Victoria Land, Antarctica: Jour. Geology, v. 73, p. 175–181.

Cailleux, A., 1956, Mares, mardelles et pingos: Paris, Acad. Sci., Comptes Rendus, v. 242, p. 1912–1914.

Christiansen, E. A., 1961, Geology and ground water resources of the Regina area, Saskatchewan: Saskatchewan Research Council Geology Div. Rept. 2, 72 p.

Clayton, L., 1962, Glacial geology of Logan and McIntosh Counties, North Dakota:

North Dakota Geol. Survey Bull. 37, 84 p.

Clayton, L., 1967, Stagnant-glacier features of the Missouri Coteau in North Dakota, in Clayton, L., and Freers, T. F., eds., Glacial geology of the Missouri Coteau and adjacent areas: North Dakota Geol. Survey Misc. Ser. 30, p. 25–46.

Clayton, L., and Cherry, J. A., 1967, Pleistocene superglacial and ice-walled lakes of west-central North America, in Clayton, L., and Freers, T. F., eds., Glacial geology of the Missouri Coteau and adjacent areas: North Dakota Geol. Survey Misc. Ser. 30, p. 47–52.

Frye, J. C., and Willman, H. B., 1958, Permafrost features near the Wisconsinan glacial margin in Illinois: Am. Jour. Sci., v. 256, p. 518–524.

―――― 1960, Classification of the Wisconsinan Stage in the Lake Michigan glacial lobe: Illinois Geol. Survey Circ. 285, 16 p.

Frye, J. C., Glass, H. D., and Willman, H. B., 1962, Stratigraphy and mineralogy of the Wisconsinan loesses of Illinois: Illinois Geol. Survey Circ. 334, 55 p.

Frye, J. C., Willman, H. B., and Black, R. F., 1965, Outline of the glacial geology of Illinois and Wisconsin, in Wright, H. E., Jr., and Frey, D. G., eds., The Quaternary of the United States: Princeton, N.J., Princeton Univ. Press, p. 43–61.

Gravenor, C. P., 1955, The origin and significance of prairie mounds: Am. Jour. Sci., v. 253, p. 475–481.

Gravenor, C. P., and Kupsch, W. O., 1959, Ice-disintegration features in western Canada: Jour. Geology, v. 67, p. 48–64.

Gross, D. L., 1970, Geology for planning in DeKalb County, Illinois: Illinois Geol. Survey Environmental Geology Notes 33, 26 p.

Holmes, G. W., Foster, H. L., and Hopkins, D. M., 1966, Distribution and age of pingos of interior Alaska, in Proceedings of Permafrost International Conference, 1963: Natl. Research Council Pub. 1287, p. 88–95.

Horberg, C. L., 1949, A possible fossil ice wedge in Bureau County, Illinois: Jour. Geology, v. 53, p. 349–359.

Kempton, J. P., 1963, Subsurface stratigraphy of the Pleistocene deposits of central northern Illinois: Illinois Geol. Survey Circ. 356, 43 p.

Kempton, J. P., and Gross, D. L., 1970, Stratigraphy of the Pleistocene deposits in northeastern Illinois, in 34th Ann. Tri-State Field Conf. Guidebook: DeKalb, Ill., Northern Illinois Univ., p. 65–72.

Kempton, J. P., and Hackett, J. E., 1968, Stratigraphy of the Woodfordian and Altonian drifts of central northern Illinois, in Bergstrom, R. E., ed., The Quaternary of Illinois: Illinois Univ. Coll. Agriculture Spec. Pub. 14, p. 27–34.

Krinsley, D. B., 1965, Birch Creek Pingo, Alaska: U.S. Geol. Survey Prof. Paper 525-C, p. 133–136.

Maarleveld, G. C., 1965, Frost mounds: Meded. Geol. Stich., new ser. no. 17, p. 3–16.

Mackay, J. R., 1963, The Mackenzie Delta area: Canada Dept. Mines Tech. Survey Geog. Mem. 8, 202 p.

McComas, M. R., Hinkley, K. C., and Kempton, J. P., 1969, Coordinated mapping of geology and soils for land-use planning: Illinois Geol. Survey Environmental Geology Notes 29, 11 p.

McGinnis, L. D., 1968, Glaciation as a possible cause of mineral deposition: Econ. Geology, v. 63, p. 390–400.

Mullenberg, W., and Gullentops, F., 1969, The age of the pingos of Belgium, *in* Péwé, T., ed., The periglacial environment, past and present: Montreal, McGill-Queens Univ. Press, p. 303-320.

Müller, F., 1963, Observations on pingos (D. A. Sinclair, trans.): Canada Natl. Research Council Tech. Trans. 1073, 177 p.

_____ 1968, Pingos, modern, *in* Fairbridge, R. W., ed., The encyclopedia of geomorphology: New York, Reinhold Corp., p. 845-847.

Patterson, T. T., 1940, The effects of frost action and solifluction around Baffin Bay and in the Cambridge district: Geol. Soc. London Quart. Jour., v. 96, p. 99-130.

Picard, K., 1961, Reste von Pingos bei Husum/Nordsee: Schriften Naturw. Ver. Schleswig-Holstein, v. 32, p. 72-77.

Piskin, K., and Bergstrom, R. E., 1967, Glacial drift in Illinois: thickness and character: Illinois Geol. Survey Circ. 416, 33 p.

Pissart, A., 1963, Les traces de pingos du Pays de Galles (Grande Bretagne) et du plateau des Hautes Fagnes (Belgique): Zeitschr. Geomorphologie, v. 7, p. 147-165.

_____ 1968, Pingos, Pleistocene, *in* Fairbridge, R. W., ed., The encyclopedia of geomorphology: New York, Reinhold Corp., p. 847-848.

Porsild, A. E., 1938, Earth mounds in unglaciated arctic northwestern America: Geog. Rev., v. 28, p. 46-58.

Quinn, J. H., 1968, Prairie mounds, *in* Fairbridge, R. W., ed., The encyclopedia of geomorphology: New York, Reinhold Corp., p. 888-890.

Rosenkrantz, A., 1942, A geological reconnaissance of the southern part of the Svartenhuk Peninsula, West Greenland: Medd. Grönland, Bd. 135, nr. 3, 72 p.

Royse, C. F., and Callender, E., 1967, A preliminary report on some ice-walled lake deposits (Pleistocene), Mountrail County, North Dakota, *in* Clayton, L., and Freers, T. F., eds., Glacial geology of the Missouri Coteau and adjacent areas: North Dakota Geol. Survey Misc. Ser. 30, p. 53-62.

Sharp, R. P., 1942, Periglacial involutions in northeastern Illinois: Jour. Geology, v. 50, p. 113-133.

Smith, H.T.U., 1949, Periglacial features in the Driftless Area of southern Wisconsin: Jour. Geology, v. 57, p. 196-215.

_____ 1962, Periglacial frost features and related phenomena in the United States: Biul. Peryglacjalny, no. 19, p. 325-342.

Svensson, H., 1964, Traces of pingo-like frost mounds: Lund Studies Geography, Ser. A, Phys. Geography, v. 30, p. 93-106.

_____ 1969, A type of circular lake in northernmost Norway: Geog. Annalar, v. 51A, p. 1-12.

Vail, R. G., 1969, Origin of the paha topography in the Garden Plain Upland, Whiteside County, Illinois (M.A. thesis): DeKalb, Ill., Northern Illinois Univ.

Wayne, W. J., 1967, Periglacial features and climatic gradient in Illinois, Indiana, and western Ohio, east-central United States, *in* Cushing, E. J., and Wright, H. E., Jr., eds., Quaternary paleoecology: New Haven, Yale Univ. Press, p. 393-414.

Wiegand, G., 1965, Fossile Pingos in Mitteleuropa: Würzburger Geog. Arb., v. 16, 152 p.

Willman, H. B., and Frye, J. C., 1970, Pleistocene stratigraphy of Illinois: Illinois Geol. Survey Bull. 94, 204 p.

Winters, H. A., 1960, Landforms associated with stagnant glacial ice: Prof. Geographer, v. 13, p. 19–23.
Wolfe, P. E., 1953, Periglacial frost-thaw basins in New Jersey: Jour. Geology, v. 61, p. 133–141.

MANUSCRIPT RECEIVED BY THE SOCIETY DECEMBER 27, 1971

GEOLOGICAL SOCIETY OF AMERICA
MEMOIR 136
© 1973

Tunnel Valleys, Glacial Surges, and Subglacial Hydrology of the Superior Lobe, Minnesota

H. E. WRIGHT, JR.

*Department of Geology and Geophysics,
University of Minnesota, Minneapolis, Minnesota 55455*

ABSTRACT

Wide stream-cut trenches, now filled with lakes, swamps, and underfit streams, transect the drumlin plain of the Superior Lobe in a subparallel pattern trending southwest, oblique to the modern regional slope and drainage. They are pictured as the products of high-velocity streams in subglacial tunnels, driven by the great hydrostatic pressure resulting from the thick mass of still-active ice. The water for such flow cannot have come from the glacier surface, because the cold upper part of the ice could not permit its penetration. It must have come rather from the base, through melting by the geothermal flux or by the frictional heat of basal ice flow. The collecting basin for such water may have extended to the center of the ice sheet in the Hudson Bay area, and the water may have been stored for thousands of years until it could break through the frozen toe of the Superior Lobe and "catastrophically" cut the tunnel valleys during its exit.

As the ice thinned to stagnation and the hydrostatic head was lost, the subglacial streams changed their habit from erosional to depositional, forming small eskers along the trenches.

After extensive wastage of the Superior Lobe at the end of the St. Croix phase, the ice readvanced to or slightly beyond the rim of the Lake Superior Basin at least three times. The later readvances, which involved the overriding of proglacial lake beds and the deposition of

251

red clayey till, were not closely synchronized with advances of adjacent ice lobes, and the pollen sequence for northeastern Minnesota shows no pattern of climatic reversals that matches ice-lobe fluctuations. Accordingly, the hypothesis is presented that nonclimatic factors may have controlled the most recent minor advances of the Superior Lobe. Subglacial meltwater may have built up beneath the then-restricted Superior Lobe and behind the dam afforded by the cold ice of the thin toe until the reduction in basal friction permitted rapid glacial flow (a surge) through the dam.

INTRODUCTION

The till plain of east-central Minnesota is distinguished by a series of long trenches extending for 150 km from the head of the Lake Superior Basin. The trenches cut through drumlins and contain small eskers. They trend obliquely to the regional slope, which controls the modern drainage to the south, and they are now marked only by lakes, swamps, and underfit streams. The pattern is coincident with the area covered by the distal part of the Superior Lobe during the St. Croix phase of Wisconsin glaciation (Fig. 1), and it is believed to have been formed by subglacial streams as the ice began to thin.

The subglacial origin of the trenches was first proposed by Cushing (in Wright and others, 1964), who subsequently visited the tunnel valleys of Denmark and realized the similarity in form and manner of origin. The tunnel valleys of Denmark were first described by Ussing (1903), and Woldstedt (1952) and others have studied many additional ones in adjacent Schleswig-Holstein, where their downstream portions in moraines are occupied by long lakes fronted by large outwash fans. Sissons (1961) attributes systems of branching channels in various areas of Scotland to stream erosion beneath stagnant ice, and Woodland (1970) relates some deeply buried valleys in East Anglia (worked out from drill holes) to the Danish tunnel valleys of the last glaciation.

The mechanics of formation of tunnel valleys has never been worked out, however, especially concerning the source of water. This paper will attempt to explain those thermal and hydrologic conditions of the Superior Lobe necessary for the formation of subglacial streams that could erode tunnel valleys and then deposit eskers. Further consequences of subglacial hydrologic conditions will be considered, in support of a hypothesis that the later, shorter advances of the ice out of the deep Lake Superior Basin may represent glacial surges.

DESCRIPTION OF TUNNEL VALLEYS

The tunnel valleys start in the low divide between the Lake Superior Basin and the Minneapolis Lowland, west of Sandstone, Pine County (Fig. 2), where five shallow subparallel trenches occur in an area about

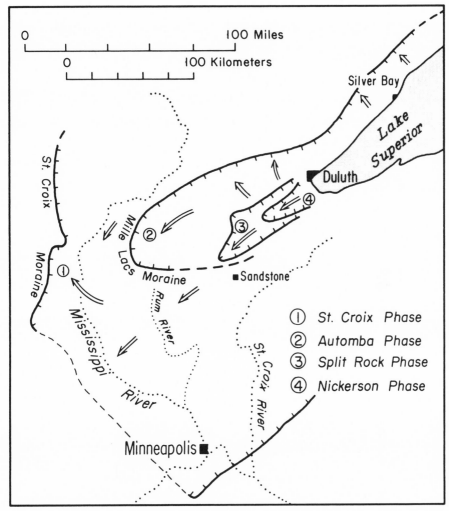

Figure 1. Sketch showing the areas covered by ice during successive phases of the Superior Lobe in Minnesota during the Wisconsin glaciation.

25 km wide, then trend southwestward for almost 150 km to the Mississippi River and the St. Croix Moraine. Although the trenches are partially obscured in the southwestern portion of the area by younger drift, as many as 12 can be recognized across a breadth of about 100 km. The pattern is accordingly fan shaped.

The tunnel valleys are generally straight or slightly curved. Some of them branch and rejoin. They range in width from less than 180 m to more than 1,000 m; perhaps an average is about 300 m. The longest continuous trench can be traced over the entire length of the pattern; others fade out to the southwest and northeast. The aggregate length of all trenches exceeds 1,000 km.

Depths of the tunnel valleys range from a few meters to a general

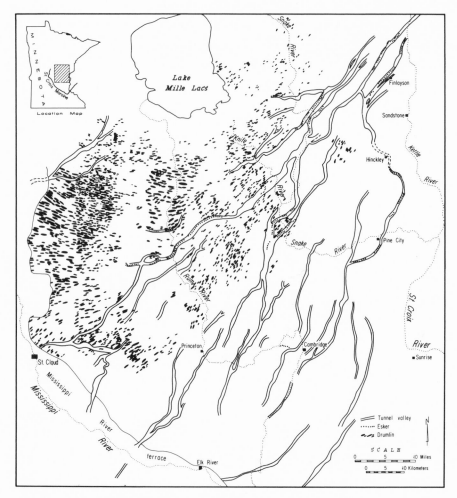

Figure 2. Map showing location of tunnel valleys, drumlins, and eskers of the Superior Lobe in east-central Minnesota.

maximum of 30 m (including lake water and lake sediment), with perhaps 10 m being a good average. The valley walls range from straight, steep, stream-cut slopes to highly irregular ice-contact slopes (Fig. 3). By far the deepest valley is at Grindstone Lake close to the divide near Sandstone. The lake surface is 15 m below the till plain, and the lake contains 45 m of water and probably at least 10 m of sediment, so the valley was cut at least 70 m into the substratum. A map of the bedrock topography in eastern Minnesota, being compiled by G. F. Lindholm and others of the U.S. Geological Survey, indicates that the Grindstone tunnel valley, as well as some of the others, extend into the bedrock.

Possibly the Grindstone Valley represents an old outlet gorge for some pre-Wisconsin proglacial lake in the Lake Superior Basin, formed in

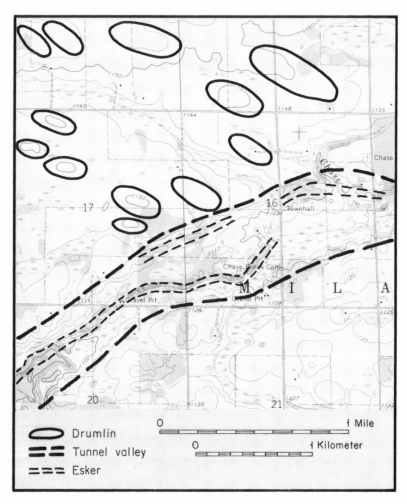

Figure 3. Section of well-developed tunnel valley and esker 5 km northwest of Milaca, Minnesota. The valley here is about 15 m deep, and the main esker is 10 m high. Note the additional small esker on the northern flank of the valley. Drumlins on the upland are outlined as ovals. (From Milaca quadrangle map of the U.S. Geological Survey. Contour interval 10 ft.)

a manner similar to the late-Wisconsin Kettle River gorge 10 km to the east, so it may not fit the integrated tunnel-valley pattern under discussion. Another valley, marked by the string of lakes leading southwest from the big bend of the St. Croix River at Sunrise, may mark a pre-Wisconsin course of the St. Croix River to join the Mississippi River north of Minneapolis. However, such old courses cannot explain the many other trenches that form the integrated pattern.

Many of the tunnel valleys contain eskers, whose crests are commonly at about the level of the till plain adjacent to the valley—an arrangement that produces the marginal troughs by which many eskers can be most easily recognized.

FORMATION OF TUNNEL VALLEYS

Relation to Form of Superior Lobe

Clues to the manner of formation of tunnel valleys and their associated eskers come from their relation to the regional slope of the bedrock and the configuration of the ice sheet in the St. Croix phase. The tunnel valleys trend subparallel to the contours of the gentle northwestern flank of the Minneapolis Lowland, whereas the postglacial streams of this region, such as the Rum and Snake Rivers, flow southward down that slope. This means that the tunnel-valley courses were determined not by the slope of the substratum but by something else, specifically the slope of the inferred ice surface.

The Superior Lobe filled the Minneapolis Lowland and flowed to the St. Croix Moraine; the ice thickness decreased gradually in the direction of flow, which was southwestward down the axis of the lowland. The hydrostatic pressure gradient was thus in the same direction, and any subglacial water must have flowed out principally in that direction.

Temperature Profiles in Ice Sheets

The existence and rate of production of water in an active Pleistocene ice lobe is difficult to determine, for various reasons. The situation is best examined first by assuming an inactive ice mass, which is comparable in certain respects to permafrost in its thermal relations. In the case of a polar glacier, the temperature near the surface (that is, at a depth of about 10 m) is not appreciably affected by seasonal fluctuations of the weather, and it approximates the mean annual atmospheric temperature. The thermal gradient from there to depth depends primarily on the geothermal flux, which is not much different in ice from the flux through rock: a gradient of about 2° C per 100 m is normal. Ice can, therefore, exist to the depth at which the pressure-melting temperature is reached. (The pressure-melting temperature decreases about 0.6° C per 1,000 m of depth in ice.) Thus, static glacial ice 100 m thick with a temperature of −5° C at the surface might have a temperature of −3° C at its base, and no water would exist. If the surface temperature were 0° C, then the gradient to the base should be slightly reversed, for it would be controlled by the pressure-melting temperature. In this case the geothermal heat would be trapped at the base of the ice, as it could not be conducted upward against the reverse gradient and would melt the ice. The geothermal flux should melt the basal ice (at a rate of about 0.5 cm per yr) and thereby produce water for subglacial streams.

In active glacial ice, the thermal gradients are complicated by the fact that heat is produced by the friction of glacial flow, especially at the base of the ice. Flow also transfers heat through the glacier. In the center of an equilibrium ice sheet, the snow is progressively buried and is thus transferred vertically downward (Robin, 1955). If the rate of burial (that is, the rate of snow accumulation) is greater than the rate of thermal conduction from the heat sources mentioned above, then cold ice moves downward so that the thermal gradient is gentle in the upper part of the ice (slight change with depth) and then steeper below.

Away from the center of an ice sheet, but still in the accumulation area, the direction of ice flow relative to the surface is not only downward but also outward and then parallel to the base; cold ice from high elevation is carried to depth and then forward, making the deep-ice gradient even steeper. Robin (1955) calculated a thermal profile for the Greenland ice sheet from known values for vertical flow (accumulation rate) and for horizontal flow. Temperature measurements subsequently made in the deep boreholes in both Greenland and Antarctica confirm his calculations, especially if slightly different values are used in the equations (Weertman, 1968; Dansgaard and Johnsen, 1969). In the Greenland borehole, the ice at the base is still subfreezing, but in Antarctica the temperature reaches the pressure-melting point at the base of the ice, and water exists there under pressure (Gow and others, 1968).

In the ablation area, the upper ice, with its gentle thermal gradient, is removed by melting, so the gradient from the new surface downward should be relatively steep, and the pressure-melting temperature should be reached at a shallower depth than is the case in the accumulation area. If the floor of the glacier rises toward the terminus, then warm ice from depth will be carried up the incline, further steepening the thermal gradient at depth.

Because of the differences between thermal profiles in the accumulation area and the ablation area on a glacier, the location of the boundary between these two areas, called the equilibrium line, is critical in attempts to reconstruct the thermal conditions at depth. The equilibrium line reflects a complex of meteorological and thermal conditions. In northwestern Greenland on the Thule Peninsula (an area of relatively heavy snowfall), it has an elevation of about 600 m above sea level (Benson, 1962), and the mean annual air temperature is about $-12°$ C (Mock and Weeks, 1965). The equilibrium line rises southward, and in west-central Greenland it has an elevation of 1,500 m, with a near-surface snow temperature of about $-15°$ C (Fig. 4).

Temperature Profiles in the Superior Lobe

In the case of the Superior Lobe in the St. Croix phase, some figures can be postulated for atmospheric temperature and ice thickness, and

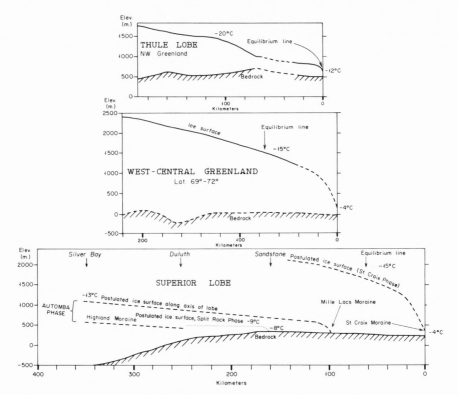

Figure 4. Surface and subsurface profiles of the Greenland ice sheet for the Thule Lobe (Clarke, 1966) and for west-central Greenland (*from* Holtzscherer, *in* Bader, 1961). Postulated profiles and spot temperatures are indicated for the Superior Lobe in the St. Croix phase, Automba phase, and Split Rock phase (Fig. 1).

the model can be evaluated to determine the gross distribution of basal meltwater and thus the possible existence of subglacial streams. Figure 4 illustrates the relations.

The subglacial topographic profile for the Superior Lobe can be reconstructed from borings, with the assumption that all drift is of Wisconsin age. The profile extends along the axis of the entire western part of Lake Superior Basin to the divide near Sandstone and thence into the Minneapolis Lowland to the St. Croix Moraine (Fig. 4). It shows especially clearly the deep trough off Silver Bay, where almost 300 m of water is underlain by glacial sediments that are proven by drilling to be more than 200 m thick (Zumberge and Gast, 1961) and are thought from echo sounding to be at least 300 m thick (Wold, *in* Farrand, 1969).

The surface profile of the ice in the St. Croix phase can be constructed with less accuracy. It seems simplest to reproduce the profile for west-central Greenland (Fig. 4). Thus, the ice might have reached a thickness of at least 2,000 m over the Sandstone Divide, and well over 3,000 m at the Silver Bay Trough.

Reconstruction of the thermal profiles must start with the surface;

that is, with the mean annual atmospheric temperature. This can be estimated for the immediate periglacial area from pollen studies that suggest the nature of the vegetation and thus the climate. For the Superior Lobe at its maximum in the St. Croix phase, periglacial sites are not sufficiently close to the moraine to provide an accurate picture, but the best indications are that closed spruce forest extended very close to the ice edge, and the belt of tundra in southern Minnesota was narrow, if it existed at all (Watts, 1967). There are scattered occurrences of fossil ice-wedge polygons in Illinois and southwestern Wisconsin, implying permafrost, but they are undated. Today in central Canada the mean annual air temperature at the northern edge of the closed coniferous forest is about −4° C (Brown, 1967; Rowe, 1959), and such a temperature may be assumed for the immediate periglacial area in the St. Croix phase. The temperature was certainly below −1° C, which is the temperature at the absolute southern limit of discontinuous permafrost in Canada (Brown, 1968), because stagnant ice, which is thermally equivalent to permafrost, persisted for several thousand years after the front of the Superior Lobe retreated from the St. Croix Moraine, even in well-drained outwash plains (Florin and Wright, 1969).

Surface temperatures of the glacier itself depend largely on elevation. Observations in Greenland show that the surface air-temperature lapse rate with elevation in much of the accumulation zone approximates the dry-adiabatic rate of −1° C per 100 m elevation, because the air is dominated by katabatic winds flowing by gravity downslope off the glacier (Mock and Weeks, 1965). In the lower parts of the accumulation zone the gradient is steeper, because percolating meltwater freezes and stores the heat. This effect should be the reverse in the ablation zone, because the meltwater tends to run off the impermeable ice surface and remove the heat from the system (Schytt, 1968). Also, the greater influence of cyclonic storms at low elevation (and, in Greenland, at a lower latitude) should reduce the temperature-elevation gradient toward the wet-adiabatic lapse rate of 0.6° C per 100 m.

Reconstruction of the temperature-elevation gradient for the surface of the Superior Lobe might make use of the relations in the main ice sheet in west-central Greenland, which is supplied by numerous cyclonic storms from the south, as was Minnesota in the Pleistocene. In this part of Greenland (lat 69° N.) the mean annual air temperature at the ice margin is about −4° C. The equilibrium line, at an elevation of 1,500 m, has a near-surface snow temperature of about −15° C, presumably within a degree or so of the mean annual surface air temperature. These two mean temperature figures give an over-all lapse rate of 0.7° C per 100 m across the ablation zone (Fig. 4).

For the Superior Lobe in the St. Croix phase, the equilibrium line might thus have had an elevation of about 1,700 m, with a surface temperature of −15° C. The postulated ice thickness is 1,400 m. The vertical temperature gradient below this position can then be extrapolated to depth, following neither the very gentle upper-ice gradient of the accumulation zone (Robin, 1955) nor the inferred oversteepened upper-ice gradient of the ablation zone, but rather (as a first approximation) deep-ice gradients such as are found in Greenland (1.7° C per 100 m; Hansen and Langway, 1966) or Antarctica (2.9° C per 100 m; Gow and others, 1968). Such gradients would bring the temperature of the Superior Lobe to the pressure-melting point at depths of 800 and 500 m, respectively, well above the base of the ice for this area. Below this point, the gradient should follow the pressure-melting temperature, which decreases with depth. Such a reversal of gradient could not exist in a steady-state glacier, however; the over-all gradient would thus be flattened to about 1° C per 100 m. At any rate, water could clearly exist at the base of the ice.

From here northward, in the accumulation area of the ice sheet, both the latitude and the surface elevation were greater, so the surface temperature was less. The thermal profile was presumably gentle or even nearly isothermal in the upper part of the ice, depending on the rate of accumulation, so that cold ice extended to greater depth than farther south. On the other hand, the ice was thicker, especially over the Lake Superior Basin, so it seems likely that the basal ice was still temperate. Still farther north, at the height of land between Lake Superior and Hudson Bay, the ice was thinner, but this area was so far removed from the moisture source that the rate of snow accumulation must have been less; the over-all thermal gradient in the ice may, therefore, have been steeper than it was over the Lake Superior Basin, so the pressure-melting temperature may still have been reached at the base of the ice.

Thus, most of the ice sheet may have been temperate at its base, whereas the toe may have been frozen. For the Superior Lobe in the St. Croix phase, the figures given above imply a frozen toe only about as wide as the terminal moraine. Although a cold ice mass precludes basal melting, it does not preclude the passage of water in distinct channels (that is, tunnel valleys) at velocities and pressures high enough to remain fluid. The thermal relations in such an environment are complex, involving the steepness of the thermal gradient from the flowing water into the cold ice adjacent to it, the hydrostatic pressure of the water, the cross section of the stream, and the frictional heat produced by the turbulent flow of the water. The best analogy is to the ground water and river systems in permafrost regions (Muller, 1947), in which Siberian rivers are described as flowing year-round on permafrost. Intrapermafrost ground-water bodies (Muller, 1947) are said to force their way through permafrost to the surface, causing icings, or to intermediate levels, causing domings (pingos, frost mounds). In stagnant debris-covered ice in Alaska, McKenzie (1969) describes the sudden burst of entrapped water. LaChapelle (1971, oral commun.) discovered bodies of entrapped water when digging a tunnel in the Blue Glacier of Washington, which is active. Apparently the ice movement itself can both create and destroy subglacial openings, and ice can change to water and back again in an intricate way, as localized pressure changes alter the freezing point. A subglacial stream, once formed, might carry enough heat to thaw the substratum in the frozen toe enough to facilitate its erosion (for frozen ground is notoriously difficult to excavate).

Postulated Sequence of Events

The Superior Lobe advanced to its terminus at the St. Croix Moraine, where it remained for hundreds of years. Debris was eroded by regelation (Weertman, 1961; Boulton, 1970) from beneath the thick, warm ice of the Lake Superior Basin, and it was transported to the frozen toe, where it became frozen to the substratum (that is, was deposited in the St. Croix Moraine). In the zone of transportation the basal ice was loaded with debris and the basal temperature was at the pressure-melting point. These conditions were favorable for the formation of drumlins, which may only require locally high concentrations of rock debris to initiate deposition (Smalley and Unwin, 1968), plus basal sliding to permit deformation of the till being deposited, and thereby the development of the till fabric that characterizes drumlins. The drumlin field starts about 40 km southwest of the Sandstone Divide and extends to the St. Croix Moraine west of the Mississippi River (Schneider, 1961). It makes a single integrated pattern (Pierz Drumlin Field) almost 100 km long, so it must have been formed during the time of glacial equilibrium when the terminus rested at the St. Croix Moraine.

Meanwhile, water could be building up beneath the thicker ice of the Lake Superior Basin. The escape of accumulated water from beneath the ice sheet could provide the immense water volume and high velocity needed to cut the tunnel valleys. The water was thus forced out over the sill near Sandstone. It collected into channels, which diverged slightly but maintained a course generally parallel to the gradient of the ice surface, cutting the tunnel valleys through the drumlin fields all the way to the ice terminus at the St. Croix Moraine. The water velocities must have been high enough to permit the tunnels to stay open against the pressure of the still-flowing ice. An alternative possibility, that the valleys are wide because they were cut by smaller streams by lateral planation, is not reasonable. The small esker streams that inherited sections of the tunnel valleys were depositing rather than eroding streams, and they are indeed minute in comparison with the tunnel valleys. Glacial streams almost always deposit, rather than erode, because under normal hydrostatic pressure their velocities are insufficient to carry the load supplied. The principal exceptions are glacial streams that have lost their load in glacial lakes. Under the hydrostatic (cryostatic) pressure provided by the ice sheet, however, the velocity is high enough not only to transport the load but to keep the tunnel open.

Surficial Source of Water

Before attempting to evaluate quantitively the process of tunnel-valley formation, a supplemental source of water must be considered. This is the water that could come from surface melting in the ablation zone. The problem exists, however, in leading the water to the base of the ice. Crevasses generally extend to depths of only 20 to 30 m, because at greater depths they are closed by glacier flow. The heat supplied by the water itself is insufficient to extend a hole to depth by melting, even if the water falls the entire depth in a glacial mill. Glen (1954) postulates that if water in a crevasse or hole is at least 100 m deep— perhaps because of the presence of a superglacial lake above a crevassed area—then the greater density of the water provides the pressure to keep the hole open; in fact, he calculates that the hole should widen at depth because of this water pressure. Such a mechanism is used by Glen to explain the sudden draining of ice-dammed lakes. In the absence of a superglacial lake, however, some other mechanism is necessary to open the ice to a depth of 100 m to 150 m to allow the critical water pressure to develop; such a mechanism might be provided by deep "tectonic" fracturing by glacial flow (Stenborg, 1969).

On the other hand, the englacial and subglacial drainage systems studied in northern Sweden by Stenborg (1969) were in temperate glaciers, whereas in the Superior Lobe the ice was cold to a depth of many hundred meters, according to the thermal gradients already discussed. Although fast-flowing water in tunnels in cold ice might maintain sufficient heat to keep from freezing, neither the Glen mechanism nor Stenborg's fracturing provides conditions for continuous flow, and any channels that might open during the melt season should tend to close by glacial flow when the source of surficial meltwater is cut off. Therefore it seems unlikely that subglacial tunnel valleys gained any water from surficial melting.

Quantitative Relations

Any attempt to quantify the relations concerned with the erosion of tunnel valleys is difficult because several factors are unknown. Stream erosion is controlled primarily by water velocity, which in turn reflects gradient, channel characteristics, sediment load, and discharge.

Gradient in this case is supplemented by the hydrostatic pressure provided by the ice mass, and this can be estimated from guesses of the ice thickness.

Channel width can be measured, but not depth, because the roof of a tunnel is of undetermined height. Channel roughness can be grossly estimated.

Sediment load is unknown, but presumably it is small enough so that the bed load cannot protect the floor from erosion. No pavement of boulders occurs in the channel bottom, as was the case with the outlet channel of glacial Lake Agassiz (Matsch and Wright, 1967), so the water velocity must have been great enough to move them until they were broken up.

Discharge might be computed from rate of meltwater production, the area draining into the tunnel valleys, and the time of storage before discharge. The geothermal flux can melt 0.5 cm of ice per year, and frictional heat of basal flow might initially provide another few centimeters (Weertman, 1969). The area from which the water might be drawn includes primarily the area covered by the Superior Lobe; that is, the Minneapolis Lowland and the Lake Superior Basin. The distance from the terminus at the St. Croix Moraine back to the height-of-land in Ontario north of Lake Superior is about 800 km, and the breadth of the lobe from the North Shore Highland of Minnesota to the Keweenaw Peninsula of northern Michigan is about 100 km. This makes an area of about 80,000 km^2. The facts that the tunnel valleys do not widen downstream and have few tributaries imply that most of the water did not come from melting ice in the Minneapolis Lowland but rather from farther north. In fact, it is postulated above that the entire southern part of the Laurentide ice sheet may have been temperate at depth, except for the frozen fringe where the ice was too thin. (A case can be made that the thickness of an ice sheet may increase until it is limited by basal melting.) Lines of equal postglacial isostatic uplift indicate that the ice sheet was thickest over Hudson Bay. Meltwater produced beneath this great area must have flowed in the direction of the hydrostatic gradient to low points in the substratum near the margin of the ice sheet. Such a low point was the bedrock divide near Sandstone in eastern Minnesota, where the tunnel valleys start. The fact that the Lake Superior Basin was eroded at least 400 m below sea level by the ice implies concentrated ice flow in this lobe, and the flow of basal meltwater may likewise have been concentrated in this direction. If the collecting area for basal meltwater is thus increased beyond the Lake Superior Basin to extend to the presumed center of the ice sheet, a distance of 1,100 km, the collecting area for basal meltwater might be 200,000 rather than 80,000 km^2.

The volume of basal water might have been greatly increased if its thickness grew to the point at which it exceeded the controlling obstacle size and began to surge. (The case for surging during later phases of the Superior Lobe is discussed later.) The frictional heat from the basal sliding of a surging glacier might produce 300 cm of basal meltwater annually (Weertman, 1969)—perhaps enough to feed the tunnel valleys without any necessity of providing storage water from a huge drainage basin. For the St. Croix phase, there is no particular evidence for surging at the terminus, but one might postulate periodic surgelike pulses of accelerated glacier flow out of the deep Lake Superior Basin whenever the basal water film built up to a sufficient thickness. Such pulses might not be transmitted all the way to the terminus with sufficient magnitude to have a record of multiple moraines, but they might supply extra basal water that could escape via tunnel valleys. In fact, if the basal water that may have initiated a push in the Lake Superior Basin became collected into tunnel valleys beyond the Sandstone Divide, then it could not maintain a surge, which requires a film rather than channeled flow (Weertman, 1969).

If the frozen toe of the ice sheet was sufficiently thick to impede the escape of subglacial meltwater, this water could effectively accumulate beneath the ice until the hydrostatic pressure became great enough (with increasing ice thickness) to extrude the water through the frozen toe (that is, to melt its way out along certain lines without losing all its heat to the cold ice that it was forced to penetrate in the frozen toe). The presence of shear zones in the toe may have aided in establishing points of extrusion. Perhaps the stresses engendered downstream by a surge might locally fracture the ice and open enough subglacial pathways to release the accumulated basal water layer and convert it to channel flow. The accumulated water could escape only through the ice, because the substratum, consisting mostly of compacted till or of discontinuously jointed crystalline bedrock, was insufficiently permeable to transmit the water.

The time available for accumulation of basal meltwater might amount to many hundreds of years, as the ice front stood at the St. Croix Moraine. During much of this time, the water did not escape, at least in discrete channels, because the drumlins that formed show no sign of contemporaneous water flow. It is even possible that basal water accumulated in deep basins throughout the thousands of years or even the tens of thousands of years of growth of the Wisconsin ice sheet. However, basal heat might have been largely removed by glacier flow during this long time, so it is not possible to judge further the extent of water storage. That water storage and sudden escape may have been involved is implied by the fact that most tunnel valleys contain eskers, formed when the water velocity was reduced so much (perhaps by exhaustion of the stored water) that the tunnels partially closed and the sediment could no longer be transported. This change to esker formation, of course, could alternatively have been caused by thinning of the ice and thus the reduction in the hydrostatic pressure.

Once the outlet points were established, channels could form and the entire series of tunnel valleys might be cut simultaneously and nearly catastrophically, somewhat in the manner of the Channeled Scablands of the Columbia Plateau (Bretz, 1969). In fact, the local crossing of tunnel valleys resembles the intersecting channels of the scablands.

If 3 cm of water were produced by basal melting and normal (nonsurging) frictional heat over an area of 200,000 km^2 and accumulated for 1,000 yrs, the water volume would amount to 6,000 km^3. This is more than the amount of water released by glacial Lake Missoula in forming the Channeled Scablands (Pardee, 1942), which are much larger erosional features than the tunnel valleys of the Superior Lobe. A similar water volume is calculated for the overflow of pluvial Lake Bonneville to form the flood features along the Snake River in southern Idaho (Malde, 1968). The volume of rock removed by the Bonneville floodwaters in the Snake River Canyon is estimated to be 1.2 km^3. If the Superior Lobe tunnel valleys have a cumulative length of 1,000 km, an average width of 0.3 km, and average depth of 10 m, the amount of material removed is 3 km^3.

These crude calculations must not be considered very seriously because the size of the drainage basin, duration of water storage, duration of cutting, and other factors are so difficult to estimate, as are the original dimensions of the tunnel valleys. On the other hand, they suggest the order of magnitude of the geologic process in comparison with cases of catastrophic stream erosion in which something is known of the parameters, including the time dimension.

It has been assumed so far that the tunnel valleys were cut simultaneously throughout their length, perhaps even catastrophically. It is possible, however, that they were not all occupied at the same time, or that they were cut progressively headward. Nonsynchronous cutting, with superposition, can also explain the occasional crossing of two valleys, plus the fact that the northern parts of some of the valleys were apparently occupied by streams at the time of the Split Rock phase of the Superior Lobe as well as the St. Croix phase. The eskers were presumably developed progressively headward as the marginal ice thinned to stagnancy; perhaps the tunnel valleys formed in the same manner just before. In this case, the cutting extended over the many hundreds of years of ice retreat, and it could not be considered as catastrophic. However, the meltwater discharge would then probably have reflected contemporaneous basal melting rather than stored water—perhaps insufficient to do the job of cutting, inasmuch as the few centimeters of basal water supplied each year could not produce a large enough water volume to maintain the large streams postulated. If supplemental water derived from surface melting cannot penetrate the cold shallow ice, and if "surge water" is not added, then it seems that the only source of enough basal water is storage, and thus "catastrophic" outflow. For these reasons it is concluded that the tunnel valleys (although perhaps not the eskers) were not formed progressively headward.

The drumlin field and the tunnel valley system can both be traced to the St. Croix Moraine, so the Superior Lobe must have remained at its maximum stand throughout the time of formation of both features, even though the subglacial condition changed from one of drumlin deposition without signs of meltwater to one of tunnel-valley formation by streams. As one explanation, the change in condition could have resulted from climatic warming great enough to shift the entire temperature profile to the warm side. This would broaden the ablation zone, and an increased rate of glacial flow would be required to maintain the terminus at the same moraine. Because the condition of tunnel-valley formation was followed immediately by the condition of esker formation, implying stagnant ice and thus decreased thickness, the preceding transition from the drumlin condition to the tunnel-valley condition may reflect the same trend of decreasing thickness and thus climatic warming.

On the other hand, as a second explanation, the drumlins may have formed when the basal meltwater back of the frozen toe was confined to a thin and discontinuous film, while within the Lake Superior Basin, where the lower part of the ice was warmer, the basal water was accumulating to greater thickness. Then, when exits were found through the frozen toe, the stored water gathered in channels and eroded the tunnel valleys.

Eskers

The subsequent conversion of a tunnel-valley stream from an eroding to a depositing habit to form an esker reflects the loss of stream velocity either through exhaustion of the stored water or through reduction in hydrostatic pressure (that is, thinning of the ice during wastage). The broad tunnels became partly plugged by ice flow as the water pressure decreased, and the esker streams were much smaller than the eroding streams that preceded them. Eskers were deposited on the floor or on the flank of the valley, or sometimes even on the plain beside the valley (Fig. 4). Gaps in the eskers probably represent nondeposition places where water velocity was locally great enough to transport the sediment. Esker crests are only locally higher than the tunnel valleys in which the eskers are located.

During the interval of intensive wastage, most of the meltwater was confined to the esker streams. The swales in the drumlin plain were only locally filled with outwash sediments. Ultimately, the exposed ice wasted completely, although stagnant ice blocks protected by a mantle of drift persisted for thousands of years (Florin and Wright, 1969), gradually melting slowly from the bottom upward by crustal heat or, if the mantle was accidentally removed, by surface melting.

SUBSEQUENT HISTORY OF SUPERIOR LOBE

Sequence of Glacial Phases

The ice beneath which the tunnel valleys and eskers formed ultimately wasted completely from the Minneapolis Lowland, except for stagnant ice blocks. It readvanced in the Automba phase with a different alignment, sending a tongue of ice westward from the head of the basin to the Lake Mille Lacs area of central Minnesota (Fig. 1). At that time the northwestern margin of the ice lobe rose only to the crest of the highland north of Lake Superior, where it formed the Highland Moraine. The

Rainy Lobe was confined to northeasternmost Minnesota, where it formed the Vermilion Moraine. Ice lobes in western Minnesota were in an unknown location, but they were small in extent compared to later time.

Retreat of the Superior Lobe into the Lake Superior Basin at the end of the Automba phase led to the formation of proglacial lakes in which red clay was deposited. Readvance of the ice over the lake beds to the rim of the basin in the Split Rock phase produced red clayey till (Fig. 1). This occurred at least 16,000 yrs ago, according to radiocarbon dates on basal organic sediments from lakes associated with its outwash. The very distant advance of the Des Moines Lobe from western Minnesota eastward across the state occurred about this time, causing the burial over much of the area of drumlins and tunnel valleys and eskers bared by the much earlier retreat of the Superior Lobe of the St. Croix phase. The differential action of eastern and western ice lobes is thus striking.

The Superior Lobe pushed up to its rim at least one more time, about 12,000 yrs ago, in the Nickerson phase (Fig. 1). Its moraine is marked in part by a frontal outwash plain and by huge eskers that represent the escape of subglacial water.

Nonsynchronous Advances of Adjacent Ice Lobes

The chronology of juxtaposed eastern and western ice movements in central and eastern Minnesota is not entirely clear, but there is enough information to suggest that adjacent ice lobes did not in all cases advance and retreat to a comparable extent at all times. Successive phases of ice advance are generally assumed to reflect distinct climate fluctuations, because the regime of glaciers is controlled primarily by the balance between snowfall in the accumulation area and wastage in the ablation area, along with the factor of glacier flow from one area to the other. The lack of synchroneity in the advances of adjacent ice lobes might be attributed to changes in the locus of maximum snow accumulation on the outer part of the ice sheet. In the case of the ice lobes that affected Minnesota, the sources of protrusion from the main ice sheet—and thus the presumed centers of snow accumulation—were, respectively, in northeastern Ontario, western Ontario, and southern Manitoba. The total distance from east to west is about 1,100 km. Leverett (1932) built the concept of westward shifting of centers of snow accumulation on the basis of the then-simple sequence of a single mid-Wisconsin advance of the Superior Lobe and other lobes to the east followed by a late-Wisconsin advance of the Des Moines and James Lobes. This concept still makes sense as a general view of the westward expansion of the Laurentide ice sheet during Wisconsin glaciation, although the role of the early advance of the Wadena Lobe from the northwest is not completely clear (nor was the significance of the now-discounted Iowan glaciation clear in Leverett's day).

A second explanation for nonsynchronous ice-lobe advance is that snow accumulation increased uniformly over the entire accumulation fringe of the ice sheet but that the distance of ice flow differed for the ice lobes involved. For example, the Superior Lobe may have extended only 800 km from its main source north of Lake Superior, whereas the Des Moines Lobe came perhaps 1,200 km from its source area in the Manitoba Lowland. If the flow rates were the same, then the maximum position of the Des Moines Lobe might postdate that of the Superior Lobe, even though both advances were impelled by the same climatic change in the accumulation area. The retreats of the two lobes might also be nonsynchronous, because the arrangement would also involve a lag for the longer ice lobe.

It is difficult to evaluate these two hypotheses, because the actual loci of increased snow accumulation cannot be easily established. Evidence presented earlier indicates that during the time involved the equilibrium line on the ice was probably close to the margin. The accumulation zone was at least partly on the lobes themselves rather than entirely on the ice sheet proper. The east-west distance between accumulation centers for the three ice lobes, and the distance of travel to the ice-lobe margins, may have been appreciably less than the figures given, in which case the time differences for maxima would be less.

These two hypotheses depend basically on climatic change, for which independent evidence may be sought in the periglacial region. One of the most useful criteria for climatic change is the pollen sequence found in lake sediments that span the time of interest. Enough sites have been studied in the Minnesota area to indicate that tundra bordered the ice margins in northeastern Minnesota from about 20,000 yrs ago until at least 11,500 yrs ago, when spruce forest began to cover the area. A reversal in the pollen sequence indicating one or more climatic fluctuations that might match the ice-margin fluctuations, although suggested in some of the early work (Wright and others, 1963), has not been subsequently confirmed (Cushing, 1967). Current studies by John Birks at the University of Minnesota of absolute pollen influx at a key site vitiate the evidence on which the earlier suggestion was made. The vegetational change was apparently unidirectional, as it has been shown to be in a similar situation in New England (Davis, 1967), and implies unidirectional climatic change. Although pollen analysis may not be sufficiently sensitive to detect climatic changes of the magnitude implied by ice-margin fluctuations, it is worth considering a nonclimatic cause for ice-margin fluctuations—the hypothesis that an individual ice lobe may advance independently in a surge.

Surging

Recent interest in surging valley glaciers and improvements in the understanding of the mechanics of glacier flow have led to suggestions that surging may have been important in Pleistocene ice sheets (Weertman,

1962). Very rapid advances of valley glaciers have been recognized at least since Tarr (1907) described the glacier surges in the Yakutat Bay region of Alaska. The great advance of the very accessible Black Rapids Glacier in the Alaska Range in 1937 focused attention on the subject. The great display of stagnant ice it produced at its terminus is in obvious contrast to the terminal portions of most Alaskan glaciers (Hance, 1937). The 1966–1968 surge of the Steele Glacier in the Yukon was reported in newspapers throughout America, and it triggered in turn two symposia on the subject (Ambrose, 1969).

Evidences for surges have now been reported for more than 200 glaciers in western North America alone (Meier and Post, 1969; Horvath and Field, 1969), but their occurrence has no obvious pattern that might reveal the mechanism. Of interest in the present problem is the good evidence for surging of ice-cap lobes on Iceland (Thorarinsson, 1969) and Spitsbergen (Schytt, 1969), where the relevant features approach the dimensions of Pleistocene ice lobes in the Great Lakes region.

Changes in flow velocity occur in two orders of magnitude. The slower one involves the progression of a kinematic wave down a glacier (Meier, 1965), started perhaps by temporary increase in accumulation either directly, through snowfall, or indirectly, such as by earthquake avalanches. The wave moves more rapidly than the ice itself. Such a wave has been followed down the Nisqually Glacier in Washington (Meier, 1965), and it can be expected to result in an advance of the ice front.

The faster type of change, more properly called a surge, apparently involves no such wave or no such change in mass budget. In some cases, only the lower reach of the glacier is affected. The ice just seems to stretch out within this reach and suddenly advances at the margin at velocities 10 to 100 times the normal velocity. Causes for such surges can be grouped in four general hypotheses, some of which are partly related. The first two involve thermal instability; the second two involve dynamic instability.

In the first hypothesis, it is known that glacial flow is highly sensitive to temperature, and that an increase in ice temperature might be enough to bring about a rapid increase in ice flow (Robin, 1969). A change in climate, addition of volcanic heat from the crust, or change in some other factor could increase the ice temperature and thus the rate of flow.

A particularly effective way in which the thermal factor can operate is through the production or the thickening of a film of water at the base of the ice (Weertman, 1962), which is the second hypothesis. Although controversy persists over whether the film is continuous and only a few millimeters thick (Weertman, 1969) or discontinuous and up to several centimeters thick (Lliboutry, 1969), there is a consensus that basal slip may be the major mechanism of ice flow as long as the thermal gradient brings the temperature at the base of the ice to the pressure-melting point. Inasmuch as the speed of ice flow results from a complex of factors, it is difficult to analyze the thermal relations quantitatively any more than in the case for the formation of tunnel valleys.

The inception of basal melting could thus initiate basal slip by decreasing the inertial friction, and the consequent ice flow could generate more heat by friction. It could also reduce some of the obstacles by erosion, thereby increasing the effectiveness of the basal film. These mechanisms could result in a surge of flow, which would be expended when the ice became so thin that the thermal relations were changed and the basal slip greatly reduced or eliminated.

The third hypothesis to explain surging depends on the presence of variations in bed profile that are large enough to affect ice thickness and the mode of flow (Robin, 1969). If, for example, an increase in accumulation causes an increase in flow down to the ablation zone, the extra ice will pile up behind an obstruction in the bed and raise the ice surface as a kind of dam, until the increase in height and the storage of ice behind the dam are enough to initiate rapid extending flow, that is, a surge.

The fourth hypothesis also postulates a dam, but it involves a different mechanism of flow (Nielsen, 1968). The terminal area of a glacier is commonly thin, drift-filled, and stagnant, and it forms a rigid dam against which the active ice piles up. The dam ultimately yields to the compressive stress and surges forward. The tensional crevasses that develop in the ice during its release produce blocks that are fluidized by the available intraglacial water, resulting in a flow velocity many times greater than that achieved by normal glacier flow. The total surge that develops produces a thin and stagnant terminus that can provide the dam necessary for a repetition of the process at a later date.

The possibility must be considered that the nonsynchroneity between the advances of the various ice lobes in Minnesota may be attributed to surging (Wright, 1969), especially because of the lack of good pollen evidence for regional climatic fluctuations. The case is best for the Split Rock and Nickerson phases of the Superior Lobe, because of favorable morphologic and thermal relations, and because a surge hypothesis is consistent with the explanation for tunnel valleys associated with the same ice lobe at an earlier phase. The hypothesis may also be proposed for certain other Wisconsin ice advances in the Great Lakes region, notably the Valders advance of the Lake Michigan Lobe (Wright, 1971).

The morphologic and thermal relations for the Superior Lobe in the Split Rock phase may be reconstructed from some of the data for the Automba phase in Figure 4. The gradient of the ice surface in the Automba phase can be postulated from the fact that the Highland Moraine (essentially a lateral moraine) along the crest of the North Shore Highland slopes from an elevation of 570 m above sea level opposite Silver Bay to 450 m opposite Duluth, a distance of 100 km. The center of the ice lobe must have been higher so that the ice could flow straight to the lateral margin and produce the fluted forms leading to the Highland

Moraine. The gradient along the axis of the lobe was probably steeper than that along the lateral margin; it may have been as steep as the gradient of the Thule Lobe in northwestern Greenland, and it is so shown in Figure 4. The narrow ice tongue (only 65 km broad) spreading the remaining distance from the Sandstone area to the terminal moraine at Mille Lacs may have had an average thickness of, say, 300 m, which is the thickness determined seismically for the outer 75 km of the Thule Lobe in northwestern Greenland (Clarke, 1966).

For the Split Rock phase itself the ice was certainly thinner at the Sandstone Divide, because it did not cross the divide nor climb the rock escarpment to the west. A thickness of 150 m is shown in Figure 4. The gradient of its north-lateral margin from the Sandstone Divide to the Duluth area, as reproduced from a series of ice-front outwash plains, is very gentle, perhaps reflecting the distension caused by surging. Between Duluth and Silver Bay the lateral margin left no clear limit, but the ice over the Silver Bay Trough was at least 1,000 m thick.

The thermal profile for the Superior Lobe in the Split Rock phase can be estimated from paleobotanical data. The periglacial area is believed from pollen studies to have featured tundra vegetation until at least 11,500 yrs ago. Plant-macrofossil analysis confirms the presence of tundra-type plants (*Dryas integrifolia, Salix herbacea, Vaccinium uliginosum,* and others) for samples in which no needles or other macro-remains of spruce, larch, or other trees are found (Wright and Watts, 1969). The northern tree line at this time was an unknown distance to the south—probably less than 75 km, but certainly more than the few kilometers that might be attributed to local cooling by katabatic winds flowing off the ice lobe.

Comparison of the location of vegetation belts today in northern Canada (Rowe, 1959) with the climatic and permafrost data (Brown, 1967) indicates that the boundary of the open coniferous forest with the continuous tundra runs close to the $-8°$ C line of mean annual air temperature, and close to the southern limit of continuous permafrost. Therefore, the tundra area bordering the Superior Lobe may have had a mean annual temperature of $-8°$ C or colder. The thin ice toe over the Sandstone Divide (150 m higher and thus 1° C colder, assuming a lapse rate of $-0.7°$ C per 100 m) must certainly have been frozen to the base, but the ice over the Silver Bay Trough must have been thick enough so that basal melting could occur there.

The thermal situation is thus favorable for surging. In the terminal area the thin, cold ice, slow-moving and frozen to the bed, acts as a dam not only to the subglacial water film but to the rapidly moving thick ice from upstream, where basal melting promotes basal slip. Geothermal and frictional heat increases the thickness of the basal water film until the ice practically floats, being anchored only on the higher obstructions on the bedrock floor. The dam mechanism, as conceived by either Robin (1969) or Nielsen (1968), could operate, but the conditions for surging are more favorable in the case of the Superior Lobe because the steep reverse slope of the bed permits a more abrupt change in flow characteristics. When the ice behind the dam piles up to a sufficiently great thickness, the dam breaks, and the ice surges forward.

When the Superior Lobe was much larger, during the St. Croix phase, basal meltwater had built up beneath the deep ice in a similar way, according to the hypothesis presented, but it escaped through subglacial tunnel valleys. At that time the frozen toe of the ice was much thicker and broader, and the dam that it produced was too strong to be overcome by the pressure of the ice behind. But when the ice became largely confined to the Lake Superior Basin, in the Split Rock and Nickerson phases, the dam was smaller and weaker. Presence of a proglacial lake at the toe may have brought about local melting of the permafrost and warming and weakening of the ice front. Tunnel valleys were probably still operating at this time in a minor way, for a series of frontal eskers and outwash fans were formed along the northwestern flank of the lobe in the Split Rock phase, representing the discharge of major subglacial streams. But presumably all of the subglacial meltwater did not escape by this route, and some may have remained to provide for the rapid basal slip required for surging.

SUMMARY AND CONCLUSIONS

The remarkable pattern of stream-cut trenches that were cut into the drumlin plain of the Superior Lobe in east-central Minnesota contains eskers in close association, and study of the three features—drumlins, trenches, and eskers—leads to the conclusion that they represent successive subglacial hydrologic conditions during ice wastage. The trenches (tunnel valleys) were formed by subglacial streams under the hydrostatic pressure caused by the great ice thickness, with the water being derived not from surface runoff but from subglacial melting from crustal heat and the friction of ice flow, perhaps after long storage behind a "dam" represented by cold ice at the toe of the ice lobe.

Thinning ice ultimately reduced the hydrostatic pressure and opened the tunnels, whereby the streams shifted from an erosional to a depositing habit, producing eskers.

The question can be raised that the tunnel valleys of Denmark (Ussing, 1903) and Schleswig-Holstein (Woldstedt, 1952) had circumstances even more favorable for the mechanism proposed here. In those studies the ice is assumed to have been temperate throughout and thus subject to downward penetration by surface meltwater. But this part of Europe had a more frigid periglacial climate than Minnesota, according to the evidence for tundra and frozen ground that reaches as far south as the Alps. Perhaps the water source for these features should be reconsidered. The morainic area is bordered up-glacier by the troughs and deeps of the western Baltic Sea east of Denmark, providing a condition favorable for deep temperate ice upstream from a frozen toe.

In a search for descriptions of features elsewhere in the Great Lakes region that may have had an origin as tunnel valleys, the submerged valley system in the eastern part of the Lake Superior Basin was encountered. These valleys were cut 100 m to 300 m into bedrock by streams flowing southward across the structure, and they end abruptly against the south coast of Lake Superior (Farrand, 1969). They may have been formed when the ice lobe that had previously fed the Valders advance of the Lake Michigan and Green Bay Lobes stood with its front along the south coast of the eastern Lake Superior Basin, with the outwash from the valleys being spread southward to make the great south-sloping sandplain north of Lake Michigan (Wright, 1971).

Later events in the history of the Superior Lobe indicate short readvances of the ice just to the southern rim of the Lake Superior Basin, not clearly matched by advances of other ice lobes. This possible nonsynchroneity in the behavior of adjacent ice lobes, along with the lack of independent evidence from pollen analysis for climatic reversals, leads to the hypothesis that ice may have advanced out of the Lake Superior Basin in a series of surges, not only in the Superior Lobe but also in the Lake Michigan Lobe. The hypothesis can be tested by sharpening the chronology of ice advance on adjacent ice lobes and by identifying criteria in the glacial drifts themselves from which surging can be inferred.

ACKNOWLEDGMENTS

The Minnesota Geological Survey supported the field studies on which this paper is based. Discussions in the field with E. J. Cushing, Roger Hooke, Ronald Shreve, Charles Matsch, Paul Conlon, and numerous students and Friends of the Pleistocene have been very helpful in some of the glaciological problems. Their searching questions sharpened the interpretations and exposed many potential loopholes.

REFERENCES CITED

Ambrose, J. W., ed., 1969, Seminar on the causes and mechanics of glacial surges, and symposium on surging glaciers: Canadian Jour. Earth Sci., v. 6, p. 807–1018.

Bader, Henri, 1961, The Greenland ice sheet: U.S. Army Materiel Command Cold Regions Research and Eng. Lab. Research Rept., I-B2, 18 p.

Benson, C. S., 1962, Stratigraphic studies in the snow and firn of the Greenland ice sheet: U.S Army Snow, Ice and Permafrost Research Establishment Research Rept. 70, 93 p.

Boulton, G. S., 1970, On the origin and transport of englacial debris in Svalbard glaciers: Jour. Glaciology, v. 9, p. 213–230.

Bretz, J. H., 1969, The Lake Missoula floods and the Channeled Scabland:

Jour. Geology, v. 77, p. 505-543.

Brown, R. J. E., 1967, Permafrost in Canada: Canada Geol. Survey Map 1246A.

—— 1968, Permafrost investigation in northern Ontario and northeastern Manitoba: Natl. Research Council Canada Div. Bldg. Research, Tech. Paper 291, 40 p..

Charlesworth, J. K., 1957, The Quaternary Era, with special reference to its glaciation: London, Edward Arnold, 2 vols., 1700 p.

Clarke, G. K. C., 1966, Seismic survey northwest Greenland, 1964: U.S. Army Materiel Command Cold Regions Research and Eng. Lab. Research Rept. 191, 19 p.

Cushing, E. J., 1967, Late-Wisconsin pollen stratigraphy and the glacial sequence in Minnesota, in Cushing, E. J., and Wright, H. E., Jr., eds., Quaternary paleoecology: New Haven, Conn., Yale Univ. Press, p. 59-88.

Dansgaard, W., and Johnsen, S. J., 1969, Comment on paper by J. Weertman, "Comparison between measured and theoretical temperature profiles of the Camp Century, Greenland, borehole": Jour. Geophys. Research, v. 74, p. 1109-1110.

Davis, M. B., 1967, Late-glacial climate in northern United States: A comparison of New England and the Great Lakes region, in Cushing, E. J., and Wright, H. E., Jr., eds., Quaternary paleoecology: New Haven, Conn., Yale Univ. Press, p. 11-44.

Farrand, W. R., 1969, The Quaternary history of Lake Superior: Internat. Assoc. Great Lakes Research Proc., 12th Conf., Ann Arbor, p. 181-197.

Florin, Maj-Britt, and Wright, H. E., Jr., 1969, Diatom evidence for the persistence of stagnant glacial ice in Minnesota: Geol. Soc. America Bull., v. 80, p. 695-704.

Glen, J. W., 1954, The stability of ice-dammed lakes and other water-filled holes in glaciers: Jour. Glaciology, v. 2, p. 316-318.

Gow, A. J., Ueda, H. T., and Garfield, D. E., 1968, Antarctic ice sheet: Preliminary results of first core hole to bedrock: Science, v. 161, p. 1011-1013.

Hance, J. H., 1937, The recent advance of Black Rapids Glacier, Alaska: Jour. Geology, v. 45, p. 775-783.

Hansen, B. L., and Langway, C. C., Jr., 1966, Deep core drilling in ice and core analysis at Camp Century, Greenland, 1961-1966: Antarctic Jour. U.S., v. 1, p. 207-208.

Horvath, E. V., and Field, W. D., 1969, References to glacier surges in North America: Canadian Jour. Earth Sci., v. 6, p. 845-852.

Leverett, Frank, 1932, Surficial geology of Minnesota and parts of adjacent areas: U.S. Geol. Survey Prof. Paper 161, 149 p.

Lliboutry, L. A., 1969, Contribution à la théorie des ondes glaciaires: Canadian Jour. Earth Sci., v. 6, p. 943-954.

Malde, H. E., 1968, The catastrophic late Pleistocene Bonneville flood in the Snake River Plain, Idaho: U.S. Geol. Survey Prof. Paper 596, 52 p.

Matsch, C. L., and Wright, H. E., Jr., 1967, The southern outlet of Lake Agassiz, in Mayer-Oakes, W. J., ed., Life, land and water: Winnipeg, Manitoba, Manitoba Univ. Press, p. 121-140.

McKenzie, G. D., 1969, Observations on a collapsing kame terrace in Glacier Bay National Monument, south-eastern Alaska: Jour. Glaciology, v. 8, p. 413-425.

Meier, M. F., 1965, Glaciers and climate, in Wright, H. E., Jr., and Frey,

D. G., eds., The Quaternary of the United States: Princeton, N.J., Princeton Univ. Press, p. 795–806.

Meier, M. F., and Post, A., 1969, What are glacier surges?: Canadian Jour. Earth Sci., v. 6, p. 807–818.

Mock, S. J., and Weeks, W. F., 1965, The distribution of ten-meter snow temperatures on the Greenland ice sheet: U.S. Army Materiel Command Cold Regions Research and Eng. Lab. Research Rept. 170, 44 p.

Muller, S. W., 1947, Permafrost or permanently frozen ground and related engineering problems: Ann Arbor, Mich., J. W. Edwards, Inc., 231 p.

Nielsen, L. E., 1968, Some hypotheses on surging glaciers: Geol. Soc. America Bull., v. 79, p. 1195–1201.

Pardee, J. T., 1942, Unusual currents in glacial Lake Missoula, Montana: Geol. Soc. America Bull., v. 53, p. 1569–1599.

Robin, G. de Q., 1955, Ice movement and temperature distribution in glaciers and ice sheets: Jour. Glaciology, v. 2, p. 523–532.

_____ 1969, Initiation of glacier surges: Canadian Jour. Earth Sci., v. 6, p. 919–928.

Rowe, J. S., 1959, Forest regions of Canada: Canada Dept. Northern Affairs and Nat. Resources Forestry Br. Bull. 123, 71 p.

Schneider, A. F., 1961, Pleistocene geology of the Randall region, central Minnesota: Minnesota Geol. Survey Bull. 40, 151 p.

Schytt, Valter, 1968, Notes on glaciological activities in Kebnekaise, Sweden, during 1966 and 1967: Geog. Annaler, v. 50A, p. 111–120.

_____ 1969, Some comments on glacier surges in eastern Svalbard: Canadian Jour. Earth Sci., v. 6, p. 867–874.

Sissons, J. B., 1961, A subglacial drainage system by the Tinto Hills, Lanarkshire: Edinburgh Geol. Soc. Trans., v. 18, p. 113–123.

Smalley, I. L., and Unwin, D. J., 1968, The formation and shape of drumlins and their distribution and orientation in drumlin fields: Jour. Glaciology, v. 7, p. 377–390.

Stenborg, Thorsten, 1969, Studies of the internal drainage of glaciers: Geog. Annaler, v. 51A, p. 13–41.

Tarr, R. S., 1907, Recent advances of glaciers in Yakutat Bay region, Alaska: Geol. Soc. America Bull., v. 18, p. 257–286.

Thorarinsson, Sigurdur, 1969, Glacier surges in Ireland, with special reference to the surges of Bruarjökull: Canadian Jour. Earth Sci., v. 6, p. 875–882.

Ussing, N. V., 1903, Om Jyllands Hedesletter og Teorierne om deres Dannelse: Copenhagen, Overs. k. danske Vid. Selsk Förh., Nr. 2, p. 99–165.

Watts, W. A., 1967, Late-glacial plant macrofossils from Minnesota, in Cushing, E. J., and Wright, H. E., Jr., eds., Quaternary paleoecology: New Haven, Conn., Yale Univ. Press, p. 89–97.

Weertman, J., 1961, Mechanism for the formation of inner moraines found near the edge of cold ice caps and ice sheets: Jour. Glaciology, v. 3, p. 965–978.

_____ 1962, Catastrophic glacier advances: Internat. Union Geodesy and Geophys., Internat. Assoc. Sci. Hydrology, Comm. Snow and Ice, Obergürgl Coloq. (1962), p. 31–39.

_____ 1968, Comparison between measured and theoretical temperature profiles of the Camp Century, Greenland, borehole: Jour. Geophys. Research, v. 73, p. 2691–2700.

_____ 1969, Water lubrication mechanism of glacier surges: Canadian Jour. Earth Sci., v. 6, p. 929–942.

Woldstedt, Paul, 1952, Die Entstehung der Seen in den ehemals vergletscherten Gebieten: Eiszeitalter u. Gegenwart, v. 2, p. 146–153.

Woodland, A. W., 1970, The buried tunnel-valleys of East Anglia: Yorkshire Geol. Soc. Proc., v. 37, p. 521–578.

Wright, H. E., Jr., 1969, Glacial fluctuations and the forest succession in the Lake Superior area: Internat. Assoc. Great Lakes Research Proc., 12th Conf., Ann Arbor, p. 397–405.

—— 1971, Retreat of the Laurentide ice sheet from 14,000 to 9,000 years ago: Quaternary Research, v. 1, p. 316–330.

Wright, H. E., Jr., and Watts, W. A., 1969, Glacial and vegetational history of northeastern Minnesota: Minnesota Geol. Survey Spec. Pub. 11, 59 p.

Wright, H. E., Jr., Winter, T. C., and Patten, H. L., 1963, Two pollen diagrams from southeastern Minnesota: Problems in the regional late-glacial and postglacial vegetational history: Geol. Soc. America Bull., v. 74, p. 1371–1396.

Wright, H. E., Jr., Cushing, E. J., and Baker, R. G., 1964, Midwest Friends of the Pleistocene, Guidebook, Eastern Minnesota (Mimeographed, 32 p.).

Zumberge, J. H., and Gast, Paul, 1961, Geological investigations in Lake Superior: Geotimes, v. 6, p. 10–13.

MANUSCRIPT RECEIVED BY THE SOCIETY DECEMBER 27, 1971

GEOLOGICAL SOCIETY OF AMERICA
MEMOIR 136
© 1973

Pleistocene-Holocene Boundary and Wisconsinan Substages, Gulf of Mexico

John H. Beard

Esso Production Research Company, Houston, Texas 77001

ABSTRACT

Late Pleistocene-Holocene climatic fluctuations and environmental conditions in the Gulf of Mexico are reflected by changes in the vertical distribution of planktonic foraminifers in the bottom sediments. The events are related closely by radiocarbon dates to continental Wisconsinan glacial-interglacial substages. The almost inverse relationship between abundances of two species, *Globorotalia menardii* (warm) and *Globorotalia inflata* (cold), allows recognition of three major episodes of climatic cooling during the Wisconsinan. Moreover, minor fluctuations of climate are reflected in detail by the combined distributional patterns of warm- versus cold-water planktonic species. Paleotemperature curves quantitatively derived from the frequency ratio of warm- versus cold-water species from the Gulf of Mexico are strikingly similar to the oxygen-isotope curve of Emiliani (1966) from the Caribbean for about the last 75,000 yrs.

Morphologic changes in the *Globorotalia menardii* group and the withdrawal of cold-water species, such as *Globorotalia inflata*, from the Gulf of Mexico characterize two climatically distinct assemblages. The older assemblage corresponds to the late Pleistocene and is characterized by *Globorotalia menardii flexuosa* (warm) and by *Globorotalia inflata* (cold). Withdrawal of *Globorotalia inflata* occurred between 4,000 and 11,000 yrs ago and corresponds closely to an incursion of abundant *Globorotalia tumida, Globorotalia ungulata,* and other warm-water species. This faunal boundary is interpreted to represent the transition from the last glacial to postglacial conditions in the Gulf of Mexico. Assuming

277

a constant average rate of deposition for pelagic sediment, the age estimated for the boundary agrees very closely with that of the radiocarbon
bracketed date of 7,000 yrs B.P. of Frye and others (1968) for the
Wisconsinan-Holocene boundary.

Paleontologic events and paleotemperature curves from the Gulf of
Mexico correlate almost exactly with those from the Caribbean and adjacent Atlantic. Widely differing opinions expressed by several authors
on Wisconsinan nomenclature in the marine section are based on correlations using different paleontological criteria and different geochemical
dating methods. Carbon-14 determinations reported in this study indicate
ages considerably younger than other published dates for what certainly
appears to be a paleontologically equivalent unit. Climatic events recognized in the Gulf of Mexico can be correlated rather precisely on the
basis of radiocarbon dates with the continental Wisconsinan glacial-interglacial substages and, therefore, development of an independent nomenclature for the marine section is neither necessary nor desirable.

INTRODUCTION

Quaternary glacial and interglacial conditions are recorded in detail
by deep-sea sediments that preserve the remains of past marine life.
Although this record is complete, at least locally, it is not clearly legible
because paleoclimatic interpretations must rely entirely on indirect evidence such as evaluation of planktonic foraminiferal assemblages. Evidence is increasingly convincing that a north-south migration of marine
planktonic organisms resulted from climatic changes related to continental
glaciation.

Vertical changes in species of planktonic foraminifers in marine sediments have been used previously to interpret conditions during glacial
and interglacial times (Phleger, 1954, 1955; Ewing and others, 1958; Ericson and others, 1963, 1964; Ericson and Wollin, 1968). Cold-water faunas
are interpreted to represent glacial stages and warm-water faunas to
represent interglacial stages. Clearly different faunal criteria must be
determined, however, for individual ocean basins and for general latitudinal belts inasmuch as the same planktonic foraminiferal assemblage is
not found from the equator to the poles. A consistent succession of
climatic events seems recognizable even though different species must
be used at different localities.

This paper presents the results of a quantitative study on the stratigraphic distribution of planktonic foraminifers in 23 deep-sea piston cores
collected during 1964 and 1966 in the Gulf of Mexico by Texas A & M
University personnel aboard the R/V Alaminos. Twelve of these cores
(Fig. 1) provide a seemingly complete record of climatic conditions for
about the last 75,000 yrs (Wisconsinan-Holocene). Cores were taken in
water depths ranging from 492 fm to 2,065 fm and are from 552 cm
to 1,160 cm in length (Table 1).

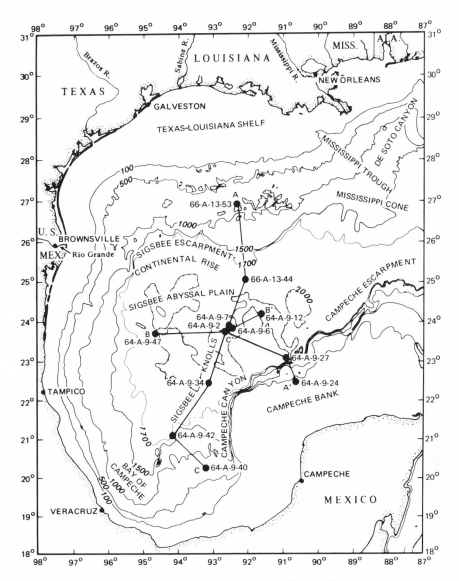

Figure 1. Bathymetric and physiographic province map, Gulf of Mexico, showing location of Texas A & M *Alaminos* cores and lines of cross sections. Depth in fathoms.

Samples (internal plugs) were cut from half-core segments at 10-cm intervals and routinely washed through a 200-mesh screen. Residues were passed through a microsplitter to obtain a representative fraction, and approximately 300 specimens were counted from the fraction greater than 177 microns to determine the coiling ratio of *Globorotalia truncatulinoides* and a frequency ratio of 17 planktonic species. A study of the most complete core showed that certain species occurred sporadically and in very low abundance. Therefore, all but nine of the more tempera-

TABLE 1. LOCATION, DEPTH OF WATER, AND LENGTH OF CORES USED IN THIS STUDY

Core	Position Lat (N.)	Long (W.)	Depth (fm)	Length (cm)
Cruise 64-A-9, R/V *Alaminos*, June 1964				
2	23°40.0'	92°34.0'	1997	910
6	23 45.5	92 26.5	1825	552
7	23 46.0	92 27.9	1895	580
12	24 08.0	91 35.0	2025	840
24	22 50.0	90 42.0	492*	756
27	23 01.0	90 50.0	1995*	761
34	22 25.0	93 06.0	—	1020
40	20 14.0	93 11.0	761	762
42	21 02.0	94 09.0	1466	1160
47	23 40.0	94 33.0	1985	935
Cruise 66-A-13, R/V *Alaminos*, September 1966				
44	25 00.0	92 00.0	1922	790
53	26 53.0	92 17.0	751	780

*Approximate.

ture-sensitive species were dropped from counts of the remaining cores, seemingly without loss of detail. A computer plotted frequency curves for each species and frequency-ratio curves for various combinations of species.

Biostratigraphic correlation between the Gulf of Mexico, Caribbean, and adjacent Atlantic is facilitated by morphologic changes in the planktonic foraminifers. Radiocarbon dates provide a means for precise correlation with the youngest glacial-interglacial stages and substages of the North American continent.

PALEOTEMPERATURE RECORD

Geographic distribution of living planktonic foraminifers clearly shows that certain species are restricted to rather narrow latitudinal belts and to cold- or warm-current systems (Phleger, 1960; Bé, 1959, 1960, 1968, 1969; Bé and Hamlin, 1967; Cifelli, 1962, 1967). The latitudinal distribution of species generally conforms to hydrologic fronts in the deep, open-ocean basin, being controlled by temperature and salinity variations (Boltovskoy, 1969). Species diversity generally increases toward the equator and, conversely, decreases toward the poles.

Modern populations of some species are subdivided into provinces that are characterized by dominance of either right- or left-coiled shells. Ericson and others (1954) demonstrated that specimens of *Globorotalia truncatulinoides* are about 93 percent right coiled in the Gulf of Mexico, whereas in colder water areas of the North Atlantic, they are dominantly left coiled. The ratio between right- and left-coiled shells of *G. truncatulinoides* has changed several times during the Pleistocene and can be used to interpret paleoclimatic conditions.

Species diversity and over-all abundance of planktonic shells, temperature-sensitive species, and coiling direction are useful indicators in the

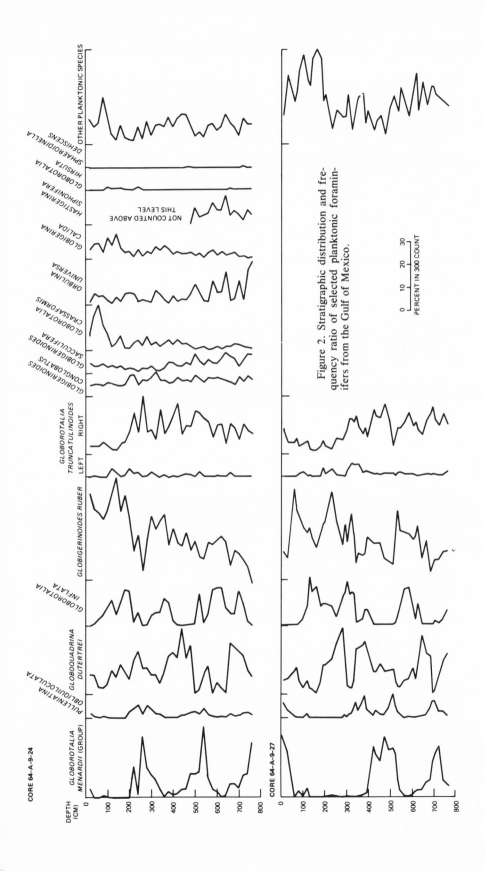

Figure 2. Stratigraphic distribution and frequency ratio of selected planktonic foraminifers from the Gulf of Mexico.

marine section to denote paleoclimatic conditions. In the Gulf of Mexico, productivity of planktonic foraminifers either was low during the glacial maxima, or the increased rate of sediment inflow during times of lowered sea level diluted the relative abundance of planktonic shells, and perhaps both were effective.

Statistical analyses show that *Globorotalia menardii* (warm) and *Globorotalia inflata* (cold) are the most temperature-sensitive species found in the Gulf of Mexico (Figs. 2, 3, 4, 5, 6, and 7). The frequency curves for these two species show an almost inverse relationship that can be used to delineate distinct glacial and interglacial episodes. Abundant *Globorotalia menardii, Globorotalia menardii flexuosa,* and *Globorotalia tumida* distinguish warm maxima. Conversely, *Globorotalia inflata* represents the maximum cold periods, being generally absent or occurring in very low numbers during the warm interglacial periods. The frequency-ratio curves between *Globorotalia menardii* and *Globorotalia inflata* (Figs. 8, 9, 10, and 11) are consistently recognizable and correlatable over much of the Gulf of Mexico, as well as the Caribbean and adjacent Atlantic. The frequency-ratio curves between members of the *Globorotalia menardii* group and *Globigerinoides ruber* consistently have similar patterns. Even though *Globigerinoides ruber* occurs commonly in the warm intervals with *Globorotalia menardii,* its abundance is much greater in the cold intervals when *Globorotalia menardii* is absent or in very low abundance.

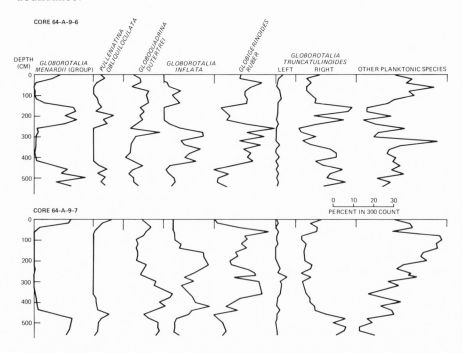

Figure 3. Stratigraphic distribution and frequency ratio of selected planktonic foraminifers from the Gulf of Mexico.

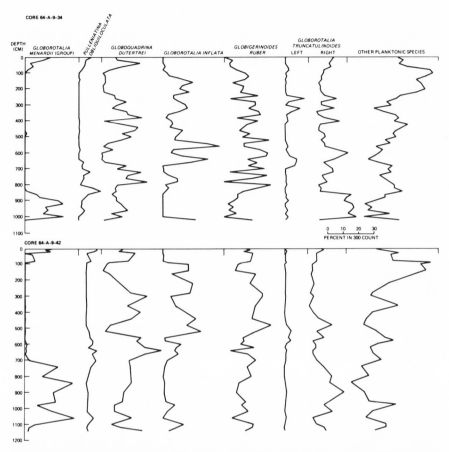

Figure 4. Stratigraphic distribution and frequency ratio of selected planktonic foraminifers from the Gulf of Mexico.

Species useful for distinguishing between glacial and interglacial events in the Gulf of Mexico are as follows (Figs. 12, 13, 14, and 15):

Warm	Cold
Globorotalia menardii	*Globorotalia inflata*
Globorotalia menardii flexuosa	*Globorotalia truncatulinoides* (left coiled)
(= *Pulvinulina tumida flexuosa*	*Globigerinoides ruber*[1]
of Koch)	
Globorotalia tumida	
Globorotalia ungulata	
Globoquadrina dutertrei	
Pulleniatina obliquiloculata	

Paleotemperature curves produced by calculating the frequency ratio of warm- versus cold-tolerant species reflect paleoclimatic conditions

---

[1] This species more abundant in cold intervals, Gulf of Mexico.

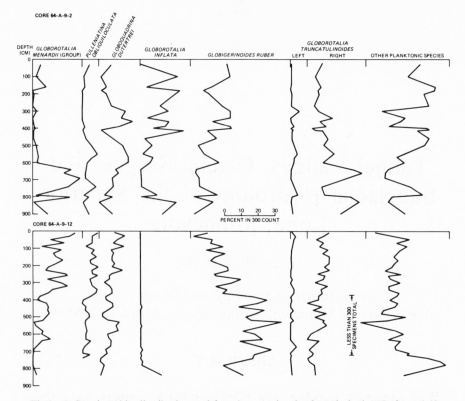

Figure 5. Stratigraphic distribution and frequency ratio of selected planktonic foraminifers from the Gulf of Mexico.

in detail. The curve derived from core 64-A-9-24 of pelagic sediment from the southern part of the Gulf of Mexico compares extremely well with the oxygen-isotope ratio curve of Emiliani (1966) from the Caribbean (Fig. 16). Significantly, both curves delineate numerous minor fluctuations that reflect in detail less severe temperature changes.

Minor temperature changes delineated by using total planktonic assemblages or oxygen-isotope ratios probably reflect paleoclimatic conditions precisely, but are more difficult to correlate reliably from core to core. As more temperature-tolerant species are added, the climate curve is in effect smoothed, and precise boundaries are more difficult to determine. Major changes determined by calculating the frequency ratio between members of the *Globorotalia menardii* group and *Globorotalia inflata*, however, are distinct, and can be correlated reliably with climatic stages described by Ericson and others (1963, 1964) and Ericson and Wollin (1968) from the Caribbean and adjacent Atlantic. Although arguments may arise concerning precise boundaries and the estimated ages for these boundaries, the major climatic events are clearly correlative between widely separated ocean basins, that is, from the Gulf of Mexico to the sub-Antarctic.

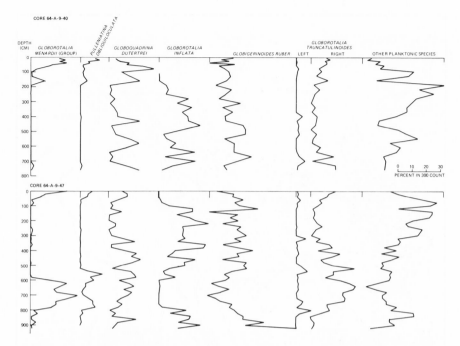

Figure 6. Stratigraphic distribution and frequency ratio of selected planktonic foraminifers from the Gulf of Mexico.

PLEISTOCENE-HOLOCENE BOUNDARY

Various faunal boundaries and the eustatic rise in sea level have been interpreted to represent climatic change from glacial conditions of the Pleistocene to postglacial conditions of the Holocene and are bracketed by radiocarbon dates that range from about 4,000 to 20,000 yrs B.P. (Morrison, 1969). Frye and others (1968, 1969) and Willman and Frye (1970), on evidence from sequences of tills and soil profiles, place the Wisconsinan-Holocene boundary above the Valderan Substage. The date indicated for the boundary is about 7,000 yrs B.P.

Gulf of Mexico

Ewing and others (1958) reported on 33 cores (6 to 12 m long) taken by Lamont Geological Observatory in 1953. The top 30 to 50 cm of each of the long cores from the abyssal plain and lower Mississippi cone is reported to be composed of foraminiferal lutite. This bed reaches its maximum thickness of 4 m on the upper cone and on the upper continental rise. The lower portion of each core consists of gray silty clay. Micropaleontological correlation indicated that the abrupt transition at the base of the foraminiferal lutite represents the Pleistocene-Recent (Holocene) boundary at about 11,000 yrs B.P.

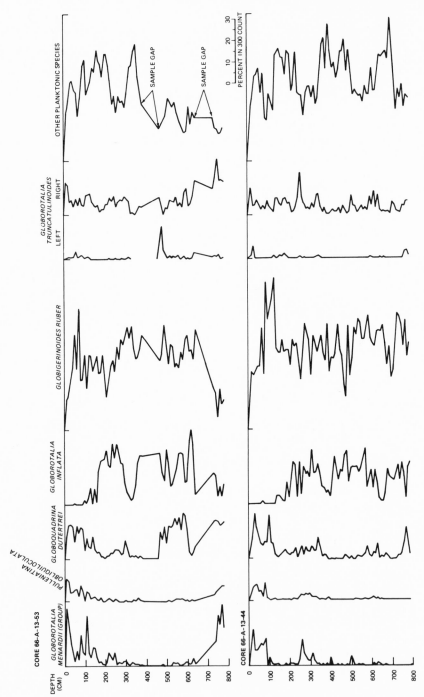

Figure 7. Stratigraphic distribution and frequency ratio of selected planktonic foraminifers from the Gulf of Mexico.

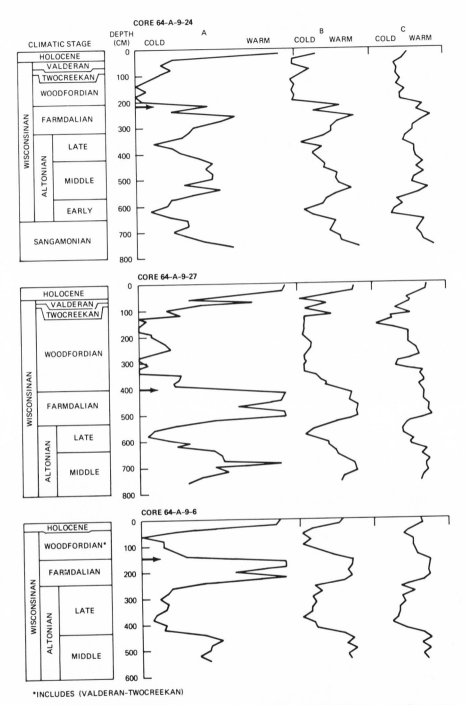

Figure 8. Paleotemperature curves and climatic stages, Gulf of Mexico. A. Frequency ratio of *Globorotalia menardii* group to *Globorotalia inflata*. B. Frequency ratio of *Globorotalia menardii* group to *Globigerinoides ruber*. C. Frequency ratio of warm-water (*Globorotalia menardii* group, *Pulleniatina obliquiloculata*, and *Globoquadrina dutertrei*) to cool-water (*Globorotalia inflata*, *Globigerinoides ruber*, and *Globorotalia truncatulinoides* [left coiled]) species. Arrow indicates last occurrence of *Globorotalia menardii flexuosa*.

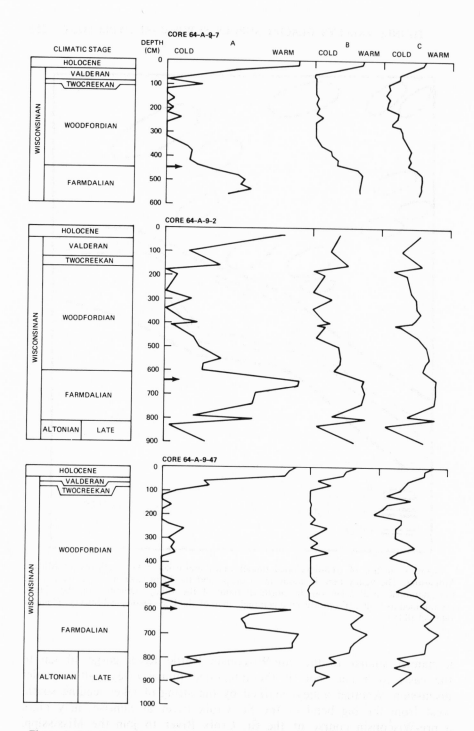

Figure 9. Paleotemperature curves and climatic stages, Gulf of Mexico. A. Frequency ratio of *Globorotalia menardii* group to *Globorotalia inflata*. B. Frequency ratio of *Globorotalia menardii* group to *Globigerinoides ruber*. C. Frequency ratio of warm-water (*Globorotalia menardii* group, *Pulleniatina obliquiloculata*, and *Globoquadrina dutertrei*) to cool-water (*Globorotalia inflata*, *Globigerinoides ruber*, and *Globorotalia truncatulinoides* [left coiled]) species. Arrow indicates last occurrence of *Globorotalia menardii flexuosa*.

TABLE 2. RADIOCARBON DATES DETERMINED ON FORAMINIFERA

Isotopes Sample no.	Sample	$-\delta C^{14}$	Age in yrs B.P.
1-4607	64-A-9-42; 710–980 cm	972 ± 4	28,750 ± 1,200
1-4609	64-A-9-24; 215–350 cm	988 ± 3	35,550 ± 2,600
1-4817	64-A-9-6; 140–240 cm	985 ± 3	33,750 ± 1,950
1-4821	64-A-9-47; 680–750 cm	989 ± 3	36,300 ± 2,500
1-4822	64-A-9-40; 30–100 cm	372 ± 13	3,740 ± 165
1-3174	66-A-13-53; 660–730 cm	970 ± 8	28,200 ± 2,400
1-3495	64-A-9-2; 50–75 cm	727 ± 19	10,430 ± 570
1-3490	64-A-9-2; 505–530 cm	922 ± 9	20,500 ± 1,000
1-3494	64-A-9-2; 705–736 cm	965 ± 4	26,900 ± 1,000
1-3497	64-A-9-2; 832–840 cm	980 ± 3	31,430 ± 1,300

Foraminiferal tests by Isotopes Inc., Westwood, New Jersey.

Morphologic changes in members of the *Globorotalia menardii* group and withdrawal of colder water species, such as *Globorotalia inflata* from the gulf of Mexico, characterize two climatically distinct assemblages. The older assemblage corresponds to the late Pleistocene and is dominated by *Globorotalia menardii flexuosa* (warm) and by *Globorotalia inflata* (cold). Radiocarbon dates from this study (Table 2) indicate that the withdrawal of *Globorotalia inflata* occurred between 4,000 and 11,000 yrs ago and corresponds closely to an incursion of abundant *Globorotalia menardii, Globorotalia tumida, Globorotalia ungulata,* and *Pulleniatina obliquiloculata*. This faunal boundary is interpreted to represent the climatic change from the last glacial to postglacial conditions in the Gulf of Mexico. In eight of the cores representing pelagic accumulations, the average thickness of the Holocene is around 40 cm (Table 3).

Thinness of the Holocene section and relatively widely spaced sample cuts (10 cm) preclude a detailed evaluation of Holocene climatic fluctuation in most of the cores studied. Evidence for minor climatic fluctuation during the Holocene, however, is found in an expanded (370 cm) section in core 64-A-9-12 (Fig. 11) taken east of the Sigsbee Knolls in a small topographic basin. Significant radiocarbon dates could not be obtained, however, because of the small amount of foraminiferal tests per volume of sediment.

TABLE 3. THICKNESS OF HOLOCENE SEDIMENTS IN CORES FROM AREAS OF PELAGIC ACCUMULATION

Core	Thickness of the Holocene (in cm)
64-A-9-2	50
64-A-9-6	30
64-A-9-7	30
64-A-9-24	40
64-A-9-27	50
64-A-9-34	40
64-A-9-42	30
64-A-9-47	50
Average	40

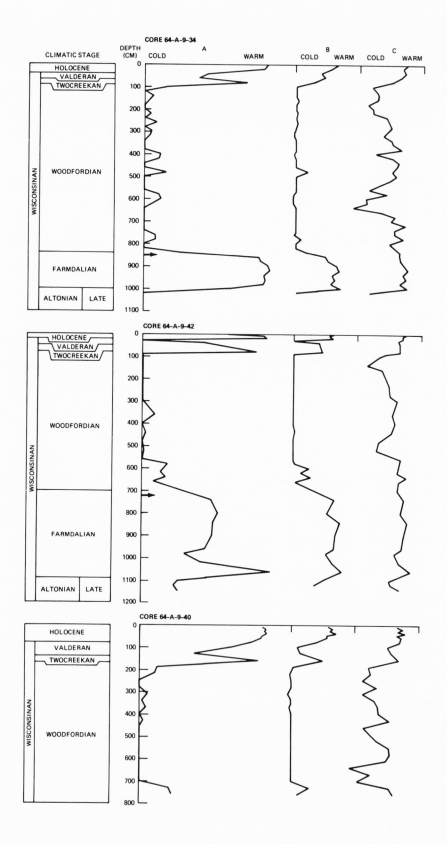

Accepting the 7,000-yr date indicated by Frye and others (1968, 1969) for the Wisconsinan-Holocene boundary, and assuming a constant average rate of sedimentation of about 50 cm per 1,000 yrs for the Holocene in core 64-A-9-12, ages for minor climatic changes are estimated. Assuming that the fluctuating percentage of warm-water species actually reflects minor climatic changes and not productivity, dissolution, or dilution, relative colder temperatures are indicated at about 2,000, 4,500, and 5,600 yrs B.P., the latter two being more significantly cold. More cores through expanded sections with radiocarbon dates are needed before serious consideration can be given to these dates. Detailed knowledge of the periodicity of Holocene climatic fluctuations, however, would substantially increase oúr ability to predict climatic trends for the immediate future. Should the trend of glaciation continue to be cyclic in nature, as it now appears to have been for the past 75,000 yrs or so, a major glacial episode is predicted to occur around the 40th or 50th century.

Evidence from the present study in the Gulf of Mexico indicates that a climatic transition prevailed from about 13,000 to 7,000 yrs B.P., the time represented by the Valderan and Twocreekan Substages of the Wisconsinan. The interval is marked by a change in distribution from about 10 percent *Globorotalia inflata* to greater than 10 percent *Globorotalia menardii* in an interval of about 60 cm in core 64-A-9-24. The lower boundary, the dramatic decrease in the abundance of *Globorotalia inflata*, seemingly occurred about 13,000 yrs ago and corresponds with the end of the cold Woodfordian climate and the beginning of the milder Twocreekan climate. The upper boundary, the dramatic increase in the abundance of the *Globorotalia menardii* group, seemingly occurred around 7,000 yrs B.P. and corresponds with the end of the mild but relatively colder Valderan climate and the beginning of the warm Holocene climate.

Allowing for some minor adjustment in the accuracy of radiocarbon dates based on foraminiferal tests and shell debris, it is clearly evident that some authors selected the Pleistocene-Holocene boundary on the basis of a faunal boundary corresponding to the Woodfordian-Twocreekan boundary at about 12,500 to 13,000 yrs B.P., whereas some selected a faunal boundary corresponding to the midpoint of the climatic transition at about 10,000 yrs B.P. by combining several faunal criteria, and others selected a faunal boundary corresponding to the Valderan-Holocene boundary. The redefinition by Frye and others (1968) that places the end of the Valderan Substage at the contact of the Cochrane Till with thin overlying and discontinuous post-Cochrane deposits in the James

─────────────

Figure 10. Paleotemperature curves and climatic stages, Gulf of Mexico. A. Frequency ratio of *Globorotalia menardii* group to *Globorotalia inflata*. B. Frequency ratio of *Globorotalia menardii* group to *Globigerinoides ruber*. C. Frequency ratio of warm-water (*Globorotalia menardii* group, *Pulleniatina obliquiloculata*, and *Globoquadrina dutertrei*) to cool-water (*Globorotalia inflata*, *Globigerinoides ruber*, and *Globorotalia truncatulinoides* [left coiled]) species. Arrow indicates last occurrence of *Globorotalia menardii flexuosa*.

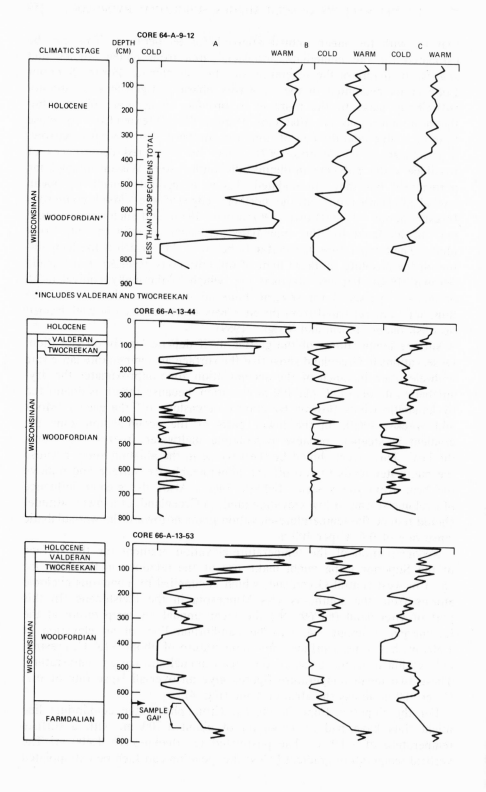

CORE 64-A-9-12

CLIMATIC STAGE

DEPTH (CM)

HOLOCENE

WISCONSINAN

WOODFORDIAN*

LESS THAN 300 SPECIMENS TOTAL

COLD   A   WARM   B   COLD   WARM   C   COLD   WARM

*INCLUDES VALDERAN AND TWOCREEKAN

CORE 66-A-13-44

HOLOCENE
VALDERAN
TWOCREEKAN

WISCONSINAN

WOODFORDIAN

CORE 66-A-13-53

HOLOCENE
VALDERAN
TWOCREEKAN

WISCONSINAN

WOODFORDIAN

FARMDALIAN

SAMPLE
GAP

Bay Lowland of Ontario, Canada, is accepted for defining the Pleistocene-Holocene boundary. This boundary has been radiocarbon dated at about 7,000 yrs B.P.

Atlantic Ocean

Ericson and others (1956) showed that a sharp faunal change in sediments of the major part of the Atlantic Ocean and adjacent seas marks a transition from a relatively cold to the present relatively mild oceanic climate. The climatic change began 13,000 to 15,000 yrs ago, and the radiocarbon-dated midpoint of the major change from glacial to postglacial conditions occurred about 11,000 yrs B.P. The major change appears to be broadly simultaneous throughout the North Atlantic and adjacent seas.

Northeastern Pacific Ocean

Duncan and others (1970) suggest that the Pleistocene-Holocene boundary in the northeastern Pacific Ocean off Oregon and Washington is marked by a change from a dominance of planktonic foraminifers below to a dominance of radiolarians above. The boundary is dated at 12,500 yrs B.P. The authors state, however, that although a rather abrupt change in the faunal composition occurred approximately 12,500 yrs B.P., the interval from about 12,500 to 7,000 yrs B.P. shows a somewhat gradual transition from the planktonic Foraminifera-rich section to the uppermost Radiolaria-rich section.

According to Bandy (1967), a marked change occurred in marine micro-faunas of the southern California area at the end of the Pleistocene about 11,000 yrs B.P. A shift from sinistral (Pleistocene) to dextral (Holocene) population of *Globigerina pachyderma* is accompanied by the appearance in Holocene sediments of dextral populations of *Globigerina subcretacea*, by an increase in numbers of orbulina chambers, and by a dramatic increase in the abundance of radiolarians.

Red Sea

Berggren and Boersma (1969) described late Pleistocene and Holocene climatic fluctuations in the Red Sea on the basis of fluctuating percentages of *Globigerinoides ruber* and *Globigerinoides sacculifera*. The youngest zone, characterized by the dominance of *G. sacculifera*, is considered

---

Figure 11. Paleotemperature curves and climatic stages, Gulf of Mexico. A. Frequency ratio of *Globorotalia menardii* group to *Globorotalia inflata*. B. Frequency ratio of *Globorotalia menardii* group to *Globigerinoides ruber*. C. Frequency ratio of warm-water (*Globorotalia menardii* group, *Pulleniatina obliquiloculata*, and *Globoquadrina dutertrei*) to cool-water (*Globorotalia inflata*, *Globigerinoides ruber*; and *Globorotalia truncatulinoides* [left coiled]) species. Arrow indicates last occurrence of *Globorotalia menardii flexuosa*.

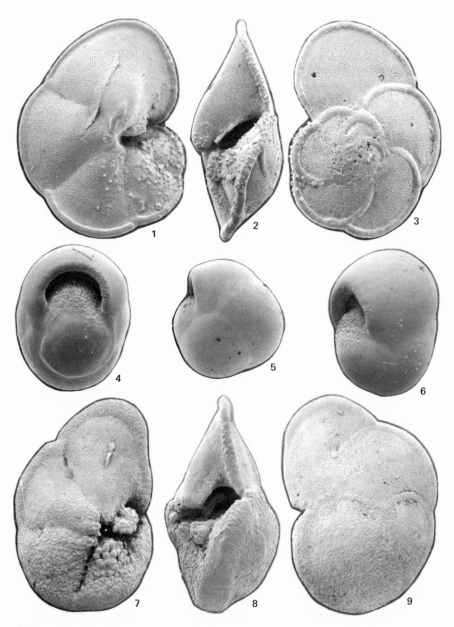

Figure 12. *Globorotalia* and *Pulleniatina* from the Holocene in Texas A & M 64-A-9 piston core number 24 at a depth of 20 to 30 cm. 1-3. *Globorotalia ungulata* Bermúdez; × 61, × 36, × 61. 4-6. *Pulleniatina obliquiloculata* (Parker and Jones) *finalis* Banner and Blow; × 28, × 28, × 28. 7-9. *Globorotalia tumida* (Brady); × 32, × 32, × 28.

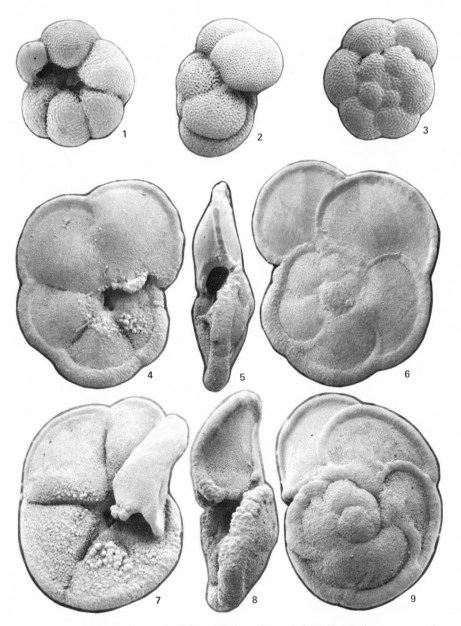

Figure 13. *Globoquadrina* and *Globorotalia* from Texas A & M 64-A-9 piston core number 24. 1–3. *Globoquadrina dutertrei* (d'Orbigny) from the Farmdalian Substage at a depth of 220 to 230 cm; × 28, × 28, × 28. 4–6. *Globorotalia menardii* (d'Orbigny) from the Holocene at a depth of 20 to 30 cm; × 28, × 25, × 25. 7–9. *Globorotalia menardii flexuosa* (Koch) from the Farmdalian Substage at a depth of 220 to 230 cm; × 28, × 25, × 35.

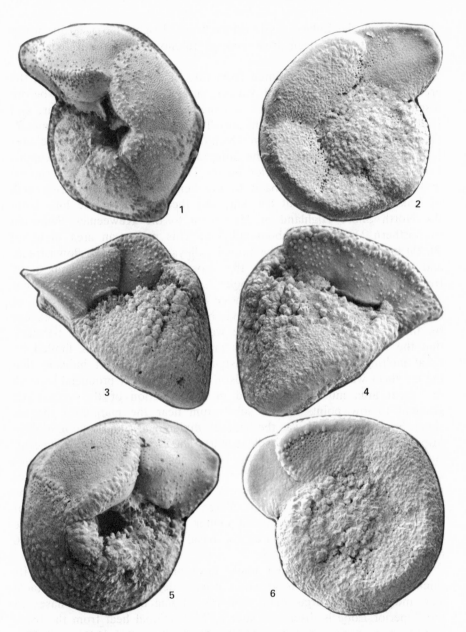

Figure 14. *Globorotalia truncatulinoides* (d'Orbigny) from Texas A & M 64-A-9 piston core number 24. 1-3. Specimens from the Holocene at a depth of 20 to 30 cm; × 50, × 50, × 50. 4-6. Specimens from the Woodfordian Substage at a depth of 160 to 170 cm; × 55, × 54, × 57.

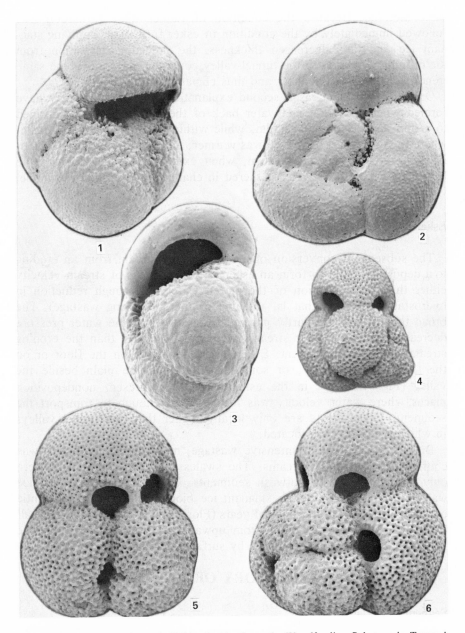

Figure 15. *Globorotalia* and *Globigerinoides* from the Woodfordian Substage in Texas A & M 64-A-9 piston core number 24 at a depth of 160 to 170 cm. 1-3. *Globorotalia inflata* (d'Orbigny); × 50, × 57, × 50. 4-6. *Globigerinoides ruber* (d'Orbigny); × 54, × 65, × 65.

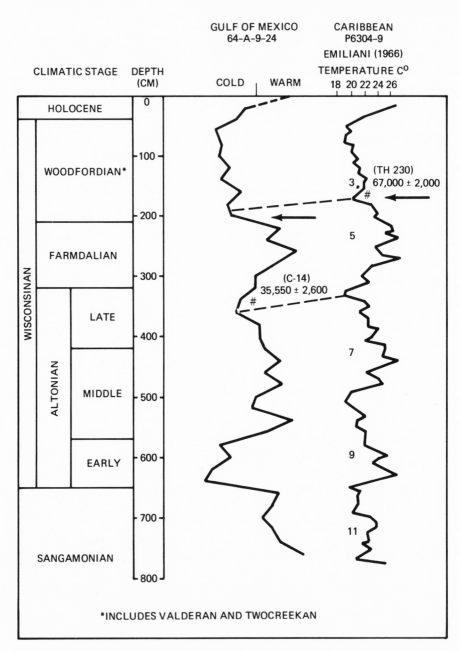

Figure 16. Comparison of paleotemperature curve from the Gulf of Mexico with oxygen-isotope curve of Emiliani (1966) from the Caribbean. Arrows indicate last (youngest) occurrence of *Globorotalia menardii flexuosa*. Numbers alongside Emiliani's curves indicate his letter stages. Depth scale same for both cores.

to represent establishment of a normal Indian Ocean fauna in the Holocene following the Würm glaciation. The base of the zone was dated at about 11,000 yrs B.P.

Indian Ocean

Frerichs (1968), from studies in the Andaman Sea and the Bay of Bengal, concluded that a marked decrease in the relative abundance of the *Globigerina rubescens* group and a significant increase in the radiolarian number marks the Pleistocene-Holocene boundary. A radiocarbon date of 8,775 ± 145 yrs B.P. was obtained from a 38-cm section of core immediately below the faunal boundary.

According to Vincent (1970), the Pleistocene-Holocene boundary in deep-sea cores from the southern Mozambique channel area of the Indian Ocean is marked by changes in the relative abundance of planktonic foraminifers. The temperate species *Globorotalia inflata* decreases markedly in relative abundance in the Holocene; it comprises about 20 percent of the planktonic foraminiferal population below the boundary and only about 3 percent or less above. Conversely, the percentage of warm-water species shows an increase in the Holocene. *Globoquadrina dutertrei* increases from about 5 percent of the population below the boundary to about 15 to 20 percent above, and *Pulleniatina obliquiloculata* increases from less than 1 percent to 10 percent. The radiolarian number increases significantly in sediments above the boundary, dated by radiocarbon as approximately 10,000 yrs B.P.

RELATIONSHIP BETWEEN PALEOCLIMATIC AND EUSTATIC SEA-LEVEL CHANGES

A striking correlation exists between climatic changes and sea-level changes determined for the Gulf of Mexico (Milliman and Emery, 1968; Ballard and Uchupi, 1970). High sea level during interglacial warm intervals is considered to be eustatic and is documented by radiocarbon age determinations using the warm-water planktonic foraminifers.

The timing of sea-level changes and paleoclimatic events coincides closely within the limitations and resolution of the radiocarbon dating system for about the past 30,000 yrs. The curves depart slightly for the older units owing in part to lack of control and finite radiocarbon dates, increased percentage of error in radiocarbon dates, and in part because of the inherent error in estimating ages by assuming a constant average rate of sedimentation. The coldest climate resulting from the last major glacial advance (Woodfordian) seemingly was from about 18,000 to 20,000 yrs B.P., and thus corresponds closely to the lowest sea-level stand of Milliman and Emery (1968). Successively and relatively higher temperatures and sea-level stands occurred with temporary glacial retreats during the pulsating withdrawal of the late Wisconsinan continental ice sheet.

A significantly warmer climate is recorded at about 7,000 yrs B.P.; this coincides closely with the latest significant rise of sea level to within a few feet of its present position.

WISCONSINAN STAGE

Paleotemperature curves reflecting fluctuating climatic conditions in the Gulf of Mexico during the Wisconsinan compare almost exactly with graphs of continental warm-cold substages of Frye and Willman (this volume, p. 135-152). Dating for the last 40,000 yrs is on the basis of radiocarbon (Fig. 17). By assuming a constant average rate of sedimentation for pelagic accumulations, ages generally can be determined for the last 75,000 yrs.

Valderan Substage

Climatic conditions during the Valderan were intermediate between those of the Woodfordian glaciation and the Holocene. *Globorotalia inflata*

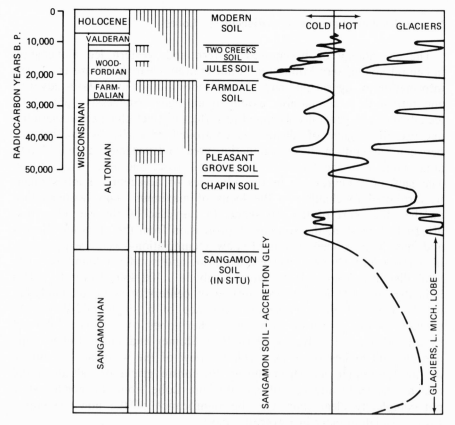

Figure 17. Climatic history as interpreted from the Lake Michigan Lobe by Frye and Willman (this volume, p. 135-152).

occurs in very low numbers, but the lack of significant numbers of the *Globorotalia menardii* group and low abundance of other warm-water species suggest a rather mild but colder climate. The exact duration of the Valderan Substage in the marine section cannot be determined from this study for lack of radiocarbon dates. Judging from the thinness of pelagic accumulation, however, it possibly was on the order of about 4,000 yrs (about 7,000 to 11,000 yrs B.P.).

Twocreekan Substage

Climatically the Twocreekan was somewhat milder than the Holocene but warmer than the Valderan. *Globorotalia inflata* usually is absent or in very low numbers, whereas members of the *Globorotalia menardii* group are rather common but in much lower abundance when compared with the Holocene. Seemingly, the duration of the Twocreekan was on the order of about 2,000 yrs. Extrapolated dates are in good agreement with those of Frye and others (1968); that is, about 11,000 to 12,500 yrs B.P.

Woodfordian Substage

The Woodfordian Substage is characterized by a dramatic increase in abundance of *Globorotalia inflata* (from less than 3 percent to about 20 percent). Members of the *Globorotalia menardii* group are essentially absent or occur in very low numbers. The fluctuating percentage of *Globorotalia inflata* and *Globigerinoides ruber* indicates numerous minor climatic changes generally, and two more significant changes, one near the base and one near the top. Willman and Frye (1970) concluded that Woodfordian glaciers left a record primarily of pulsating retreat. According to these authors, stratigraphic evidence indicates that more than 30 pulses occurred in a span of about 8,000 yrs. One readvance of the glacial front was as much as 100 mi. Rhythmic fluctuations are indicated by the fact that many of the moraines attained a rather uniform height of 50 ft and a width of 1 to 2 mi.

Radiocarbon dates from the present study indicate an age greater than 20,500 ± 1,000 yrs B.P., but less than 26,900 ± 1,000 yrs B.P. for the base of the Woodfordian Substage. These dates bracket the 22,000 yr date proposed by Frye and others (1968, 1969) and Frye and Willman (this volume, p. 135-152).

Farmdalian Substage

The Farmdalian Substage is characterized by an abundant assemblage of *Globorotalia menardii flexuosa*, and *Pulleniatina obliquiloculata*. The *Globorotalia menardii* group constitutes about 15 to 30 percent and *Pulleniatina obliquiloculata* about 10 percent of the planktonic foraminiferal assemblage. *Globorotalia inflata* is not present or occurs in very low numbers (less than 6 percent). Indications from the percentage of the *Globorotalia menardii* group are that the Farmdalian was a warm episode approaching that of the Holocene. This differs slightly from Frye and Willman (this volume, p. 135–152) in that they show the Farmdalian to be slightly milder than the Holocene.

Radiocarbon dates from the Farmdalian Substage range from 26,900 ± 1,000 yrs B.P. to 36,300 ± 2,500 yrs B.P. The average time span estimated for the Farmdalian in five cores penetrating the base is about 10,000 radiocarbon yrs (Table 4).

Accepting 22,000 yrs B.P. for the Woodfordian-Farmdalian boundary, and assuming a constant average rate of sedimentation of about 10 cm per 1,000 yrs, the estimated age for the lower boundary of the Farmdalian in core 64-A-9-24 is about 32,000 radiocarbon yrs B.P. This date is in good agreement with the average estimated time span, but slightly older than the 28,000 yr date indicated by Frye and others (1968).

Altonian Substage

The Altonian Substage consists of alternating occurrences of cold-water faunas dominated by *Globorotalia inflata*, and warm-water faunas dominated by *Globorotalia menardii flexuosa, Pulleniatina obliquiloculata,* and *Globoquadrina dutertrei.* Core 64-A-9-24 taken on Campeche Bank is the most complete core studied. Because of the seemingly constant rate of pelagic sedimentation in the area, it is ideally suited as a model to determine the age of individual climatic changes affecting the Gulf of Mexico during the last 75,000 yrs. The Altonian Substage here is subdivided informally into three climatic parts—a lower and upper cold separated by a warm episode, here referred to as the Middle Altonian. Willman and Frye (1970) do not subdivide Altonian time, but they show three glacial advances in the Early Altonian and two glacial advances in the Late Altonian. The Early and Late Altonian tills are separated by the Pleasant Grove and Chapin Soils. An equally, if not more, complex Altonian is recognized in the Gulf of Mexico. In general, the tripartite subdivision is strikingly similar to major climatic changes suggested by Frye and Willman (this volume) for the Lake Michigan Lobe.

TABLE 4. THICKNESS, AVERAGE RATE OF SEDIMENTATION, AND DURATION OF FARMDALIAN SUBSTAGE

Core	Farmdalian thickness (cm)	Average rate of sedimentation (cm/000)	Duration radiocarbon yrs
64-A-9-6	100	07	14,300
64-A-9-27	130	18	7,200
64-A-9-24	110	10	11,000
64-A-9-47	190	27	7,400
64-A-9-42	390	32	12,200
Average	184	19	10,420

Significant radiocarbon dates are not available from the Altonian in the Gulf of Mexico. Ages estimated on the basis of sedimentation rates indicate that the upper boundary is at about 32,000 yrs B.P., the upper-middle boundary is at about 42,000 yrs B.P., and the lower-middle boundary is at about 57,000 yrs B.P.

The accuracy of estimated ages for individual glacial-interglacial boundaries of the Altonian are limited by having to assume a constant average rate of sedimentation for both the glacial and interglacial episodes. The inaccuracy of estimated ages must increase in the older part of the section as thickness per unit volume of sediment is somewhat reduced owing to compaction.

Marine faunal evidence indicates that Wisconsinan-Holocene climatic episodes were cyclic in nature, averaging about 10,000 yrs per major pulse (Table 5).

SANGAMONIAN STAGE

Faunally the Sangamonian is similar to the Farmdalian and Middle Altonian. Evidence from the climatic curve of core 64-A-9-24 indicates that the Sangamonian was a very warm interglacial stage with temperatures exceeding those of the Wisconsinan interglacials and perhaps also those of the Holocene.

Although glacial-interglacial boundaries in marine sediments are chosen somewhat arbitrarily, a pronounced warm interval at about 75,000 yrs B.P. should represent a main warm episode of the Sangamonian Stage. A significantly warm climate, however, continued on to about 65,000 yrs B.P. The dramatic increase at this time of *Globorotalia inflata* marks the beginning of the significantly cooler Wisconsinan climate.

FORAMINIFERAL EVENTS

Biostratigraphic subdivision of the marine Wisconsinan-Holocene is facilitated by planktonic foraminiferal events recognized as extinction levels,

TABLE 5. ESTIMATED DURATION OF THE HOLOCENE AND THE WISCONSINAN SUBSTAGES FROM CORE 64-A-9-24

Stages and Substages	Time span (from core 64-A-9-24) (1,000 yrs)
Holocene	7.0
Valderan-Twocreekan	5.5
Wisconsinan	
Woodfordian	9.5
Farmdalian	10.0
Late Altonian	10.0
Middle Altonian	15.0
Early Altonian	8.0
Average	9.3

initial appearances, joint occurrences, and relative abundance. The following planktonic events are recognized in the Gulf of Mexico. (1) Abundant *Globorotalia tumida* s.s. and *Globorotalia ungulata* (Holocene-Twocreekan). (2) Last (youngest) occurrence of *Globorotalia inflata* (Valderan). (3) Last (youngest) occurrence of *Globorotalia inflata* together with *Globorotalia tumida* s.s. and *Globorotalia ungulata* (Valderan-Twocreekan). (4) Last (youngest) occurrence of abundant (about 20 percent) *Globorotalia inflata* (Woodfordian). (5) Last (youngest) occurrence of *Globorotalia menardii flexuosa* and *Globoquadrina hexagona* (Farmdalian). However, according to Bé and McIntyre (1970), *Globorotalia menardii flexuosa* is living in the northern Indian Ocean.

RATES OF SEDIMENTATION

Depositional rates differ widely from core to core depending on the general position in an ocean basin and local bottom topography. Rates vary from about 7 cm per 1,000 yrs in the vicinity of the Sigsbee Knolls to as much as 38 cm per 1,000 yrs on the Sigsbee abyssal plain (Table 6). Rates are higher in the vicinity of Campeche Canyon and increase markedly as the Mississippi cone is approached. Clastic deposition is highest (as much as 50 cm per 1,000 yrs for core 64-A-9-12) south and east of the Sigsbee Knolls and north of the Campeche escarpment where carbonaceous debris and quartz grains are common constituents of the sediments.

It is apparent from these data that a constant rate of sedimentation cannot be assumed for the central Gulf of Mexico as a whole. However, in a local area of pelagic sedimentation, a uniform rate of deposition would be expected. Thus, extrapolated ages based on rates of sedimentation from a single core of pelagic sediments are considered accurate for a limited time span.

CORRELATION

Paleontologic events and paleoclimatic curves from the Gulf of Mexico correlate exactly with those from the Caribbean and adjacent Atlantic (Fig.

TABLE 6. THICKNESS AND AVERAGE RATES OF SEDIMENTATION FOR POST-FARMDALIAN TIME

Core	Thickness (cm)	Average rate (cm/1,000 yrs)
64-A-9-2	610	28
64-A-9-6	140	07
64-A-9-7	450	20
64-A-9-24	210	10
64-A-9-27	400	18
64-A-9-34	840	38
64-A-9-42	700	32
64-A-9-47	590	27
64-A-9-53	640	29

18). In areas of pelagic accumulation, the entire section from Illinoian to Holocene can be penetrated by 1,200-cm cores. In all three regions, *Globorotalia menardii flexuosa* does not occur in abundance in sediments younger than the Farmdalian Substage of the Wisconsinan, and *Globorotalia truncatulinoides* (90 percent left coiled) is restricted to the Illinoian and older Pleistocene sections (Beard, 1969).

Paleoclimatic curves from the Gulf of Mexico, moreover, can be corre-lated confidently with paleoclimatic curves from the sub-Antarctic of the South Pacific, even though the curves are developed on different faunal criteria (Fig. 19). Paleoclimatic curves reflecting changes in plank-

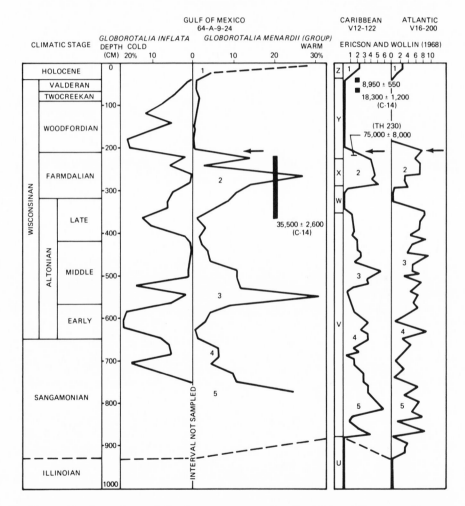

Figure 18. Comparison of the frequency curve for the *Globorotalia menardii* group from the Gulf of Mexico with frequency curves of the *Globorotalia menardii* group of Ericson and Wollin (1968) from the Caribbean and adjacent Atlantic. Arrows indicate last (youngest) occurrence of *Globorotalia menardii flexuosa*. Depth scale is the same for all three cores. Numbers refer to correlative temperature maxima. Bar indicates interval sampled for C-14 analyses. Date shown reflects oldest sediment sampled.

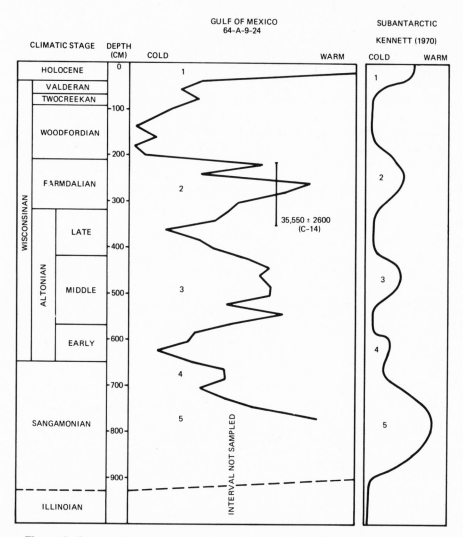

Figure 19. Comparison of paleoclimatic curve from the Gulf of Mexico with generalized paleoclimatic curve of Kennett (1970) from sub-Antarctic deep-sea cores. Numbers refer to correlative temperature maxima.

tonic marine organisms are strikingly similar to the generalized curve suggested by Frye and Willman (this volume, p. 135–152) for the Lake Michigan Lobe and can be correlated reliably by utilizing radiocarbon dates.

Correlation of climatic stages recognized in deep-sea sediments from the Gulf of Mexico are shown on a series of cross sections (Figs. 20, 21, and 22). The distinct vertical changes in the frequency ratio of the *Globorotalia menardii* group and *Globorotalia inflata* are correlated confidently peak-for-peak between adjacent cores, and the major units are recognizable from the northern continental slope across the Gulf of Mexico to Campeche Bank.

A very persistent zone of volcanic glass shards (Fig. 23), first reported

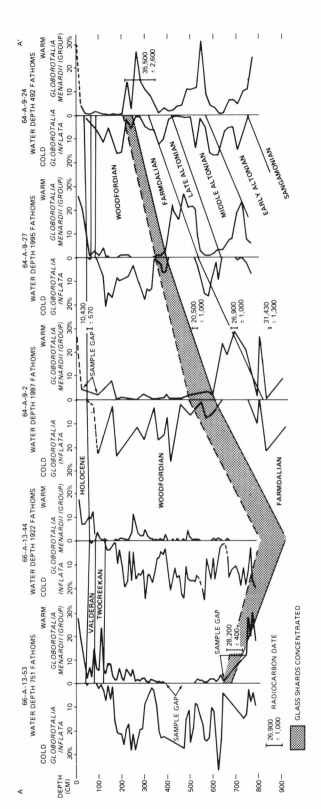

Figure 20. Stratigraphic cross section A-A' showing correlations from the continental slope, northern Gulf of Mexico, to Campeche Bank, southern Gulf of Mexico.

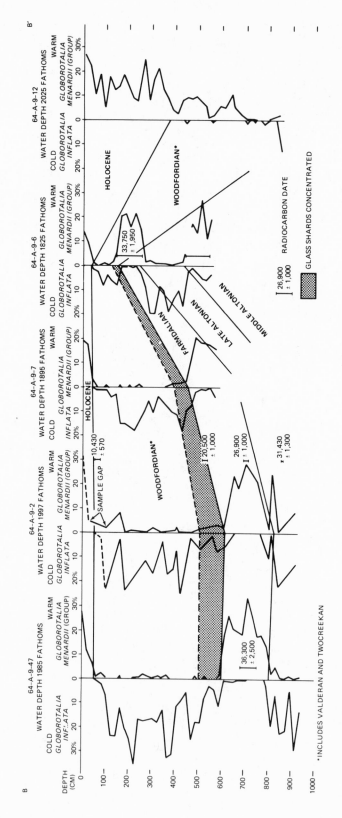

Figure 21. Stratigraphic cross section B-B′ showing west-east correlations across the Sigsbee Knolls.

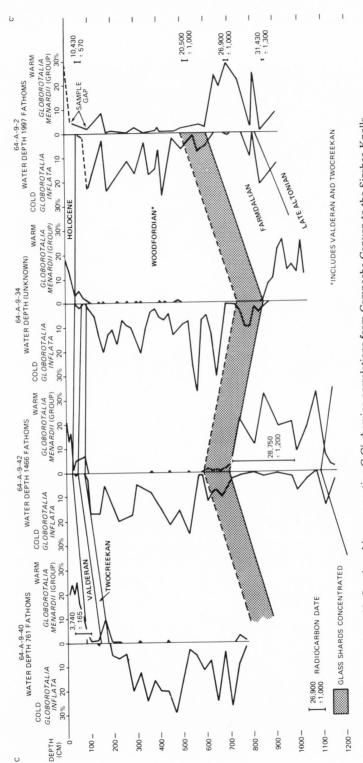

Figure 22. Stratigraphic cross section C-C' showing correlations from Campeche Canyon to the Sigsbee Knolls.

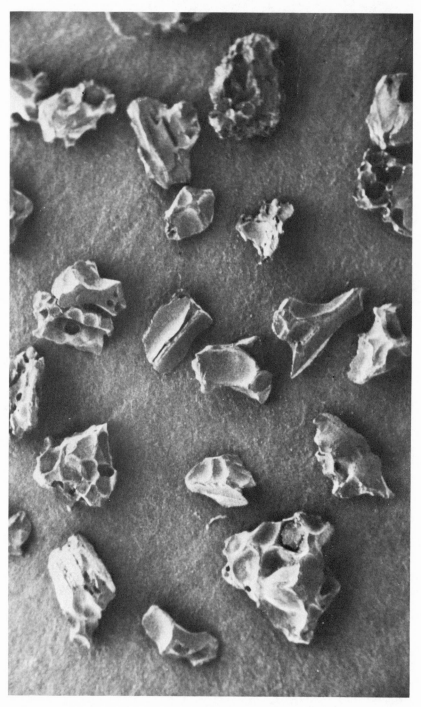

Figure 23. Volcanic glass shards from early Woodfordian in core 64-A-9-a at a depth of 500–510 cm; × 179.

by Ewing and others (1958), occurs in the Gulf of Mexico near the base of the Woodfordian in all the cores penetrating this unit and substantiates the paleontological correlation. In some of the cores ash occurs near the top of the Farmdalian (Fig. 24). Significantly, volcanic glass shards are concentrated in the Peoria Loess of Kansas (Swineford and Frye, 1951). The Peoria Loess was depositing during Early Woodfordian time and, therefore, the shards in it possibly settled from the same ash fall that contributed glass to sediments in the Gulf of Mexico.

Correlation between the Caribbean, Atlantic, and Gulf of Mexico is facilitated by the last (youngest) occurrence of *Globorotalia menardii flexuosa* and *Globoquadrina hexagona*, which do not occur in abundance above the Farmdalian Substage or the middle Wisconsinan of other authors. Various opinions, however, have been expressed on the "absolute" age of the Wisconsinan interglacial herein referred to as the Farmdalian. Ericson and others (1963, 1964) and Ericson and Wollin (1968) refer this warm interval to the middle Wisconsinan or "X" (warm) climatic stage. In a core (V12-122) from the Caribbean, as well as a core (V16-200) from the equatorial Atlantic, the top of the "X" climatic stage is at a depth of 200 cm. *Globorotalia menardii flexuosa* and *Globoquadrina hexagona* do not occur in abundance above this level. A thorium (230) date from slightly above the "X" climatic stage indicates an age of 75,000 ± 8,000 yrs B.P.

Emiliani (1966) indicates that the last (youngest) occurrence of *Globorotalia menardii flexuosa* is at a depth of about 200 cm in core P6304-9 from the Caribbean and shows a thorium date of 67,000 ± 2,000 yrs immediately above this level. Emiliani (1955) equated the last (youngest) interval with *Globorotalia menardii flexuosa* (his climatic stage 5) with the Sangamonian Stage.

In core 64-A-9-24 from the Gulf of Mexico the last (youngest) occurrence of abundant *Globorotalia menardii flexuosa* and *Globoquadrina hexagona* is at a depth of 210 cm. As previously discussed, this faunal event corresponds to the climatic boundary that is bracketed by radiocarbon dates of 20,500 ± 1,000 yrs B.P. and 26,900 ± 1,000 yrs B.P. The estimated age for the boundary is about 22,000 yrs B.P. which, therefore, corresponds to the Woodfordian-Farmdalian boundary of Frye and others (1968).

Thus, what appears to be a paleontologic equivalent horizon in three widely separated ocean basins is dated by different geochemical methods at 75,000, 67,000, and 22,000 yrs B.P., and has been referred to the Sangamonian, middle Wisconsinan, and Farmdalian, respectively. Obviously, the different geochemical dating methods are in error relative to one another, or the climatic and paleontologic events are not time-synchronous in the different ocean basins. Radiocarbon dates, however, obtained from a mixture of foraminiferal test and other shell debris contained in marine sediments must be used with caution. In addition

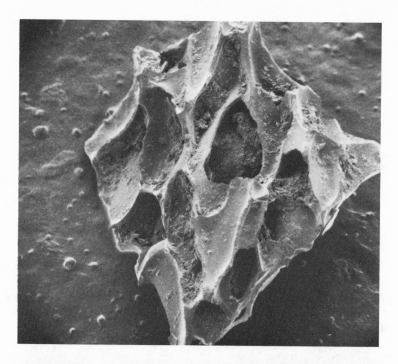

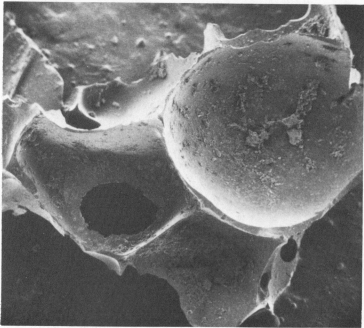

Figure 24. Volcanic glass shards from the Farmdalian in core 64-A-9-42 at a depth of 740 to 750 cm; × 18.

to possible instrument error or contamination, reworking of older shell material into younger sediments sometimes occurs. Some of the radiocarbon dates shown on Table 2 and cross sections (Figs. 20, 21, 22) are too old to be correlated as absolute ages inasmuch as the ages do not consistently agree with paleotemperature and paleontologic correlations. The older dates are readily explained by contamination or reworking. On the other hand, successive radiocarbon dates from a single core (that is, 64-A-9-2) seemingly are relatively accurate, inasmuch as the dates are from Foraminifera-rich samples representing short intervals of time and are increasingly older from top to bottom.

Significantly, the average radiocarbon date from the warm Farmdalian ("X" climatic stage of Ericson and Wollin, 1968) is about 31,000 yrs B.P., which is considerably younger than the Th-230 dates shown at the top of this unit by Ericson and Wollin (1968). The thorium (230) method seems to produce dates considerably older than carbon-14 for what appears to be a time-equivalent unit. Unfortunately, thorium (230) analyses were not made on samples from the Gulf of Mexico so a direct comparison could not be made.

The climatic curve drawn on the stratigraphic distribution of the *Globorotalia menardii* group in core 64-A-9-24 from the Gulf of Mexico compares very closely with the climatic curve from Caribbean core V12-122 of Ericson and Wollin (1968). Although similarities exist between the climatic curves from core 64-A-9-24 and from the equatorial Atlantic core (V16-200) of Ericson and others (1964), peak-for-peak correlations are less confidently made below the Late Altonian. Seemingly, the effects of glaciation were less severe in the equatorial Atlantic, and the climatic curve is somewhat smoothed as compared with curves from the Caribbean or Gulf of Mexico.

The differences in geochemical dating have led to widely differing opinions as to how the climatic stages of the Pleistocene in the marine environment correlate with climatic stages of the North American continent. Radiocarbon dates from this study indicate a close correlation with the continental climatic stages and substages of Frye and others (1968, 1969) and Willman and Frye (1970) for about the latest 40,000 yrs.

Although the climatic curve of Ericson and Wollin (1968) from the Caribbean matches almost exactly with the curve from this study of the Gulf of Mexico, the correlation of those authors departs somewhat from the usage here because of the different geochemical dates. Part of what Ericson and Wollin (1968) considered Sangamonian or climatic stage "V" is here correlated with the Early and Middle Altonian inasmuch as their climatic curve clearly indicates a cold interval within the "V" climatic stage that corresponds with a similarly cold interval in the Gulf of Mexico. The "W" climatic stage or early Wisconsinan of Ericson and Wollin (1968) is correlated with the late Altonian, and their "X" climatic stage or middle Wisconsinan correlates with the Farmdalian Substage. The

"Y" climatic stage or late Wisconsinan seems to correlate with the Wood-fordian, Twocreekan, and Valderan Substages combined, and the "Z" or postglacial stage of Ericson and Wollin (1968) probably correlates with the Holocene.

CONCLUSIONS

1. Changes in the vertical distribution of planktonic foraminifers in sea-bottom cores reflect paleoclimatic conditions precisely. Three major glacial episodes are recognized during about the last 75,000 yrs (Wisconsinan) in the Gulf of Mexico.

2. Paleotemperature curves quantitatively derived from the frequency ratio of warm- versus cold-water species are strikingly similar to the oxygen-isotope curve of Emiliani (1966) from the Caribbean, which indicates a temperature change of about 6° C between glacial and interglacial episodes.

3. Climatic events recognized in the Gulf of Mexico are correlated closely on the basis of radiocarbon dates with the continental Wisconsinan glacial-interglacial substages of Frye and others (1968) and, therefore, development of an independent nomenclature for the marine section is neither necessary nor desirable.

4. Morphologic changes in the *Globorotalia menardii* (group) and withdrawal of colder water species such as *Globorotalia inflata* facilitate interregional correlations and biostratigraphic subdivision of the Wisconsinan and Holocene Stages.

5. The major warming trend of the Holocene began about 7,000 yrs B.P. Older faunal changes, ranging in age from about 13,000 to 7,000 yrs B.P. and regarded by some to correspond to the Pleistocene-Holocene boundary, represent faunal changes reflecting a transitional climate that includes the Valderan and Twocreekan Substages. The dramatic decrease in abundance of the cold-water species *Globorotalia inflata* occurred about 13,000 yrs ago, or near the Woodfordian-Twocreekan boundary. The similarly dramatic increase in the abundance of warm-water species, such as the *Globoratalia menardii* group, on the other hand, occurred about 7,000 yrs ago at the Pleistocene-Holocene boundary.

6. Should the trend of glaciation continue to be cyclic in nature, as it now appears to have been for the past 75,000 yrs or so, a major glacial episode is predicted to occur around the 40th or 50th century.

ACKNOWLEDGMENTS

The author thanks Arnold H. Bouma and W. R. Bryant of Texas A & M University for the core samples (collection of the cores taken in 1964 by Texas A & M University was sponsored by ONR Contract N000 14-68-A-0308-0002 and NSF Grant GA-1296); John C. Frye and H. B. Willman, Illinois Geological Survey, for advice and for unpublished data; E. W. Borden, Humble Oil and Refining Company, for help with the computer program; my colleagues at Esso Production Research Company, especially L. A. Smith, J. L. Lamb, and J. F. van Sant, for advice on biostratigraphic and paleoclimatic concepts; R. D. Hockett for scanning-electron-microscope illustrations; and Russell M. Jeffords for reading of the manuscript and helpful advice. The author is especially grateful to Gene R. Kellough for constructive advice and for help with the tedious work of determining the quantitative distribution of planktonic foraminiferal tests; and to Esso Production Research Company and the Humble Oil and Refining Company for permission to publish these data.

REFERENCES CITED

Ballard, R. C., and Uchupi, Elazar, 1970, Morphology and Quaternary history of the continental shelf of the Gulf Coast of the United States: Marine Sci. Bull., v. 20, p. 547–559.

Bandy, O. L., 1967, Foraminiferal definition of the boundaries of the Pleistocene in southern California, U.S.A., in Progress in oceanography, v. 4: New York, Pergamon Press, p. 27–48.

Bé, A. W. H., 1959, Ecology of Recent planktonic Foraminifera; Pt. 1—Areal distribution in the western North Atlantic: Micropaleontology, v. 5, p. 77–100.

―― 1960, Ecology of Recent planktonic Foraminifera; Pt. 2—Bathymetric and seasonal distributions in the Sargasso Sea off Bermuda: Micropaleontology, v. 6, p. 373–392.

―― 1968, Shell porosity of Recent planktonic Foraminifera as a climatic index: Science, v. 161, p. 881–884.

―― 1969, Planktonic Foraminifera, in Distribution of selected groups of marine invertebrates in waters south of 35° latitude: Antarctic Map Folio Ser., Am. Geog. Soc. Folio 11, p. 9–12.

Bé, A. W. H., and Hamlin, W. H., 1967, Ecology of Recent planktonic Foraminifera; Pt. 3—Distribution in the North Atlantic during the summer of 1962: Micropaleontology, v. 13, p. 87–106.

Bé, A. W. H., and McIntyre, Andrew, 1970, *Globorotalia menardii flexuosa* (Koch); an extinct foraminiferal subspecies living in the northern Indian Ocean: Deep-Sea Research, v. 17, p. 595–601.

Beard, J. H., 1969, Pleistocene paleotemperature record based on planktonic foraminifers, Gulf of Mexico: Gulf Coast Assoc. Geol. Socs. Trans., v. 19, p. 535–553.

Berggren, W. A., and Boersma, Anne, 1969, Late Pleistocene and Holocene planktonic Foraminifera from the Red Sea, in Degens, E. T., and Ross, D. A., eds., Hot brines and recent heavy metal deposits in the Red Sea: New York, Springer-Verlag, Inc., p. 282–298.

Boltovskoy, Esteban, 1969, Living planktonic Foraminifera at the 90° E meridian from the equator to antarctic: Micropaleontology, v. 15, p. 237–255.

Cifelli, Richard, 1962, Some dynamic aspects of the distribution of planktonic Foraminifera in the North Atlantic: Jour. Marine Research, v. 20, p. 201–213.

―― 1967, Distributional analysis of North Atlantic Foraminifera collected in 1961 during cruises 17 and 21 of the R/V *Chain*: Cushman Found., Foram. Research Contr., v. 18, p. 118–127.

Duncan, J. R., Fowler, G. A., and Kulm, L. D., 1970, Planktonic foraminiferan-radiolarian ratios and Holocene-late Pleistocene deep-sea stratigraphy off Oregon: Geol. Soc. America Bull., v. 81, p. 561-566.

Emiliani, Cesare, 1955, Pleistocene temperatures: Jour. Geology, v. 63, p. 538-578.

—— 1966, Paleotemperature analysis of the Caribbean cores P6304-8 and P6304-9 and a generalized temperature curve for the last 425,000 years: Jour. Geology, v. 74, p. 109-126.

Ericson, D. B., and Wollin, Goesta, 1968, Pleistocene climates and chronology in deep-sea sediments: Science, v. 162, p. 1227-1234.

Ericson, D. B., Wollin, Goesta, and Wollin, Janet, 1954, Coiling direction of *Globorotalia truncatulinoides* in deep-sea cores: Deep-Sea Research, v. 2, p. 152-158.

Ericson, D. B., Broecker, W. S., Kulp, J. L., and Wollin, G., 1956, Late Pleistocene climates and deep-sea sediments: Science, v. 124, p. 385-389.

Ericson, D. B., Ewing, W. M., and Wollin, Goesta, 1963, Pliocene-Pleistocene boundary in deep-sea sediments: Science, v. 139, p. 727-737.

—— 1964, The Pleistocene epoch in deep-sea sediments: Science, v. 146, p. 723-732.

Ewing, W. M., Ericson, D. B., and Heezen, B. C., 1958, Sediments and topography of the Gulf of Mexico, *in* Weeks, E., ed., Habitat of oil: Tulsa, Okla., Am. Assoc. Petroleum Geologists, p. 995-1053.

Frerichs, W. E., 1968, Pleistocene-Recent boundary and Wisconsin glacial biostratigraphy in the northern Indian Ocean: Science, v. 159, p. 1456-1458.

Frye, J. C., and Willman, H. B., 1973, Wisconsinan climatic history interpreted from Lake Michigan Lobe deposits and soils: Geol. Soc. America Mem. 136, p. 135-152.

Frye, J. C., Glass, H. D., Kempton, J. C., and Willman, H. B., 1969, Glacial tills of northwestern Illinois: Illinois Geol. Survey Circ. 437, 47 p.

Frye, J. C., Willman, H. B., Rubin, Meyer, and Black, R. F., 1968, Definition of Wisconsinan Stage: U.S. Geol. Survey Bull. 1274-E, 22 p.

Kennett, J. P., 1970, Pleistocene paleoclimates and foraminiferal biostratigraphy in subantarctic deep-sea cores: Deep-Sea Research, v. 17, p. 125-140.

Milliman, J. D., and Emery, K. O., 1968, Sea levels during the past 35,000 years: Science, v. 162, p. 1121-1123.

Morrison, R. B., 1969, The Pleistocene-Holocene boundary; an evaluation of the various criteria used for determining it on a provincial basis, and suggestions for establishing it world-wide: Geologie en Mijnbouw, v. 48, p. 363-371.

Phleger, F. B., 1954, Foraminifera and deep-sea research: Deep-Sea Research, v. 2, p. 1-23.

—— 1955, Foraminiferal faunas in cores offshore from the Mississippi delta: Deep-Sea Research, v. 3 (Supp., Papers in Marine Biology and Oceanography), p. 45-57.

—— 1960, Ecology and distribution of Recent foraminifers; studies of planktonic foraminifers (CH 5): Baltimore, John Hopkins Press, p. 213-253.

Swineford, Ada, and Frye, J. C., 1951, Petrography of the Peoria Loess in Kansas: Jour. Geology, v. 59, p. 306-332.

Vincent, Edith, 1970, Pleistocene-Holocene boundary in southwestern Indian Ocean [abs.]: Am. Assoc. Petroleum Geologists Bull., v. 54, p. 558-559.

Willman, H. B., and Frye, J. C., 1970, Pleistocene stratigraphy of Illinois: Illinois Geol. Survey Bull. 94, 204 p.

MANUSCRIPT RECEIVED BY THE SOCIETY DECEMBER 27, 1971

GEOLOGICAL SOCIETY OF AMERICA
MEMOIR 136
© 1973

Climatic Fluctuations During the Late Pleistocene

C. C. Langway, Jr.

*U.S. Army Cold Regions Research and Engineering
Laboratory, Hanover, New Hampshire 03755*

W. Dansgaard

S. J. Johnsen

H. Clausen

University of Copenhagen, Copenhagen, Denmark

ABSTRACT

The oxygen-isotope ratio in polar snow is determined mainly by the temperature of formation of the precipitating clouds. A continuous core 1,390 m long through the ice sheet at Camp Century, Greenland, reveals a climatic record, inferred from those ratios, spanning possibly the last 100,000 yrs. The depth-age relationship of the core is calculated from present ice-flow patterns and simple assumptions; the paleoclimatic data are interpreted from the analysis of oxygen-isotope–ratio measurements on nearly 7,000 individual samples cut from the core. The ice-core record reveals that the Wisconsin Stage started 73,000 yrs B.P. Many perturbations of the oxygen-isotope ratios are observed within the Wisconsin Stage that agree with climatic oscillations dated by radioactive methods. An 11‰ shift in the 0 isotope data shows that the Wisconsin Stage ended very rapidly, within a 2,500 yr interval, at about 13,000 yrs B.P. Spectral analyses of the data show oscillations with periods of 78, 181, 400, and 2,400 yrs.

ISOTOPIC VARIATIONS

The stable oxygen-isotope-ratio $(0^{18}/0^{16})$ variations in polar snows are determined by the temperature of formation of the precipitating clouds (Dansgaard, 1954; Picciotto and others, 1960). The oxygen-18 content, $\delta(0^{18})$, increases in summer deposits and decreases in winter deposits (Epstein, 1956) producing an isotopic concentration variation in deposited snows that is measurable by mass spectroscopic means (Epstein and Mayeda, 1953; Dansgaard, 1961). The isotopic concentration variation provides a means of delimiting annual accumulation cycles and a reliable estimate of past net accumulation rates. This technique has been used successfully on the Greenland Ice Sheet (Epstein and Sharp, 1959; Benson, 1962; Langway, 1962, 1967, 1970) and, with more difficulty because of the lower average annual precipitation, on the Antarctic Ice Sheet (Gonfiantini and Picciotto, 1959; Gonfiantini, 1965; Sharp and Epstein, 1962; Gow and others, this volume, p. 323-326).

ICE-CORE STUDY

A unique opportunity for testing this concept for studying paleoclimates was presented in 1966, for the first time in history, when the U.S. Army Cold Regions Research and Engineering Laboratory successfully core drilled through the northern Greenland Ice Sheet at Camp Century (Garfield and Ueda, 1967) and recovered a 1,390-m long, 12-cm diameter ice core (Hansen and Langway, 1966).

A depth-age relationship was calculated for the core by developing a physical model that incorporated and satisfied all of the known physical parameters of the Camp Century location and the flow theories of ice (Dansgaard and Johnsen, 1969). Nearly 7,000 small vertical sample increments were cut from the core in continuous sequence and analyzed for 0^{18}.

Although useful for interpreting accumulation in the upper firn layers, the stable isotope technique becomes less reliable with greater depths because various physical processes (for example, molecular diffusion) tend to diminish the isotopic gradients and set a limit on the applicability of the method to a few thousand years. Nevertheless, Dansgaard and others (1969) have shown that long-term variations in the $\delta(0^{18})$ are preserved for many tens of thousands of years. This means that the average isotopic composition of a vertical section of a deep ice core reflects the average climatic conditions existing at the surface during the time the original deposit formed. An increase in $\delta(0^{18})$ along the core corresponds to warming climate; decreasing $\delta(0^{18})$ corresponds to cooling climate.

INTERPRETATION OF RESULTS

Results of the $\delta(0^{18})$ measurements for the entire core are shown in Figure 1. The continuous δ-profile is given in 200-yr increments plotted

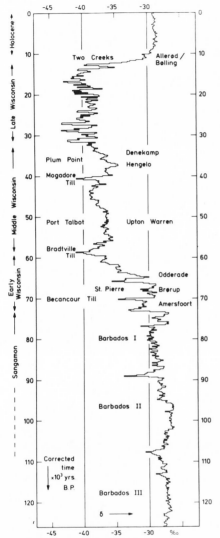

Figure 1. (From Dansgaard and others, 1970a) δ in 200-yr intervals plotted on the corrected time scale. Tentative interpretations in European and American terminology are shown to the right and to the left. To the outer left is proposed a division of the Wisconsin glaciation in accordance with the characteristic features of the δ curve.

against time and provided with both European and American stratigraphic terms. A more detailed plot of the last 800 yrs (0 to 780 yrs B.P.) shows that most of the climatic changes known from historical sources are reflected as δ-variations in the Camp Century ice core (Johnsen and others, 1970). These include the warm periods around the 1930s and the middle of the 18th and 16th centuries. Cold periods are shown around the 1820s and the "little ice age" between 1740 and 1600.

The $\delta(0^{18})$ data from the Holocene (past 10,000 yrs B.P.; Dansgaard and others, 1970a) show that the last millenium was colder than the preceding one. The post-Wisconsin climatic optimum is depicted from 4,100 to 8,000 yrs B.P. with two extreme maxima near 5,000 yrs B.P. and 6,000 yrs B.P. The Alleroed/Boelling shown together on Figure 1 are very distinct in a detailed plot and are recognized at 11,900 to 10,900 yrs B.P. and 12,400 to 12,100 yrs B.P., respectively (in our time scale). These dates compare very favorably with European pollen records (Iversen, 1954) and varve chronologies (Tauber, 1970).

The main features of the curve suggest a division of the Wisconsin Stage into three intervals as shown to the left of the figure: (1) early (73,000 to 59,000 yrs B.P.), comprising the main cooling from the Sangamon Interglacial; (2) middle (59,000 to 32,000 yrs B.P.), with generally more stable conditions; and (3) late (32,000 to 13,000 yrs B.P.), comprising the coldest part and further characterized by greatly varying conditions. The climatic transition to the Holocene is much more abrupt than the inception of the Wisconsin Stage from the Sangamon Interglacial. The δ-record compares favorably with the Pleistocene glacial stratigraphy in the Ontario and Erie Basins (Goldthwait and others, 1965) as is shown in Figure 1.

The pre-Wisconsin record compares well with the high sea-level stands, Barbados I, II, and III, which were dated by Pa^{230} and Th^{230} at 82,000, 103,000, and 122,000 yrs B.P. (Veeh, 1966; Broecker and others, 1968).

Spectral analyses based on the Fourier integral were carried out on the continuous δ-record. A separate detailed study on the past 800 yrs (Johnsen and others, 1970) showed two dominant peaks that correspond to periods of 78 and 181 yrs. The spectral analysis of the data between 1,000 and 10,000 yrs B.P. shows climatic oscillations with periods of approximately 400 to 2,400 yrs. It is postulated that these oscillations reflect solar-activity variations (Dansgaard and others, 1970b).

REFERENCES CITED

Benson, C., 1962, Stratigraphic studies in the snow and firn of northwest Green-land, 1952, 1953, and 1954: U.S. Army Snow, Ice, and Permafrost Establish-ment Research Rept. 26, 93 p.

Broecker, W. S., Thurber, D. L., Goodard, J., Ku, T. L., Matthews, R. K., and Mesolella, K. J., 1968, Milankovitch hypothesis supported by precise dating of coral reefs and deep-sea sediments: Science, v. 159, p. 297–299.

Dansgaard, W., 1954, The 0^{18} abundance in fresh water: Geochim. et Cosmochim. Acta, v. 6, p 241–260.

—— 1961, The isotopic composition of natural waters: Medd. Grönland, Bd. 165, Nr. 2, 120 p.

Dansgaard, W., and Johnsen, S. J., 1969, A flow model and a time scale for the ice core from Camp Century, Greenland: Jour. Glaciology, v. 8, p. 215–223.

Dansgaard, W., Johnsen, S. J., Clausen, H., and Langway, C. C., Jr., 1970a, Climatic record revealed by the Camp Century ice core, in Late Cenozoic glacial ages: Symposium Proc., New Haven, Conn., Yale Univ.

—— 1970b, Ice cores and paleoclimatology, in Radiocarbon variations and absolute chronology: XII Nobel Symposium Proc., Uppsala, Sweden, p. 337–351.

Dansgaard, W., Johnsen, S. J., Møller, J., and Langway, C. C., Jr., 1969, One thousand centuries of climatic record from Camp Century on the Green-land ice sheet: Science, v. 166, p. 377–381.

Epstein, S., 1956, Variations of the $0^{18}/0^{16}$ ratios of fresh water and ice: Natl. Acad. Sci.—Natl. Research Council Nuclear Sci. Ser. Rept. 19, p. 20–28.

Epstein, S., and Mayeda, T., 1953, Variations of 0^{18} content of water from natural sources: Geochim. et Cosmochim. Acta, v. 4, p. 213–224.

Epstein, S., and Sharp, R. P., 1959, Oxygen-isotope studies: Am. Geophys. Union Trans., v. 40, p. 81–84.

Garfield, D., and Ueda, H., 1967, Drilling through the Greenland ice sheet: U.S. Army Materiel Command Cold Regions Research and Eng. Lab. Tech. Rept., 14 p.

Goldthwait, R. P., Dreimanis, A., Forsyth, J. L., Karrow, P. F., and White, G. W., 1965, Pleistocene deposits of the Erie Lobe, in Wright, H. E., Jr., and Frey, D. G., eds., The Quaternary of the United States: Princeton, N.J., Princeton Univ. Press, p. 85–97.

Gonfiatini, R., 1965, Some results on oxygen isotope stratigraphy in the deep drilling at King Boudouin Station, Antarctica: Jour. Geophys. Research, v. 70, p. 1815-1819.

Gonfiantini, R., and Picciotto, E., 1959, Oxygen isotope variations in Antarctica snow samples: Nature, v. 184, p. 1557-1558.

Gow, A. J., Epstein, S., and Sharp, R. P., 1973, Climatological implications of stable isotope variations in deep ice cores from Byrd Station, Antartica: Geol. Soc. America Mem. 136, p. 323-326.

Hansen, B. L., and Langway, C. C., Jr., 1966, Deep core drilling in ice and core analyses at Camp Century, Greenland, 1961-1966: Antarctic Jour. U.S., v. 1, p. 207-208.

Iversen, J., 1954, The late glacial flora of Denmark and its relation to climate and soil: Danmarks Geol. Undersøgelse, Ser. II, no. 80.

Johnsen, S. J., Dansgaard, W., Clausen, H., and Langway, C. C., Jr., 1970, Climatic oscillations 1200-2000 A.D.: Nature, v. 227, p. 482-483.

Langway, C. C., Jr., 1962, Some physical and chemical investigations of a 411 meter deep Greenland ice core and their relationship to accumulation: Internat. Assoc. Sci. Hydrology Pub. 58, p. 101-118.

____ 1967, Stratigraphic analysis of a deep ice core from Greenland: U.S. Army Materiel Command Cold Regions Research and Eng. Lab. Research Rept. 77, 130 p.

____ 1970, Stratigraphic analysis of a deep ice core from Greenland: Geol. Soc. America Spec. Paper 125, 186 p.

Picciotto, E., DeMaere, X., and Friedman, I., 1960, Isotopic composition and temperature of formation of Antarctic snows: Nature, v. 187, p. 857-859.

Sharp, R. P., and Epstein, S., 1962, Comments on annual rates of accumulation in West Antarctica: Internat. Assoc. Sci. Hydrology Pub. 58, p. 273-285.

Tauber, H., 1970, The Scandinavian varve chronology and C^{14} dating, in Radiocarbon variations and absolute chronology: XII Nobel Symposium Proc., Uppsala, Sweden, p. 171-196.

Veeh, H. H., 1966, $Th^{230}/U^{234}/U^{238}$ ages of Pleistocene high sea-level stand: Jour. Geophys. Research, v. 71, p. 3379-3386.

MANUSCRIPT RECEIVED BY THE SOCIETY DECEMBER 27, 1971

GEOLOGICAL SOCIETY OF AMERICA
MEMOIR 136
© 1973

Climatological Implications of Stable Isotope Variations in Deep Ice Cores from Byrd Station, Antarctica

ANTHONY J. GOW

U.S. Army Cold Regions Research and Engineering Laboratory, Hanover, New Hampshire 03755

SAMUEL EPSTEIN

ROBERT P. SHARP

Division of Geological and Planetary Sciences, California Institute of Technology, Pasadena, California 91109

ABSTRACT

Oxygen- and hydrogen-isotope analyses of ice cores from a hole drilled 2,164 m through the Antarctic Ice Sheet suggest that the Wisconsin cold interval began about 75,000 yrs B.P., reached its climax about 17,000 yrs B.P., and terminated about 11,000 yrs B.P.

RESULTS AND DISCUSSION

Oxygen- and hydrogen-isotope analyses of ice cores from the 2,164 m hole drilled through the Antarctic Ice Sheet at Byrd Station (Gow and others, 1968) reflect temperature variations estimated to extend back more than 75,000 yrs B.P. Age of the ice was estimated on the basis of a constant rate of snow accumulation of 12 g cm^{-2} yr^{-1} water equiva-

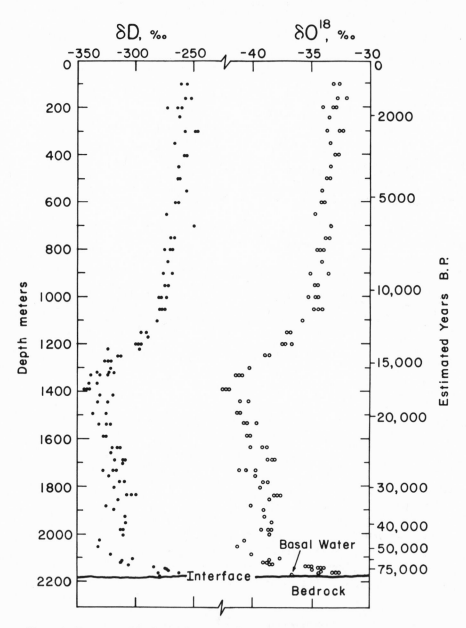

Figure 1. Oxygen and hydrogen isotope ratios, with respect to Standard Mean Ocean Water, in ice samples from the deep drill hole at Byrd Station, Antarctica. Data plotted linearly against depth (left) and nonlinearly against estimated age (right). The basal water sample was obtained from the bottom of the ice sheet, which is at the pressure melting point.

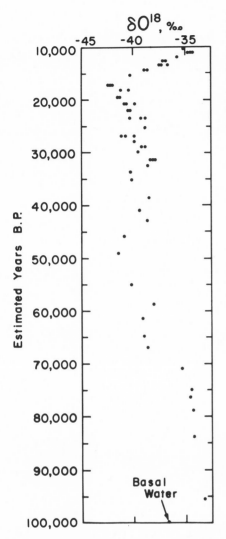

Figure 2. Oxygen-isotope ratios of Byrd core samples from below 1,000 m, plotted on a linear time scale.

lent with due allowance also being made for thinning of the ice sheet by plastic flow. $\delta(O^{18})$ and δD data from more than 100 samples taken at intervals of 33 to 62 m at depths ranging between 99 and 2,162 m are plotted against depth in Figure 1; $\delta(O^{18})$ values of samples from below 1,000 m are plotted in linear fashion against estimated age in Figure 2.

The isotope data are interpreted as follows. The Wisconsin cold interval in Antarctica was initiated about 75,000 yrs B.P., reached its climax about 17,000 yrs B.P., and terminated about 11,000 yrs B.P. Inflections on the curve suggest three intra-Wisconsin warmer phases peaked at about 25,000, 31,000, and 39,000 yrs B.P. and intervening colder intervals at about 27,000, 34,000, and 46,000 yrs B.P. These interpretations indicate that major climatic changes in Antarctica and the Northern Hemisphere have been essentially synchronous. The likelihood of synchronism of climatic variations in the two hemispheres is further strenghthened by the similarity between the Byrd Station data and results obtained on deep cores from the Greenland Ice Sheet at Camp Century (Dansgaard and others, 1969; Langway and others, this volume, p. 317–321).

The warmer phases recorded in the Byrd Station cores are all appreciably colder than pre- or post-Wisconsin times, from which it is concluded that the Northern Hemisphere ice sheets did not disappear at any time during the Wisconsin and that eustatic rises of sea level all fell short of present-day sea level. Analyses of cores from the lowermost part of the ice sheet indicate a pre-Wisconsin climate even warmer than the present and probably reflect true interglacial conditions. Temperatures in Antarctica during the Wisconsin were possibly as much as 7° to 8° C colder than they are today.

REFERENCES CITED

Dansgaard, W., Johnsen, S. J., Moller, J., and Langway, C. C., Jr., 1969, One thousand centuries of climatic record from Camp Century on the Greenland Ice Sheet: Science, v. 166, p. 377–381.

Gow, A. J., Ueda, H. T., and Garfield, D. E., 1968, Antarctic Ice Sheet: Preliminary results of first core hole to bedrock: Science, v. 161, p. 1011–1013.

Langway, C. C., Jr., Dansgaard, W., Johnsen, S. J., and Clausen, H., 1973, Climatic fluctuations during the late Pleistocene: Geol. Soc. America Mem. 136, p. 317–321.

MANUSCRIPT RECEIVED BY THE SOCIETY DECEMBER 27, 1971

Index

Aberdeen Readvance, 125
Adirondack Mountains, 48, 49, 50, 58, 73, 81, 83, 191
Africa, 128
Agassiz phase, 158
Alabama, 279
Alaska, 260, 269
Alberta, 222, 239
Alborn phase, 164
Alexandria Moraine, 158, 160, 170, 172, 179, 181
Algona Moraine, 160, 172, 181
Almond drift, 90, 122
Almond Moraine, 64, 73
Alps, 121, 125-126, 129, 272
Altamount Moraine, 13, 15, 16, 160, 171, 172
Altonian, 3, 8, 18, 19, 20, 21, 22, 23, 25, 80, 81, 85, 86, 87, 88, 137, 137-139, 140, 144, 145, 148, 149, 232, 287, 288, 290, 298, 300, 302, 303, 305, 306, 307-309, 313
    Early, 287, 298, 302, 303, 305, 306, 307, 313
    Middle, 287, 298, 302, 303, 305, 306, 307, 308, 313
    Late, 287, 288, 290, 298, 302, 303, 305, 306, 307, 308, 309, 313
America, *see* North America
Anoka Sandplain, 154, 155, 156, 176, 177, 178-179, 181, 183
Antarctica, 257, 260, 284, 306, 318, 323-325
Antarctic ice sheet, 42, 127
Antelope Moraine, 15, 172
Arctic, 127, 146
Argyle Till Member, 139, 149
Arlington Moraine, 231, 246
Ashtabula Till, 94
Atlantic, 12, 14, 15, 51, 128, 277, 280, 282, 284, 293, 305, 311, 313
Augusta Moraine, 47, 53, 55, 56, 59, 63, 64, 65
Automba Drumlin Field, 158, 161, 163
Automba phase, 158, 161-162, 163, 164, 179, 253, 258, 266, 267, 270

Baltic, 121, 124, 273
Barbados, 319, 320
Becancour Till, 80, 319
Belgium, 121, 126
Bemis Moraine, 158, 160, 170, 171, 181
Big Sioux River, 160, 170, 171
Big Stone Moraine, 182
Binghamton drift, 122
Block Island, 65
Bloomington Drift, 90
Bloomington Morainic System, 18, 150, 231, 246
Bloomington Outwash, 93
Bologoye/Edrovo, 121, 124
Boston Drift, 90
Boston marine clay, 39
Boston Till, 77, 85, 91, 92
Boulder pavement, 111, 167, 168, 169, 170, 262
Boundary Mountains, 191, 192, 205, 206, 214
Bradtville Till, 83, 84, 85, 113, 114, 319
Brainerd Drumlin Field, 159
Brainerd Sublobe, 158
Brandenburg-Frankfurt Moraine, 121, 124
Brandenburg maximum, 125
Bridport readvance, 48, 59-61, 63, 64, 66
Burlington drift, 64
Bush Creek Interstadial, 85
Buzzard's Bay Moraine, 39, 40

Caesar Till, 85, 90, 91, 92, 93
California, 293
Cambridge readvance, 39, 41
Camden Moraine, 93
Canada, 4, 7, 12, 14, 28, 161, 166, 182, 189-217, 221-227, 259, 271, 293
Canadian Shield, 74, 155
Canning Till, 83, 84
Cape Cod Moraine, 15
Capron Till Member, 139, 149
Carbon, 8, 14, 25, 37, 38, 39, 40, 41, 42, 52, 53, 58, 61, 62, 63, 65, 66, 78, 79, 80, 81, 82, 84, 85, 86, 87, 88, 89, 90, 91, 92, 94, 95, 115, 120, 121, 122, 123, 124, 125,

126, 127, 128, 137, 139, 141, 142, 144, 150, 161, 164, 165, 180, 181, 182, 238, 267, 277, 278, 280, 285, 289, 291, 299, 300, 301, 302, 303, 305, 306, 307–309, 311, 313, 314, 317
Carbonate leaching, 77, 83, 85
Caribbean, 277, 278, 280, 282, 284, 298, 305, 311, 313, 314
Carlingford Readvance, 125
Carroll Outwash, 93
Cartersburg Till Member, 93
Cary, 3, 21, 23, 25, 64, 74, 81, 95, 111, 117, 118, 123, 161
Cary-Port Huron Interstade, 90, 95–96
Cary Till, 95, 118
Catfish Creek Drift, 90, 109, 110–114
Catfish Creek Till, 93, 107, 108, 110, 111, 113, 114, 115, 116, 117, 118, 122
Catskill Plateau, 48, 50
Catskills, 65
Center Grove Till, 91
Central Interior, see Central Lowlands
Central Lowlands, 8, 23, 24, 25, 27
Cerro Gordo Moraine, 90
Champaign Drift, 93
Champaign Morainic System, 18
Champlain Lobe, 50
Champlain Lowland, 48
Champlain Sea, 61
Champlain Valley, 48, 49, 50, 57, 59, 61, 62, 63
Channeled Scablands, 264
Chapin Soil, 139, 140, 145, 149, 300, 302
Charlestown Moraine, 39, 40
Chaudière Hills, 191, 192
Chaudière Till, 196, 197, 215, 216
Chaudière Valley, 197, 198, 201, 203, 204, 206, 216, 217
Cherry River Moraine, 66
Chippewa Lobe, 157, 159
Chisago Lobe, 155
"Classical" Wisconsin, 21, 22, 23, 25, 88
Climate, 17, 43, 63, 78, 81, 82, 83, 88, 89, 107, 128, 230, 246, 259, 269
Climatic fluctuations, 37, 38, 42, 120, 124, 127, 128, 148, 150, 162, 252, 267, 268, 270, 277, 278, 282, 284, 285, 288, 289, 290, 291, 292, 293, 298, 299, 300, 301, 302, 303, 305, 306, 307–309, 311, 313, 314, 317–320
Climatic history, 135–150, 266, 268
Climatic implications, 88, 120, 193, 271, 273, 323–325
Climatic optimum, 319
C line, 121, 124
Coastal Plain, 5
Cochrane Till, 291
Columbia Plateau, 264
Connecticut, 12, 14, 37, 39, 40, 51, 55, 64, 65, 120
Connecticut Valley, 39

Connecticut Valley Lobe, 48, 50, 51, 53, 65
Connersville Interstadial, 80, 82, 91
Cores, 26, 27, 121, 127, 129, 137, 139, 235, 236, 237, 278, 279, 280–288, 289, 290, 291, 292, 294–298, 299, 302, 303, 304, 305, 306, 307–309, 310, 311, 312, 313, 314
Coteau des Prairies, 160, 168, 170, 239
Crevasse filling, 148, 172
Cromwell Moraine, 160, 161, 162
Cryoturbations, see Involutions
Cuba Moraine, 91, 120, 121, 125
Culvers Gap Moraine, 47, 53, 55, 56, 63, 64, 65

Dakotas, 14, 15, 155, 171
Darby Till, 85, 90, 92, 93
Decatur Sublobe, 72, 73, 74, 80, 90
Deep-sea sediments, 28, 123, 127, 278, 284, 293, 303, 306, 311
Defiance Moraine, 94
DeKalb Mounds, 229–247
Delaware Valley, 52, 53, 65
Delta, 42, 57, 163, 178, 181
Denmark, 124, 252, 272, 273
Denville re-entrant, 56, 57
Des Moines Lobe, 123, 153–183, 222, 267, 268
Des Moines River, 155, 181
D line, 121, 124
Don Formations, 83
Drift, 3–28, 50, 63, 64, 65, 75, 77, 80, 83, 84, 85, 86, 118, 122, 125, 142, 145, 154, 155, 157, 159, 160, 161, 163, 166, 167, 168, 169, 170, 171, 172, 173, 174, 175, 179, 180, 183, 222, 231, 232, 241, 253, 266
Driftless Area, 12, 14, 138, 140, 147
Drolet Lentil, 197, 198, 199, 200, 204, 205, 206, 207, 208, 209, 210, 211, 212, 214, 216
Drumlin Readvance, 121, 125
Drumlins, 50, 63, 73, 96, 153, 154, 159, 161, 162, 163, 166, 172, 174, 180, 181, 251, 252, 254, 255, 261, 264, 265, 266, 267, 272
Dunwich Till, 84, 86, 87, 113, 114

Early Wisconsin, 3, 8, 18, 20, 21, 24, 71, 72, 77, 80–86, 88, 110, 120, 153, 168, 196, 319
East Anglia, 252
East Iowa Formation, 16
East Iowan, 22
East Jylland Stadial, 124
East White Sublobe, 90
East Wisconsin Formation, 16
England, 125
Environmental conditions, 28, 49, 52–53, 55, 277
Equilibrium line, 257, 258, 259

Erie Basin, 18, 72, 74, 76, 79, 83, 86, 87, 89,
    93, 94, 95, 96, 108, 118, 121, 320
Erie Interstade, 80, 90, 93, 94, 107–129, 141
Erie Lobe, 6, 8, 20, 26, 71–96, 107, 116, 118,
    119, 120, 122, 136, 138, 221, 222, 223,
    225, 226, 227
Erie-Ontario Lobe, 47
Erratics, 3, 4, 5, 6, 24, 194, 200, 202, 203
Esker, 154, 160, 162, 163, 164, 172, 173,
    174, 240, 252, 254, 255, 256, 261, 264,
    265, 266, 267, 272
Europe, 16, 17, 18, 121, 122, 124–126, 272
Eustatic, 121, 127, 128, 129, 285, 299, 325
Evans Creek Stade, 121, 123

Farmdale, 21
Farmdale Loess, 21
Farmdale Soil, 139, 140, 300
Farmdalian, 3, 10, 23, 50, 52, 80, 81, 86,
    137, 139–140, 144, 149, 287, 288, 290,
    292, 295, 298, 300, 301, 302, 303, 304,
    305, 306, 307–309, 311, 312, 313
Farmersville Moraine, 93, 94
Fayette Drift, 90
Fayette Till, 91
Forest bed, 10, 11, 15, 91
Foraminifers, 127, 277, 278, 279, 280–288,
    289, 290, 291, 292, 293, 294–298, 299,
    300, 301, 302, 303, 304, 305, 306,
    307–309, 311, 313, 314
  North-south migration, 278
Ft. Covington Till, 81
Fort Wayne Morainic System, 18
Fossils, 27, 42, 78, 82, 83, 88, 89, 96, 128,
    135, 137, 143, 144–145, 147, 271
France, 4, 121, 125, 129
Frankfurt maximum, 125
Fraser, 120, 123
Fresh Pond Moraine, 39, 40

Gahanna Drift, 83, 84
Gahanna Till, 85
Galt Moraine, 96
Garfield Heights Till, 85
Gary Moraine, 15
Gayhurst Formation, 197
Georgian Bay Lobe, 72, 73, 81, 83, 84, 87,
    89, 90, 221, 223, 225, 226, 227
Germany, 121, 124, 125
Glacial dispersal, 189–217
Gowanda Moraine, 64, 65
Grand River Lobe, 19, 20
Grand River Sublobe, 73, 74, 84, 90
Grantsburg Sublobe, 123, 154, 155, 158, 159,
    160, 165, 171, 174–175, 176, 177, 178,
    179, 180, 181
Great Britain, 6, 16, 27, 121, 125, 129
Great Lakes, 20, 38, 42, 43, 48, 72, 74, 76,
    81, 154, 157, 174, 269, 270, 273
Great Plains, 8, 240
Green Bay Lobe, 73, 80, 138, 157, 273

Greenland, 121, 126, 129, 168, 242, 257, 258,
    259, 260, 271, 317, 318, 325
Green Mountains, 48, 50
Green River Lobe, 149
Gulf of Mexico, 128, 277–315
Gyttja, 115, 121, 122, 144

Halton Till, 90, 95
Hamburg Moraine, 64, 66
Harbor Hill Moraine, 39, 40, 47, 50, 51, 52,
    53, 59, 63, 64, 120, 122
Hartwell Moraine, 91
Hayesville Till, 90
Hempstead Gravel, 51, 52
Herod Gravel, 51
Hewitt phase, 158, 159
Highland Front Moraine, 38, 40, 43, 48, 59,
    61, 62, 63, 64
Highland Moraine, 158, 160, 161, 162, 179,
    180, 266, 270, 271
Hinuera, 121, 126
Hiram Till, 19, 90, 92, 94, 108
Holocene, 21, 144, 277–315, 319
Hudson Bay, 183, 251, 260, 263
Hudson-Champlain Lobe, 47–66
Hudson-Champlain Lowland, 48
Hudson-Champlain Valley, 47, 59, 64
Hudson Highlands, 48, 49, 55, 56
Hudson Lobe, 50
Hudson River, 39, 50, 51, 55, 58, 60, 61, 93
Hudson Valley, 47, 48, 49, 50, 55, 57, 58,
    59, 61, 62, 63, 122
Hudson Valley Lowland, 48, 56
Humboldt Moraine, 160, 172
Huron Basin, 71, 72, 76, 94, 95, 96, 118
Huron Lobe, 8, 20, 71–96, 119, 122, 138,
    221, 223, 225, 226, 227
Hydrogen isotope, 323–325
Hydrology, 251–273

Ice advance, 8, 12, 15, 16, 17, 24, 25, 26,
    37–43, 50, 53, 55, 56, 57, 59, 63, 71, 72,
    89, 90, 93, 116, 119, 120, 123, 136, 137,
    140, 141, 142, 143, 145–146, 149, 150,
    154, 155, 157, 161, 162, 164, 165, 168,
    170, 171, 172, 179, 180, 181, 183, 189,
    216, 217, 230, 246, 251, 252, 261, 266,
    267, 268, 270, 273
Iceberg origin, 3, 4, 6, 15, 16, 24
Ice core, 317, 318, 319, 323–325
Iceland, 269
Ice retreat, 12, 16, 17, 24, 25, 26, 37–43, 53,
    55, 56, 57, 59, 61–63, 71, 72, 90, 93,
    118–120, 122, 123, 124, 136, 137, 139,
    140, 141, 142, 143, 145–146, 149, 150,
    161, 162, 163, 164, 165, 179, 180, 183,
    246, 259, 267, 268
Ice-sheet origin, 3, 5, 6, 24, 27
"Ice stand on the necks," 51, 52, 59, 63, 64,
    65
Ice-wedge casts, 89, 147, 244, 259

Idaho, 265
Illinoian, 18, 19, 20, 25, 77, 135, 136, 137,
    138, 140, 144, 145, 147, 149, 232, 305,
    306
Illinois, 7, 8, 10, 11, 12, 14, 16, 18, 19, 21,
    22, 23, 26, 85, 89, 90, 135, 136, 137,
    138, 139, 140, 141, 142, 144, 146, 147,
    149, 229–247, 259
Illinois Glacial Lobe, 18
Illinois Lobe, 20
Illinois Valley, 137, 139, 145
Indiana, 8, 10, 11, 12, 14, 17, 20, 72, 74, 77,
    80, 81, 82, 83, 85, 86, 88, 89, 90, 91, 94,
    118, 119, 120, 122, 138, 244
Indian Ocean, 299, 304
Ingersoll Moraine, 90
Involutions, 82, 84, 89, 147, 244
Iowa, 8, 9, 11, 13, 14, 16, 18, 22, 26, 123,
    138, 154, 155, 156, 160, 165, 167, 168,
    171, 172, 181
Iowan, 3, 16, 18, 19, 20, 21, 22, 24, 25, 168,
    267
Ireland, 125
Isostatic changes, 38, 39, 76, 127, 128
Isotope, 28, 135, 323–325
Itasca Moraine, 158, 159, 160, 167, 179

James Lobe, 160, 170, 171, 267, 291
Johnville Till, 196, 197
Jubileean Substage, 137, 140, 148
Jules soil, 140, 141, 142, 150, 300

Kansan, 19, 138, 155
Kansan Formation, 16
Kansas, 14, 23, 311
Kattegatt area, 121, 124
Kennard Outwash, 93
Kennebunk glacial advance, 42
Kent drift, 65
Kent Moraine, 64
Kent Till, 19, 90, 94, 108, 122
Kentucky, 18
Kettle Moraine, 3, 13, 14, 15, 16
Killbuck Sublobe, 73, 74, 84, 90, 94
Killwangen Phase, 121, 125
Kingston Outwash, 93
Kühlung Interstadial, 121, 124
Kumara, 121, 126

Labrador, 83
Labradorean Ice Sheet, 80, 85
Lac-aux Araignées Basin, 191, 192
Lac-Mégantic, 189, 190, 192, 195, 217
Lagro Formation, 90
Lake Agassiz, 154, 158, 160, 165, 180,
    182–183, 263
Lake Aitkin, 158, 161, 164, 179, 180, 182
Lake Albany, 48, 55, 56–58, 59
Lake Algonquin, 96
Lake Arkona, 80, 95
Lake Bonneville, 123, 265

Lake Border Moraine, 142
Lake Champlain, 49, 60, 61
Lake Coveville, 48, 59, 61, 63
Lake Duluth, 165
Lake Erie, 4, 12, 14, 72, 73, 76, 86, 87, 88,
    94, 107, 108, 109, 110, 111, 114, 117,
    118, 125, 126, 136, 137, 225
Lake Erie Basin, 79, 81, 86
Lake Escarpment Moraine, 64, 65, 94, 95,
    122
Lake Fort Ann, 48, 59, 61
Lake George, 49, 57, 60, 61
Lake Grantsburg, 154, 163, 175–178, 180,
    181
Lake Huron, 12, 14, 72, 73, 74, 86, 93, 136,
    137, 225
Lake Iroquois, 81, 87, 96
Lake Leverett, 107, 108, 111, 112, 117, 119
Lake Maumee, 89, 95, 118, 119
Lake Michigan, 12, 14, 23, 86, 136, 137, 138,
    141, 150
Lake Michigan Basin, 52, 149, 165
Lake Michigan Lobe, 3, 7, 20, 22, 26, 72,
    73, 80, 83, 86, 121, 123, 135–150, 157,
    161, 222, 270, 273, 300, 302, 306
Lake Missoula, 264
Lake Nemadji, 158, 165
Lake Ontario, 12, 14, 73, 74, 76, 79, 81, 83,
    84, 86, 87, 90, 225
Lake Quaker Springs, 48, 59, 61, 63
Lake Saginaw, 95
Lake Scarborough, 81
Lake Simcoe Lobe, 73
Lake Superior, 12, 14, 154, 156, 160, 161,
    162, 163, 171, 179, 180, 183, 253, 260,
    263, 268
Lake Superior Basin, 153, 157, 159, 161,
    162, 163, 164, 175, 177, 180, 251, 252,
    254, 258, 260, 261, 263, 266, 267, 272,
    273
Lake Till, 90
Lake Tillson, 57
Lake Upham, 158, 161, 164, 179, 180
Lake Warren, 90, 108, 118, 119, 319
Lake Warrensburg, 58
Lake Whittlesey, 90, 95, 96
Lammermuir Readvance, 125
Lascaux Interstadial, 121, 125, 126
Late Midlandian, 121, 125
Late Pleistocene, 317–320
Late Sangamon Loess, 21
Late Valdaj, 124
Late Weichselian, 124, 125, 129
Late Wisconsin, 3, 8, 18, 19, 21, 24, 37–43,
    52, 71, 72, 78, 80, 81, 82, 84, 85, 86, 87,
    88–96, 107, 110, 111, 117, 120, 122, 123,
    124, 126, 127, 128, 129, 221, 222, 223,
    254, 267, 299, 314, 319
Late Würm, 125
Laugerie Interstadial, 125
Laurentide ice sheet, 37–43, 82, 83, 88, 89,

95, 123, 125, 126, 129, 154, 170, 263
Lavery Till, 19, 94, 122
Lennoxville Till, 189, 196, 197-217
Liman Substage, 137, 140, 148
"Little ice age," 319
Little Megantic Mountains, 189, 191, 192, 195, 197, 198, 201, 202, 203, 204, 205, 206, 210, 211, 214, 215, 216, 217
Lockbourne gravels, 85
Loess, 13, 20, 21, 22, 23, 72, 75, 81, 84, 87, 88, 91, 93, 123, 137, 139, 141, 142, 144, 145, 146-147, 149, 150, 167, 229, 236, 238, 240, 246
Long Island, 37, 39, 47, 49, 50-53, 56, 61, 62, 63, 65
Louisiana, 279
Lucan Moraine, 94
Luzerne readvance, 48, 56-58, 62, 63, 64, 65

Mad River Valley Train, 93
Maine, 12, 37, 38, 40, 41, 42, 43, 191, 192, 195, 198, 199, 202, 203, 207, 208, 209, 210, 211, 212, 215, 216
Main Glacial Advance, 83-86
"Main Morainic System," 16
"Main" Wisconsin, 88
Malden Till, 141, 231, 238
Manhasset Formation, 51
Manitoba, 154, 155, 156, 165, 166, 168, 170, 267, 268
Mankato, 3, 21, 25
Marine sedimentation, 135
Marine transgression, 37, 42, 61
Markhem Member, 139, 149
Marks Moraine, 165
Marseilles Morainic System, 18, 121, 123, 142
Marshall Moraine, 172
Martha's Vineyard, 37, 39, 53, 120
Massachusetts, 3, 12, 14, 37, 40, 50
Massawippi Formation, 197
McDonough Loess Member, 139, 149
Meadowcliffe Till, 84, 87
Meadow Loess, 139
Megantic area, 190, 196, 206, 215
Melvin Loess, 88
Mexico, 279
Miami drift, 91
Miami Lobe, 8, 20
Miami Sublobe, 73, 74, 80, 84, 85, 90, 91, 93, 94
Michigan, 20, 65, 72, 85, 95, 108, 118, 143, 157, 165, 263
Middletown readvance, 39, 40, 41, 64, 65
Middle Wisconsin, 3, 8, 21, 24, 71, 72, 77, 78, 80, 81, 84, 86-88, 110, 161, 168, 311, 313, 319
Midwest, 4, 6, 10, 17, 47, 48
Millbrook Till, 85, 87
Mille Lacs Moraine, 158, 160, 162, 179, 180, 253, 258

Minneapolis Lowland, 153, 154, 155, 157, 162, 171, 172, 174, 177, 181, 252, 256, 258, 263, 266
Minnesota, 8, 10, 12, 13, 14, 21, 123, 153, 154, 155, 156, 157, 158, 159, 160, 161, 162, 165, 166, 167, 168, 170, 171, 172, 175, 176, 178, 179, 181, 182, 251-273
Minnesota River, 153, 154, 155, 156, 158, 160, 167, 168, 169, 170, 171, 172, 176, 177, 181, 182
Mississinawa Morainic System, 18, 94
Mississippi, 279
Mississippi River, 12, 14, 136, 148, 154, 156, 159, 160, 163, 174, 175, 176, 177, 179, 180, 182, 253, 254, 255, 261, 285
Mississippi Valley, 20, 139, 182
Mitchell Moraine, 94
Mogadore Till, 87, 319
Mohawk Lowland, 107, 118, 122
Mohawk Valley, 64, 65, 66
Mohican Substage, 137, 140, 148
Montana, 8, 12, 123
Montauk Till, 51, 65
Monteregian Hills, 193
Moraine, 13, 14, 15, 16, 18, 19, 20, 24, 25, 26, 37, 38, 39, 40, 41, 42, 43, 47, 50, 51, 52, 53, 55, 56, 59, 61, 65, 76, 90, 91, 94, 95, 96, 122, 136, 141, 142, 145, 146, 148, 150, 159, 162, 163, 164, 170, 171, 172, 174, 175, 178, 179, 180, 181, 182, 190, 203, 230, 234, 245, 252, 259, 260, 264, 266, 267, 270, 271, 272
Morphology, 76
Morton Loess, 22, 141, 142
Mount Megantic, 192, 193, 196, 206, 212
Mt. Olive Moraines, 90, 91, 120

Nantucket Moraine, 15, 39, 40
Narragansett Valley Lobe, 50
Navarre Till, 90, 108
Nebraska, 12, 14, 156
Nebraskan, 138, 155
New Brunswick, 12, 40, 121, 122
New England, 37-43, 47, 48, 50, 53, 55, 58, 61, 64, 65, 120, 121, 190, 191, 197, 203, 206, 268
New England Upland, 48
Newfoundland, 128
New Hampshire, 12, 40, 191, 192, 195, 196, 207, 210, 211, 212, 215, 216
New Hampton Moraine, 55
New Holland Till, 90, 94
New Jersey, 9, 12, 14, 47, 48, 49, 51, 52, 54, 55, 56, 57, 62
New Paris Interstadial, 80, 88
New York, 5, 6, 8, 12, 14, 18, 37, 39, 40, 48, 50, 57, 59, 60, 61, 64, 65, 66, 72, 74, 77, 81, 85, 87, 94, 121, 122, 191
New Zealand, 121, 126, 129
Nickerson Moraine, 158, 160, 164

Nickerson phase, 158, 162–165, 180, 253, 267, 270, 272
North America, 4, 5, 6, 7, 8, 9, 15, 16, 17, 120–124, 128, 145, 269, 280, 313
North Dakota, 8, 123, 154, 156, 160, 166, 168, 239
North Sea, 124

O^{18}, 126–127, 318–319, 324–325
Ocean level, see Sea level
Ohio, 3, 4, 6, 8, 10, 11, 12, 13, 14, 18, 19, 23, 65, 72, 74, 77, 78, 79, 80, 81, 83, 85, 86, 87, 88, 89, 90, 91, 92, 93, 94, 107, 108, 118, 119, 120, 122, 123
Ohio Basin, 18
Ohio River, 14, 76, 86, 88, 94
Ohio Valley, 3, 4, 5, 88
Olean Drift, 72, 73, 83, 84, 85, 122
Ontario, 4, 18, 26, 74, 77, 80, 82, 84, 87, 90, 93, 94, 107, 108, 110, 117, 118, 119, 156, 165, 221–227, 263, 267, 293
Ontario Basin, 4, 72, 76, 79, 81, 83, 95, 96, 107, 118, 121, 122, 320
Ontario Lobe, 8, 48, 71–96, 108, 118, 122, 221, 223, 225, 226, 227
Oregon, 123, 127, 293
Organic, 8, 9, 10, 15, 16, 19, 21, 25, 39, 41, 77, 78, 81, 84, 86, 87, 88, 91, 120, 139, 144, 149, 161, 180, 236, 267
Otto Peat Bed, 82, 83, 84
Outwash, 3, 4, 6, 10, 11, 51, 52, 53, 65, 76, 85, 91, 93, 122, 137, 139, 142, 146, 150, 159, 163, 164, 172, 178, 180, 182, 183, 229, 230, 232, 238, 239, 240, 252, 266, 267, 271, 273
Oxygen isotope, 126–127, 277, 284, 298, 314, 317–320, 323–325

Pacific Ocean, 293
Paleoclimate, 317, 318
Paleotemperature curves, 127, 140, 147, 277, 278, 280–285, 287, 288, 291, 292, 293, 300, 304, 307–309, 313, 314, 315
Paleontologic studies, 27, 28, 71, 77, 82, 87, 149, 277–315
Paleosol, 8, 77, 81, 87, 88, 94, 122
Palynology, 49, 77, 78, 80, 82, 88, 89
Paris Moraine, 96
Patrician Superior Lobe, 83
Peat, 9, 12, 16, 21, 23, 39, 47, 52, 65, 78, 79, 81, 83, 84, 85, 88, 89, 111, 114, 115, 139, 164, 180
Pellets Island Moraine, 47, 53, 55, 56, 63, 64, 65
Pennsylvania, 8, 12, 14, 15, 18, 19, 23, 57, 72, 74, 80, 85, 87, 94, 121, 122
Peoria Loess, 72, 141, 142, 146, 311
Peorian, 18, 22
Periglacial, 136, 143, 147, 149, 230, 246, 259, 271, 272

Permafrost, 90, 147, 150, 168, 230, 241, 244, 245, 246, 256, 259, 260, 271, 272
Pierz Drumlin Field, 159, 261
Pike soil, 137, 140, 148
Pine City Moraine, 158, 180
Pine City-Split Rock phase, 158
Pinedale, 120, 123
Pineo Ridge Moraine, 37, 43
Pineo Ridge readvance, 37, 38, 40, 42, 43
Pingo, 229–247, 260
Pinnacle Hills Moraine, 64
Plano Silt Member, 139, 149
Pleasant Grove Soil, 139, 140, 149, 300, 302
Pleistocene-Holocene boundary, 277–315
Plum Point Interstadial, 80, 81, 84, 86, 87, 88, 319
Pollen, 27, 39, 48, 49, 52, 53, 55, 58, 62, 82, 87, 88, 123, 125, 126, 135, 144–145, 252, 259, 268, 270, 271, 273, 319
Pommeranian Moraine, 121, 124
Pommeranian phase, 125
Pond Ridge Moraine, 40, 41
Portage Uplands, 191, 192
Port Huron, 20, 38, 42, 43, 64, 65, 74, 81, 90, 95, 96
Port Huron Moraine, 95, 143
Port Stanley Drift, 90, 109, 110, 111
Port Stanley Till, 94, 107, 108, 110, 111, 112, 116–118, 122
Port Talbot Interstadial, 80, 81, 84, 86, 87, 88, 113, 114, 319
Powell Moraine, 73, 90, 94, 95, 119, 120, 121, 123
Prairie Coteau, 171, 172, 181

Quebec, 38, 40, 41, 42, 43, 60, 61, 66, 74, 77, 81, 83, 189–217

Rainy Lobe, 157, 158, 159, 161, 162, 179, 182, 267
Red Lakes Lowland, 155, 179, 180, 181
Red River, 154, 155, 158, 160, 166, 170, 181, 182
Red Sea, 293
Reesville Moraine, 90, 93, 94, 120, 121, 125
Rhode Island, 12, 14, 40
Richland Loess, 142, 236, 238
River Warren, 182, 183
Robein Silt, 139
Rocky Fork Till, 83
Rocky Mountains, 123
Rogers Lake, 37, 39, 40, 41, 120
Rondout (Minisink) Valley, 49, 65
Ronkonkoma-Culvers Gap Moraine, 48
Ronkonkoma Moraine, 37, 39, 40, 47, 50, 51, 52, 53, 62, 63, 64, 65, 120, 122
Rosendale readvance, 48, 55, 56–58, 63, 64, 65
Roslyn Till, 47, 51, 52, 53
Roxana Silt, 21, 23, 139, 149

Saginaw Lobe, 20, 72, 73, 85, 88, 90, 138, 222
St. Croix Moraine, 153, 154, 157, 158, 159, 160, 161, 162, 167, 172, 173, 174, 175, 179, 181, 253, 256, 258, 259, 261, 263, 264, 265
St. Croix phase, 157-161, 162, 163, 251, 252, 253, 256, 257, 258, 259, 260, 263, 265, 267, 272
St. Croix River, 154, 158, 159, 160, 165, 174, 176, 177, 178, 179, 180, 181, 253, 254, 255
St. John's Morainic System, 18, 94
St. Lawrence Basin, 4
St. Lawrence Lowland, 62, 71, 80, 81, 82, 83, 192
St. Lawrence River, 12, 73, 191
St. Lawrence Valley, 38, 43, 61, 77, 87, 95
St. Louis River, 154, 158, 164, 165, 175, 179, 180, 182
St. Louis Sublobe, 154, 158, 164, 179-180, 182
St. Pierre Interstade, 71, 81, 82-83
Salamonie Morainic System, 18, 94
Sandwich Moraine, 39, 40
Sangamonian, 10, 15, 20, 21, 25, 50, 81, 82, 135, 140, 144, 145, 147, 149, 287, 298, 300, 303, 305, 306, 307, 311, 313, 319
Sangamon soil, 21, 82, 140, 148, 149, 300
Saskatchewan, 23, 26, 27, 168, 239
Scandinavia, 124
Scandinavian ice sheet, 124-125, 129
Scarborough Formation, 82, 83, 84
Schlieren Phase, 121, 125
Scioto Lobe, 8, 19, 20, 91
Scioto River, 93
Scioto Sublobe, 72, 73, 74, 80, 84, 85, 90, 91, 93, 94
Scotland, 17, 125, 252
Scottish Readvance, 125
Seaforth Moraine, 94, 122
Sea level, 38, 39, 41, 42, 124, 127, 128, 135, 285, 300, 320
Sedimentation rate, 178, 278, 291, 299, 302, 303, 304
Seminary Till, 87
Shabbona Moraine, 231, 246
"Shelbyville drift," 21, 22, 91
Sidney Interstadial, 8, 80, 84, 86
Sidney, Ohio, 8, 72, 85, 88, 89
Snake River Canyon, 265
Soil, 25, 26, 77, 82, 85, 88, 91, 94, 125, 135-150, 180, 212, 222, 230, 232, 234, 285, 300
South America, 126, 129
South Carolina, 128
South Dakota, 8, 123, 156, 160, 166, 168, 170, 171
Southwold Drift, 84
Southwold Till, 87, 113, 114, 118
Spain, 125, 129

Spitzbergen, 269
Split Rock phase, 162-165, 180, 181, 182, 253, 258, 265, 267, 270, 271, 272
Stone line, 167
Stone pavement, see Boulder pavement
Striae, 10, 50, 72, 73, 91
Striations, 5, 12, 16, 167, 168, 169, 190, 192, 196, 198, 202, 214
Submarine moraines, 41
Sunnybrook Till, 83, 84
Superior Lobe, 123, 153-183, 251-273
Superior Lowland, 155
Surges, 38, 146, 154, 251-273
Susacá Interstadial, 126
Sussex Moraine, 47, 53, 55, 56, 59, 63, 64, 65
Sweden, 17, 262

Tazewell, 3, 21, 23, 25, 111
Temperature variations, 82, 83, 88, 126, 127, 135, 140, 145, 147, 149, 323-325
Terminal Moraine, 47, 48, 55, 56, 57, 63
Terrace, 183, 259
  Atlantic, 128
  Mississippi River, 159, 179
  Ohio River, 4, 5
Texas, 279, 294, 295, 296, 297
Thomson Moraine, 164, 165
Thorium dates, 311, 313, 320
Thorncliffe Formation, 87
Thule Lobe, 258, 271
Till, 4, 15, 16, 19, 21, 22, 24, 25, 26, 50, 63, 65, 74-76, 77, 78, 79, 80, 81, 83, 84, 85, 86, 87, 90, 91, 92, 94, 107, 108, 110, 111, 115, 117, 118, 119, 120, 122, 123, 137, 141, 142, 144, 145, 146, 147, 148, 149, 150, 157, 159, 163, 164, 166, 167, 168, 169, 173, 174, 175, 177, 178, 179, 180, 181, 182, 189-217, 221-227, 229, 240, 252, 254, 256, 261, 267
  Carbonates, 50, 74, 75, 85, 111, 112, 113, 117, 118, 144, 222, 238, 243, 245, 285
  Fabric, 72, 74, 75, 83, 87, 95, 166, 174, 190, 198, 261
  Lithology, 26, 72, 74, 75, 83, 87, 91, 92, 94, 95, 108, 111, 112, 113, 114, 141, 168, 169, 174, 189-217, 222, 223, 232
  Texture, 26, 74, 75, 76, 87, 95, 108, 112, 113, 167, 190, 212-214, 222
  Trace elements, 75, 189-217, 221-227
Tinley Moraine, 142
Tiskilwa Till Member, 141, 231, 232, 237, 238, 240
Titusville Till, 84, 85, 86, 87
Toimi drumlins, 180
Trafalgar Formation, 90
Tundra, 52, 53, 55, 58, 62, 63, 120, 259, 268, 271, 272
Tunnel valleys, 154, 159, 162, 163, 172, 174, 177, 179, 251-273
Tursac Oscillation, 125

Twocreekan, 3, 23, 81, 88, 96, 137, 150, 287,
    288, 290, 291, 292, 298, 300, 301, 303,
    304, 305, 306, 307–309, 314, 319
Two Creeks buried forest, 7, 21, 143
Two Creeks soil, 140, 300
Tymcht Till, 90

Ula Interstadial, 121, 124
Union City Moraine, 90, 94, 95, 119, 120,
    121, 123
United States, 3–28
Upper Leaside Till, 90, 95
Upper Melvin Loess, 93
Utica Valley Train, 93

Valderan, 23, 81, 89, 96, 137, 140, 150, 238,
    285, 287, 288, 290, 291, 292, 298, 300,
    301, 303, 305, 306, 307–309, 314
Valders, 21, 72, 143, 270, 273
Valders Till, 95, 150, 164
Valley Heads Drift, 90, 122
Valley Heads Moraine, 64, 65, 66, 73, 94,
    121
Valparaiso Morainic System, 15, 18, 121,
    123, 142
Varve, 78, 84, 89, 110, 111, 114, 116, 117,
    119, 176, 197, 237, 319
Vashon Stade, 121, 123, 124
Vepsovo Stadial, 121, 124
Vermilion Moraine, 158, 160, 161, 162, 179,
    267
Vermont, 10, 12, 40, 48, 49, 50, 59, 60, 61,
    65, 191, 207, 210, 211, 212, 216
Vineyard Moraine, 39, 40
Volcanic glass shards, 306, 307–309, 310,
    311, 312

Wabash Morainic System, 18, 94, 123
Wabash Valley, 72, 88
Wadena Drumlin Field, 158, 159, 160, 166,
    170, 172, 179, 181
Wadena Lobe, 158, 159, 160, 167, 169, 170,
    172, 173, 179, 267

Wallkill Moraine, 47, 48, 53, 55, 56, 62, 63,
    64, 65, 122
Wallkill Valley, 47, 48, 49, 50, 53–56, 57, 58,
    61, 62, 63, 65
Washington, 121, 123, 260, 269, 293
Waterloo-Auburn Moraine, 64, 66
Wayne Moraine, 94
Wedron Formation, 141, 231
Weichselian, 107, 125, 126, 127
Wentworth Till, 90, 95
White Mountains, 83, 191
White River Sublobe, 72, 73, 74, 80, 90, 93,
    94
Whitewater Basin, 91
Whitewater Drift, 84, 85, 86
Whitewater Till, 83, 85
Winnebago Drift, 23
Winnebago Formation, 139, 149
Winnebago Till, 21, 22
Winnipeg Lowland, 155, 166
Wisconsin, 7, 8, 11, 12, 13, 14, 15, 16, 22,
    138, 143, 147, 154, 156, 159, 165, 171,
    176, 181, 244, 259
Wisconsin Formation, 16
Wood, 10, 78, 79, 84, 89, 90, 92, 123, 146,
    181
Woodfordian, 3, 8, 15, 18, 19, 20, 22, 23, 47,
    48, 50, 52, 53, 55, 56, 57, 59, 63, 64, 65,
    80, 81, 85, 88, 120, 137, 138, 140,
    141–143, 144, 146, 148, 149, 150,
    229–247, 287, 288, 290, 291, 292, 296,
    297, 298, 300, 301, 303, 304, 305, 306,
    307–309, 310, 311, 314
Wright Moraine, 160, 161, 162
Würm, 83, 299
Wyoming Till, 95

Xenia Moraine, 91

Yarmouthian, 140, 145, 147
Yarmouth Soil, 140, 147, 148

Zürich Phase, 121, 125